Elliott Coues

Fur-bearing animals

A monograph of North American Mustelidae

Elliott Coues

Fur-bearing animals
A monograph of North American Mustelidae

ISBN/EAN: 9783337238230

Printed in Europe, USA, Canada, Australia, Japan

Cover: Foto ©berggeist007 / pixelio.de

More available books at **www.hansebooks.com**

DEPARTMENT OF THE INTERIOR.
UNITED STATES GEOLOGICAL SURVEY OF THE TERRITORIES.
F. V. HAYDEN, U. S. GEOLOGIST.

MISCELLANEOUS PUBLICATIONS, No. 8.

FUR-BEARING ANIMALS:

A MONOGRAPH

OF

NORTH AMERICAN MUSTELIDÆ,

IN WHICH AN ACCOUNT OF

THE WOLVERENE, THE MARTENS OR SABLES, THE ERMINE, THE MINK
AND VARIOUS OTHER KINDS OF WEASELS, SEVERAL SPECIES OF
SKUNKS, THE BADGER, THE LAND AND SEA OTTERS, AND
NUMEROUS EXOTIC ALLIES OF THESE ANIMALS,

IS CONTRIBUTED TO THE

HISTORY OF NORTH AMERICAN MAMMALS,

BY

ELLIOTT COUES,

CAPTAIN AND ASSISTANT SURGEON UNITED STATES ARMY,
SECRETARY AND NATURALIST OF THE SURVEY.

ILLUSTRATED WITH SIXTY FIGURES ON TWENTY PLATES

WASHINGTON:
GOVERNMENT PRINTING OFFICE.
1877.

J. Watts de Peyster:
LL. D.

MASTER OF ARTS, COLUMBIA COLLEGE, OF NEW YORK, 1872.
ROSE HILL, in the TOWNSHIP of RED HOOK, near TIVOLI P. O., DUCHESS CO., N. Y.

January, 1886.

JUDGE ADVOCATE, with the rank of MAJOR 1849,
COLONEL N. Y. S. I., 1846; assigned for "*Meritorious Conduct*" 1849;
BRIGADIER GENERAL for "*Important Service*" (first appointment—in N. Y. State—to that rank, hitherto elective) 1851, M. F. S. N. Y.
ADJUTANT GENERAL, S. N. Y., 1855.
BREVET MAJOR-GENERAL, S. N. Y., for "*Meritorious Services*,"
[first and only General officer receiving such an honor (the highest) from S. N. Y., and the only officer *thus* brevetted (Major General) in the United States,]
by "*Special Act*," or "*Concurrent Resolution*," *New York State Legislature*, April, 1866.
In the State Military Roster, "*Legislative Manual*" and "*Civil List*," as fifth Major-General, by Brevet, of the five Major-Generals, S. N. Y.)

LAWS OF NEW YORK, Vol. 2.—89th Session, 1866, Page 2142.

Concurrent Resolution requesting the Governor to Confer upon Brigadier-General J. WATTS DE PUYSTER [de Peyster] the brevet rank of Major" [General] in the National Guard of New York.

Resolved (If the Senate concur), That, it being a grateful duty to acknowledge, in a suitable manner the services of a distinguished citizen of this State, rendered to the National Guard and to the United States prior to and during the Rebellion, the Governor be and he is hereby authorized and requested to confer upon Brigadier General J WATTS DE PUYSTER [de Peyster] the brevet rank of *Major-General* in the National Guard of New York, for meritorious services, which mark of honor shall be stated in the Commission conferred.

STATE OF NEW YORK, in Assembly, April 9th, 1866.
The foregoing Resolution was duly passed. By order of the Assembly,
J. B. CUSHMAN, Clerk.

STATE OF NEW YORK, in Senate, April 20th, 1866
The foregoing Resolution was duly passed. By order of the Senate,
* So in original. JAS. TERWILLIGER, Clerk.

MILITARY AGENT, S. N. Y. (in Europe), 1851-'53.
HONORARY MEMBER, SECOND CLASS, of the MILITARY ORDER of the LOYAL LEGION of the U. S.
FIRST HONORARY MEMBER Third (Army of the Potomac) Corps Union.
HONORARY MEMBER of the THIRD CORPS RE-UNION ASSOCIATION, and MEMBER of COMMITTEE OF ARRANGEMENTS for the Grand Meeting of the Corps upon the Battlefield of Gettysburg.
HONORARY MEMBER of the CLARENDON HISTORICAL SOCIETY of Edinburgh, Scotland, and of the NEW BRUNSWICK (Canada) HISTORICAL SOCIETY (of St. John).
ASSOCIATE MEMBER of the MILITARY SERVICE INSTITUTION OF THE UNITED STATES.
MEMBER—10th June, 1872, DIRECTOR—of the GETTYSBURG BATTLEFIELD MEMORIAL ASSOCIATION, and VICE PRESIDENT of the SARATOGA [Battlefield] MONUMENT ASSOCIATION
MEMBER of the NETHERLANDISH LITERARY ASSOCIATION
(*Maatschappij der Nederlandsche Letterkunde*), at Leyden, Holland.
RECIPIENT, 1856, of *Three Silver Medals* from H. R. M. OSCAR, King of Sweden and Norway, &c., for a Military Biography of LEONARD TORSTENSON, Field Marshal. Generalissimo; of a *Gold Medal* in 1852, from WASHINGTON HUNT (Governor S. N. Y.), for "Efforts to Improve
the Military System of New York." &c., &c.; and Suggestions for a
Paid Fire Department with Steam Fire Engines, &c., &c.;
of a *Gold Medal*, only similar distinction ever ordered and directed, and conferred by the supreme military authority of the State of New York, by a Special Order, dated September 8th, 1851, of WASHINGTON HUNT, Governor and Commander in Chief of the Military Forces (S. N. Y.), authorized
to be worn in attest of "*Zeal, Devotion and Meritorious Service*!"
of a *Gold Medal*, in 1832, from the FIELD and STAFF OFFICERS of his Command, 9th Brig., 3d Div., N. Y. S. Troops, "In testimony of their Esteem and Appreciation of his Efforts towards
the Establishment of an efficient Militia," &c.;
in 1870, of a Magnificent *Badge, Medal* and *Clasps*, voted at the Annual Meeting of the Third Corps (Army of the Potomac) Union, held at Boston, Mass., Thursday, May 5th, 1870, when
A Resolution was adopted to present a *Gold Medal*, of the value of $500, to Gen. J. WATTS DE PEYSTER, of New York, as a testimonial of the appreciation by the Corps of his eminent services in placing upon record the true history of its achievements, and in defending its commanders and their men from written abuse and misrepresentation;
and of several other Badges, Medals, &c., for services in connection with the military service of New York State.

HONORARY MEMBER
of the NEW JERSEY and of the MINNESOTA and MONTANA HISTORICAL SOCIETIES, and of the PHRE-
NOKOSMIAN SOCIETY of PENNSYLVANIA COLLEGE. *Gettysburg*; of the PHILOSOPHIAN SOCIETY. *Missionary Institute*, Selin's Grove; of the DIAGNOTHEAN LITERARY SOCIETY of *Franklin and Marshall College*, Lancaster; of the EUTERPEAN SOCIETY, *Muhlenburg College*, Allentown, Penn., and of the GASMAN LITERARY and PHI
LOLOGIAN SOCIETIES of *Lycuem College, Nebraska City*.
HONORARY MEMBER of the LYCEUM SOCIETY, *Cazenovia, Madison Co., N. Y.*; and HONORARY MEMBER FOR LIFE of the AMERICAN RIFLE ASSOCIATION, to whom Gen. DE PEYSTER
presented the most original, exquisite and unique *Gold Badge* and *Clasp*, to
be shot for at the Annual Tests of Marksmanship.
HONORARY MEMBER of the NEW YORK BURNS CLUB.
BURNS was a member of the *Dumfries Volunteers*, of which Col. ARENT SCHUYLER DE PEYSTER, 8th or King's Foot, B. A., was Colonel; to whom the "National Bard of Scotland" addressed
just before his death, in 1796, "POEM ON LIFE"), and
LIFE MEMBER of the ST. NICHOLAS CLUB of NEW YORK
of which city JOHANNES DE PEYSTER—first of the name in the New World—was Schepen, 655; Alderman. 1666; Burgomaster, 1673; Deputy Mayor, 1677. Mayoralty offered and refused).

MEMBER
of the NEW YORK, of the RHODE ISLAND (Newport), and of the PENNSYLVANIA HISTORICAL SOCIETIES; HOLLAND SOCIETY OF NEW YORK; AMERICAN HISTORICAL ASSOCIATION,
and of the MILITARY ASSOCIATION of the STATE of NEW YORK,
and of the CENTURY CLUB, New York City.

LIFE MEMBER
of the HISTORICAL SOCIETY of MICHIGAN, of the NEW YORK GALLERY OF FINE ARTS, of the ALUMNI ASSOCIATION OF COLUMBIA COLLEGE, in the City of New York, and Director of the NEW YORK INSTITUTION for the INSTRUCTION of the DEAF and DUMB. AMERICAN NUMIS-
MATIC and ARCHAEOLOGICAL SOCIETY of NEW Y'K, and of
the HUGUENOT SOCIETY OF AMERICA, N.
LIFE MEMBER or FELLOW of the AMERICAN GEOGRAPHICAL SOCIETY; of the ROYAL HISTORICAL SOCIETY OF GREAT BRITAIN: PATRON of the ASSOCIATION for the BENEFIT
of COLORED ORPHANS, and of the NEW YORK DISPENSARY;
LIFE DIRECTOR of the AMERICAN TRACT and LIFE MEMBER of the AMERICAN BIBLE SOCIETY, N. Y.

CORRESPONDING MEMBER
of the STATE HISTORICAL SOCIETIES of MAINE, of VERMONT, of RHODE ISLAND (Providence), of CONNECTICUT, and of WISCONSIN; of the LONG ISLAND, of the BUFFALO, and of ONEIDA COUNTY (S. N Y.) HISTORICAL SOCIETIES; of the NEW ENGLAND HISTORICO-
GENEALOGICAL SOCIETY of the QUEBEC LITERARY and HISTORI-
CAL SOCIETY: of the NUMISMATIC and ANTIQUARIAN
SOCIETY of PHILADELPHIA, Pennsylvania;
&c., &c., &c.

PREFATORY NOTE.

U. S. GEOLOGICAL AND GEOGRAPHICAL
SURVEY OF THE TERRITORIES,
Washington, D. C., July 21, 1877.

This treatise on Fur-bearing Animals of North America, prepared by Dr. Elliott Coues, Assistant Surgeon United States Army, at present on duty with the Survey, is published as a specimen fasciculus of a systematic History of North American Mammals, upon which the author has been long engaged.

In the forthcoming work, which will be published by the Survey as soon as it can be prepared for the press, it is proposed to treat the Mammals of North America, living and extinct, in the same comprehensive and thorough manner in which the single family of the *Mustelidæ* has been elaborated.

The form of the final work, however, will necessarily be modified, in order to bring the whole matter within reasonable compass, as well as to adapt it more perfectly to the wants of the general public, which it is designed to meet. The technical and critical portions of the treatise will be condensed as far as may be deemed compatible with its distinctively scientific character, while the aspects of the subject which are of more general interest, such as the life-histories of the species and the economic or other practical relations which animals sustain toward man, will be presented in ample detail.

Other considerations have also had weight with me in deciding to publish this Monograph of the *Mustelidæ* in advance of the general "History", and as a separate volume. This family of Mammals is one of special interest and importance, from an economic point of view, as all the species furnish valuable peltries, some of which, like Sable, Ermine, and Otter, are in great demand; while their pursuit is an extensive and important branch of our national industries.

It is believed that the Monograph satisfactorily reflects the present state of our knowledge of these animals, and forms a desirable contribution to the literature of the general subject. The *Mustelidæ*, like most other families of North American Mammals, have not been systematically revised for many years, during which much new material, hitherto unused, has become available for the purposes of science; while the steady and rapid progress of scientific inquiry has rendered it necessary to reopen and discuss many questions in a new light. The same principles and methods of study which the author has successfully applied to the elucidation of the *Rodentia* of North America have been brought to bear upon the investigation of the *Mustelidæ*.

The Memoir is based upon specimens secured by the Survey under my direction, together with all the material contained in the National Museum, for the opportunity of examining which the Survey acknowledges, in this as in other instances, its indebtedness to the Smithsonian Institution.

The illustrations of the present volume, with few exceptions,* were engraved by Mr. H. H. Nichols, of Washington, from photographs on wood made under Dr. Coues's direction by Mr. T. W. Smillie, of Washington. This method of natural history illustration may still be regarded in the light of an experiment; but the cuts may be considered fine specimens of the engraver's art, when it is remembered that photography gives no lines to be followed by the graver. Though showing less detail, particularly of the under surfaces of the skulls, than might have been secured by hand-drawing, the cuts possess the merit of absolute accuracy of contour.

This opportunity is taken to reprint, by permission, a Circular relating to the proposed "History", which was addressed by Dr. Coues to the Medical Staff of the Army, of which he is a member. The Circular is sufficiently explicit to require no comment; but I may here express my high appreciation of the courtesy with which the wishes of the Survey have been met by the Surgeon General of the Army.

F. V. HAYDEN,
United States Geologist.

---

* The several figures on the electrotype plate VI were kindly loaned by Mr. E. A. Samuels, of Boston, from the Massachusetts Agricultural Report for 1861. The figures on plate XII were drawn on wood by Mr. S. W. Keen, of Washington, from photographs furnished by Mr. H. W. Parker, Agricultural College, Amherst, Mass.

[Reprinted.]

WAR DEPARTMENT,
SURGEON GENERAL'S OFFICE,
WASHINGTON, *March* 31, 1877.

CIRCULAR ORDERS, }
    No. 1. }

The attention of the Medical Officers of the Army is particularly invited to the following communication addressed to them by Assistant Surgeon ELLIOTT COUES, U. S. Army.

It is hoped that their assistance and co-operation will be cheerfully given for the reasons stated and in the manner indicated by Dr. COUES.

BY ORDER OF THE SURGEON GENERAL:

                         C. H. CRANE,
       Assistant Surgeon General, U. S. Army.

OFFICE OF U. S. GEOLOGICAL AND GEOGRAPHICAL SURVEY,
Washington, D. C., March 13, 1877.

TO THE MEDICAL OFFICERS OF THE ARMY:

Medical Officers of the Army, and others who may be interested in the matter, are respectfully and earnestly invited to coöperate with the undersigned in the preparation of a work entitled *"History of North American Mammals,"* to be published by the Government.

It is now twenty years since the last general work upon the Quadrupeds of this country appeared. The progress of our knowledge during this period renders the demand for a new treatise imperative. It is proposed to make the forthcoming *"History"* a standard scientific treatise, covering the whole ground, and fully exhibiting the present state of our knowledge of the subject. The plan of the work may be briefly indicated; its scope includes—

1. The Classification of North American Mammals according to the latest and most approved views of leading therologists, including diagnoses of the orders, families, genera and species.
2. The most acceptable Nomenclature of each species and variety, with extensive Synonymy.
3. The elaborate technical Description of each species and variety, including much anatomical detail, especially respecting the skull and teeth.
4. The Geographical Distribution of the species—an important matter, concerning which much remains to be learned.
5. The "Life-histories" of the species, or an account, as full and complete as it can be made, of their *habits.* This is also a matter requiring much further study.
6. The Bibliography of the subject.

While the strictly scientific character of the work will be maintained, the "life-histories," being of general interest, will be divested as far as possible of technicalities, and treated with a free hand, in popular style. The author has long been engaged in gathering material for this work, already far advanced, and hopes to publish at no distant day. His resources and facilities for the preparation of the descriptive and other technical portions of the treatise have been ample; but he has still, in common with other naturalists, much to learn respecting the Geographical Distribution and Habits of North American Mammals. To these points, therefore, special attention is invited, with the expectation that much important and valuable information may be secured with the assistance of Medical and other Officers of the Army, many of whom enjoy unusual facilities for acquiring a knowledge of this subject, and whose individual experiences, in many cases, represent a fund of information not yet on scientific record, but which, it is hoped, may now be made fully available.

The Geographical Distribution of animals can be thoroughly worked out only by means of observations made at very many different places. To this end it is desirable that lists should be prepared of the various species found in any given locality, noting their relative abundance or scarcity, times of appearance and disappearance, nature of their customary resorts, and other pertinent particulars. A sufficient number of such reports, from various stations, would greatly increase our knowledge, and render it more precise. It is believed that the "History of the Post," as already prepared by Medical Officers, usually includes information of this kind, which, by the permission of the Surgeon General, is made available for the present purpose.

As a rule, the habits of *larger* "game" animals, such as are ordinarily objects of the chase for pleasure or profit, and of all those which sustain obvious economic relations with man, as furnishing food or furs, or as committing depredations upon crops or live stock, are the best known; yet there is much to be learned even respecting these. The habits of many of the *smaller*, insignificant or obscure species are almost entirely unknown. Full and accurate information respecting the habits of the numerous species of Hares, Squirrels, Shrews, Moles, Mice, Rats, Bats, Weasels, Gophers, &c., is particularly desired. The Bats offer a peculiarly inviting and little explored field of research. Among points to which attention may be directed, in any case, are the following:

Date and duration of the rut.—Period of gestation.—Usual time of reproduction.—Number of young produced.—Duration of lactation.—Care of the young, by one or both parents.—State of monogamy or polygamy.—Times of disappearance and re-appearance of such animals as are migratory, and of such as hybernate.—Completeness or interruption of torpidity.—Times of changing pelage, of acquiring, shedding and renewing horns.—Habits connected with these processes.—Habits peculiar to the breeding and rutting seasons.—Construction of nests, burrows, or other artificial retreats.—Natural resorts at different seasons.—Nature of food at various seasons; mode of procuring it; laying-up of supplies; quantity required.—Various cries, of what indicative.—Natural means of offense and defense, and how employed.—General disposition, traits, characteristics.—Methods of capturing or destroying, of taming or domesticating.—Economic relations with man; how injurious or beneficial, to what extent, used for what purposes, yielding what products of value.

Other points will doubtless suggest themselves to the observer. Anatomical notes of careful dissections of soft parts, particularly of the digestive and reproductive organs, are valuable. Anecdotal records of personal experiences possess at least the interest which attaches to originality, and are very acceptable. Persons are frequently deterred from communicating their observations for fear that what they have to offer may not be wanted. This is generally a mistake. In the first place, duplication of data serves the important purpose of corroborating and confirming the accuracy of reports furnished, and in all cases of seasonal phenomena, which of course vary with latitude, the same observations may be profitably repeated at different stations. Secondly, persons who write books are generally supposed to know more than they really do.

Specimens of common and well-known animals, especially if bulky, are of course less desirable than those of rare and obscure species; but specimens of any species secured beyond the ordinary geographical range, or illustrating unusual conditions, such as albinism, melanism, or malformations, or representing embryonic stages of growth, are always in demand. Small dry parcels may be conveniently mailed direct to the undersigned; large packages should be sent in accordance with Circular Orders, No. 2, War Department, Surgeon General's Office, April 13, 1875, (copy herewith [*]), or by express, if the Quartermaster's Department cannot furnish transpor-

[*] WAR DEPARTMENT,
SURGEON GENERAL'S OFFICE,
Washington, April 13, 1875.

CIRCULAR ORDERS,
No. 2.

The following General Order from the Adjutant General's Office is published for the information of Medical Officers:

GENERAL ORDERS
No. 49.

WAR DEPARTMENT,
ADJUTANT GENERAL'S OFFICE.
Washington, April 6, 1875.

The Quartermaster's Department is authorized to transport to the Medical Museum at Washington such objects as may be turned over to its officers for that purpose at any military post or station by the officers of the Medical Department.

BY ORDER OF THE SECRETARY OF WAR:

E. D. TOWNSEND,
Adjutant General.

Medical Officers in turning over packages to the Quartermaster's Department for transportation will take receipts in duplicate, and will forward one of the receipts to the Surgeon General. All packages for the Museum should be plainly marked "Surgeon General, U. S. A., Washington, D. C.," with "Army Medical Museum" inscribed in the lower left hand corner.

BY ORDER OF THE SURGEON GENERAL:

C. H. CRANE,
Assistant Surgeon General U. S. Army.

tation. Specimens, after examination by the undersigned for the purposes of the work in hand, will be deposited, in the name of the donor, in the Army Medical Museum, or in the National Museum.

Printed instructions for collecting and preserving specimens will be furnished on application to the Smithsonian Institution. Medical Officers receiving this Circular are requested to bring it to the notice of others who may be interested in the matter, and are cordially invited to open correspondence with the writer upon the subject.

It is proper to add, that, for all information or specimens furnished, full credit will be given in every instance, both in the text of the treatise in which such material is utilized, and in the records and publications of the Museum in which it is finally deposited; and that the author will regard coöperation in this work as a personal favor, to be fully appreciated and gratefully acknowledged.

ELLIOTT COUES,
Assistant Surgeon, U. S. Army.

SUMMARY OF CONTENTS.

CHAPTER I.

THE FAMILY MUSTELIDÆ.

General considerations—Systematic position and relations of the *Mustelidæ*—Division into subfamilies—Schedule of the North American genera—Their differential characters—Diagnoses of the North American subfamilies—The anal glands of *Mustelinæ*—The fossil North American species of *Mustelinæ*—Derivation and signification of names applied to *Mustelidæ*........................... 1

CHAPTER II.

SUBFAMILY MUSTELINÆ: THE WOLVERENE.

The genus *Gulo*—Generic characters—*Gulo luscus*, the Wolverene—Synonymy—Habitat—Specific characters—Description of external characters—Measurements—Anal glands—Description of the skull and teeth—Measurements of skulls, European and American—Nomenclature of the species—Relation of the European and American animal—General history, geographical distribution, and habits of the species—Its distribution in the Old World............... 32

CHAPTER III.

MUSTELINÆ—Continued: THE MARTENS.

The genus *Mustela*—Generic characters, &c.—Analysis of North American species—*Mustela pennanti*, the Pekan or Pennant's Marten—Synonymy—Habitat—Specific characters—Description of external characters—Dimensions—Skull and vertebræ—General history, habits, and geographical distribution—Interpolated matter relating to exotic species of *Mustela*—*M. martes*—Synonymy—Description of its skull and teeth—*M. foina*—Synonymy—Notes on its characters—*M. zibellina*—Synonymy—Measurements of skulls of the three species—Comparative diagnoses of *M. martes, americana,* and *foina*—*Mustela americana*, the American Sable or Marten—Synonymy—Description and discussion of the species—Table of measurements—Geographical variation in the skull—General history and habits of the species................. 59

CHAPTER IV.

MUSTELINÆ—Continued: THE WEASELS.

The genus *Putorius*—Generic characters and remarks—Division of the genus into subgenera—Analysis of the North American species—The subgenus *Gale*—*Putorius vulgaris*, the Common Weasel—Synonymy—Habitat—Specific characters—General characters and relationships of the species—Geographical distribution—Habits—*Putorius erminea*, the Stoat or Ermine—Synonymy—Habitat—Specific characters—Discussion of specific characters and relationships—Table of measurements—Note on the skull and teeth—Description of external characters—Conditions of the change of color—General history and habits of the species—Its distribution in the Old World—*Putorius longicauda*, the Long-tailed Weasel—Synonymy—Habitat—Specific characters—Description—Measurements—General account of the species—*Putorius brasiliensis frenatus*, the Bridled Weasel—Synonymy—Habitat—Specific characters—General account of the species 97

CHAPTER V.

MUSTELINÆ—Continued: THE AMERICAN FERRET.

The subgenus *Cynomyonax*—Subgeneric characters—*Putorius (Cynomyonax) nigripes*, the American or Black-footed Ferret—Synonymy—Specific characters—Habitat—General account of the species—ADDENDUM: On the species of the subgenus *Putorius*—*P. fœtidus*, the Polecat or Fitch—Synonymy—Description—*P. fœtidus* var. *furo*, the Ferret—Synonymy—Remarks—Ferret breeding and handling—*P. fœtidus* var. *eversmanni*, the Siberian Polecat—Synonymy—Remarks—*P. sarmaticus*, the Spotted Polecat—Synonymy and remarks 147

CHAPTER VI.

MUSTELINÆ—Continued: THE MINK.

The subgenus *Lutreola*—Subgeneric characters and remarks—*Putorius vison*, the American Mink—Synonymy—Habitat—Specific characters—Description of external characters—Measurements—Variation in external characters—Variation in the skull—Comparison with the European Mink—Notice of allied Old World species, *P. lutreola* and *P. sibiricus*—General history and habits of the Mink—"Minkeries" 158

CHAPTER VII.

SUBFAMILY MEPHITINÆ: THE SKUNKS.

General considerations—Cranial and dental characters—The anal armature—Division of the subfamily into genera—Note on fossil North American species—The genus *Mephitis*—*Mephitis mephitica*, the Common Skunk—Synonymy—Habitat—Specific characters—Description of external characters—Description of the skull and teeth—Variation in the skull with special reference to geographical distribution—Anatomy and physiology of the anal glands and properties of the secretion—Geographical distribution and habits of the Skunk—History of the species—ADDENDUM: On hydrophobia from Skunk-bite, the so-called "rabies mephitica"............ 187

CHAPTER VIII.

MEPHITINÆ—Continued: SKUNKS.

The genus *Mephitis*, continued—*Mephitis macrura*, the Long-tailed Mexican Skunk—Synonymy—Habitat—Specific characters—Description—The subgenus *Spilogale*—*Mephitis* (*Spilogale*) *putorius*, the Little Striped Skunk—Synonymy—Habitat—Specific characters—Description of external characters—Description of the skull and teeth—History of the species—The genus *Conepatus*—*Conepatus mapurito*, the White-backed Skunk—Synonymy—Habitat—Specific characters—Description of external characters—Description of the skull and teeth—Description of the anal glands—Geographical distribution and habits... 236

CHAPTER IX.

SUBFAMILY MELINÆ: THE BADGERS.

The genus *Taxidea*—Generic characters and comparison with *Meles*—*Taxidea americana*, the American Badger—Synonymy—Habitat—Specific characters—Description of external characters—Description of the skull and teeth—Geographical variation in the skull—History of the American Badger—Its geographical distribution—Habits—*Taxidea americana berlandieri*, the Mexican Badger—Synonymy—Habitat—Subspecific character—General remarks—ADDENDUM: Description of the perineal glands of the European Badger, *Meles vulgaris* 261

CHAPTER X.

Subfamily LUTRINÆ: The Otters.

General considerations—The genus *Lutra*—Generic characters and remarks—The North American Otter, *Lutra canadensis*—Synonymy—Habitat—Specific characters—Description of external characters—Description of the skull and teeth—Variation in the skull—History of the species—Geographical distribution—Habits of Otters—Extinct species of North American Otter............ 293

CHAPTER XI.

Subfamily ENHYDRINÆ: Sea Otter.

General considerations—The genus *Enhydris*—Generic characters—*Enhydris lutris*, the Sea Otter—Synonymy—Habitat—Specific characters—Description of external characters—Description of the skull and teeth—History of the species—"The Sea Otter and its hunting"—The habits of the Sea Otter..................... 325

LIST OF ILLUSTRATIONS.

PLATE I.—GULO LUSCUS.
    Skull from above, below, and in profile. (*Reduced.*)

II.—MUSTELA PENNANTI.
    Skull from above, below, and in profile. (*Natural size.*)

III.—MUSTELA MARTES.
    Skull from above, below, and in profile. (*Natural size.*)

IV.—MUSTELA FOINA.
    Skull from above, below, and in profile. (*Natural size.*)

V.—MUSTELA AMERICANA.
    Skull from above, below, and in profile. (*Natural size.*)

VI.—PUTORIUS ERMINEA *and* P. VULGARIS.
    Figs. 1, 3, 5, 6, *P. erminea*, heads and tails. Figs. 2, 4, *P. vulgaris*, head and tail. (*Natural size.*)

VII.—PUTORIUS NIGRIPES.
    Skull from above, below, and in profile. (*Natural size.*)

VIII.—PUTORIUS FOETIDUS.
    Skull from above, below, and in profile. (*Natural size.*)

IX.—PUTORIUS VISON.
    Skull from above, below, and in profile. (*Natural size.*)

X.—MEPHITIS MEPHITICA.
    Skull of ordinary characters from above, below, and in profile. (*Natural size.*)

XI.—MEPHITIS MEPHITICA.
    Skull of large size from above, below, and in profile. (*Natural size.*)

XII.—MEPHITIS (SPILOGALE) PUTORIUS.
    Two skins, to show the peculiar markings. (*Much reduced.*)

XIII.—MEPHITIS (SPILOGALE) PUTORIUS.
    Large old skull from above, below, and in profile. (*Natural size.*)

PLATE XIV.—MEPHITIS (SPILOGALE) PUTORIUS.
    Small young skull from above, below, and in profile. (*Natural size.*)

XV.—CONEPATUS MAPURITO.
    Skull from above, below, and in profile. (*Natural size.*)

XVI.—TAXIDEA AMERICANA.
    Skull from above, below, and in profile. (*Reduced.*)

XVII.—LUTRA CANADENSIS.
    Skull from above, below, and in profile. (*Natural size.*)

XVIII.—LUTRA VULGARIS.
    Skull from above, below, and in profile. (*Natural size.*)

XIX.—ENHYDRIS LUTRIS.
    Skull from above and below. (*Reduced.*)

XX.—ENHYDRIS LUTRIS.
    Skull in profile. (*Reduced.*) Palate and teeth. (*Natural size.*)

HISTORY

OF

NORTH AMERICAN MUSTELIDÆ.

CHAPTER I.

THE FAMILY MUSTELIDÆ.

General considerations—Systematic position and relations of the *Mustelidæ*—Division into subfamilies—Schedule of the North American genera—Their differential characters—Diagnoses of the North American subfamilies—The anal glands of *Mustelinæ*—The fossil North American species of *Mustelinæ*—Derivation and signification of names applied to *Mustelidæ*.

THIS is a large, important, and well-defined family of Carnivorous Mammals, embracing the Weasels and Martens, as its typical representatives, the Skunks, Badgers, Otters, and a few other less familiar animals.

Representatives of the family exist in most portions of the globe, excepting the Australian region, home of the Marsupials and Monotremes. The group reaches its highest development in the Northern Hemisphere, or Arctogæa, where both the genera and the species are most numerous and diversified. Some twenty genera are recognized by modern authors; of these, the genus *Putorius*, including the true Weasels, has the most extensive geographical distribution in both hemispheres, and contains by far the largest number of species. In one sense, it is to be considered as the typical genus of the family. Many of the other genera consist of but a single species, and some of them are the sole representatives of the subfamilies to which they respectively belong.

The economic importance of the family may be estimated from the very high commercial value which fashion has set upon the fur of several of the species, such as the Ermine, Sable, Nutria, and Sea Otter; and various other pelts, only less valuable than these, are furnished by members of this family. These animals sustain other relations toward man, by no means to be overlooked. The serious obstacles which the Wolverene offers to the pursuit of the more valuable fur-bearing animals of British America is set forth in following pages; while the destructiveness of such species as the Mink and various kinds of Weasels is well known to the poulterer. The Skunks are infamous for the quality, familiar to every one, which places them among the most offensive and revolting of animals; they are, moreover, capable of causing one of the most dreadful diseases to which the human race is exposed. The cruel sport which Badgers have afforded from time immemorial has given a verb to the English language; while the legitimate pursuit of various *Mustelidæ* is an important and wide-spread branch of human industry.* The scientific interest with which the zoölogist, as simply such, may regard this family of animals yields to those practical considerations of every-day life which render the history of the *Mustelidæ* so important.

The definition of the family is strict. The zoölogical characters by which it is distinguished from other Carnivorous Mammals are well marked; and few if any naturalists of repute differ in their views respecting the limitation of the group. The systematic position of the family in the Carnivorous series seems to be also settled by very general consent. Singular as it may seem, when, without considering intermediate forms, we compare for instance the diminutive, slender-bodied, and nimble Weasels with the great, heavy-bodied, and comparatively slothful Bears, the closest affinities of the Musteline series are with the Ursine; the next nearest are with the Canine; and the family *Mustelidæ* may properly stand between the *Canidæ* on

---

* During the century, 1769-1868, the Hudson's Bay Company sold at auction in London, besides *many million* other pelts, the following of *Mustelidæ*:— 1,240,511 sables, 674,027 otters, 68,694 wolverenes, 1,507,240 minks, 218,653 skunks, 275,302 badgers, 5,349 sea otters. In 1868 alone, the company sold (among many thousand others) 106,254 sables, 73,473 minks, 14,966 otters, 6,298 skunks, 1,104 wolverenes, 1,551 badgers, 123 sea otters; besides which there were also sold in London, in the autumn of the same year, about 4,500 sables, 22,000 otters, &c. Another company, the Canadian, sold in London, during the years 1763 to 1839, the following: 2,931,383 sables, 29,110 wolverenes, 895,832 otters, 1,080,780 minks.—(DROSTE-HÜLSHOFF, *Der Zoologische Garten*, 1869, p. 317.)

the one hand and the *Ursidæ* on the other. In order to give a clear idea of the position and relationships of the *Mustelidæ*, the following characters* of the higher groups of Mammals under which the family comes are given:—

> Mammals having a brain with the cerebral hemispheres connected by a more or less well-developed corpus callosum and a reduced anterior commissure. Vagina a single tube, but sometimes with a partial septum. Young retained within the womb till of considerable size and nearly perfect development, and deriving its nourishment from the mother through the intervention of a "placenta" (developed from the allantois) till birth. Scrotum never in front of penis. . . (Subclass) **Monodelphia**.
>
> Brain with a relatively large cerebrum, behind overlapping much or all of the cerebellum, and in front much or all of the olfactory lobes; corpus callosum (attypically) continued horizontally backwards to or beyond the vertical of the hippocampal suture, developing in front a well-defined recurved rostrum.
> (Super-order) EDUCABILIA.
>
> Posterior members and pelvis well developed (in antithesis with the Cetaceans and Sirenians). Proximal segments of both fore and hind limbs (upper arm and thigh) more or less enclosed in the general integument of the trunk (in antithesis with the order *Primates*). Clavicles rudimentary or wanting. Scaphoid and lunar bones of the wrist consolidated into one (scapholunar) carpal. Digits clawed (not hoofed). Teeth of three kinds, all enamelled; incisors $\frac{3-3}{3-3}$ (exceptionally fewer); canines specialized and robust; one or more molars in each jaw usually sectorial. Brain without calcarine sulcus. Placenta deciduate, zonary. (= The Carnivora or "beasts of prey" of ordinary language.) . . . . (Order) FERÆ.
>
> Body elevated and adapted for progression on land by approximately equal development, freedom, and mobility of fore and hind limbs. Tail free from common integument of body. Ears well developed. Functional digits terminating in claws. Digits of neither fore nor hind feet webbed to the ends (excepting the hind feet of *Enhydra*); inner digits of fore feet not produced beyond the rest; inner digits of hind feet seldom thus produced, but often reduced or atrophied. (All these expressions in antithesis to the *Pinnipedia*, or suborder of the Seals.) . . . . (Suborder) FISSIPEDIA.

---

* For which I am principally indebted to Dr. Theo. Gill. (Smithsonian Miscellaneous Collections. | —230— | Arrangement | of the | Families of Mammals. | With analytical tables. | Prepared for the Smithsonian Institution. | By Theodore Gill, M. D., Ph. D. | [Seal.] | Washington: | Published by the Smithsonian Institution. | November, 1872. | [8vo. pp. i-vi, 1-98.])

Skull with the paroccipital process not closely applied to the auditory bulla; the mastoid process prominent and projecting outwards or downwards behind the external auditory meatus; external auditory meatus diversiform. Intestinal canal with no cæcum. Prostate gland not salient, being contained in the thickened walls of the urethra. Skull with the carotid canal distinct, and more or less in advance of the foramen lacerum posticum; condyloid foramen alone distinct from the foramen lacerum posticum; glenoid foramen generally well defined. Os penis very large. Cowper's glands not developed.

(Super-family) ARCTOIDEA.

True molars of upper jaw one (M. $\frac{1}{2}$: rarely—in *Mellivorinæ*—$\frac{1}{1}$); last premolar of upper jaw sectorial (rarely—in *Enhydrinæ*—with blunt tubercles). . . . . (Family) *Mustelidæ*.

DIVISION OF THE MUSTELIDÆ INTO SUBFAMILIES.

Having thus, by a process of gradual elimination of the characters of other groups, reached a family, *Mustelidæ*, we may proceed to inquire of what subdivisions the family itself is susceptible. Authors—even throwing the older writers out of consideration—differ greatly in their methods of reckoning the subfamilies and genera, the number of subfamilies recognized varying from three or four to eight. According to my present understanding of the subject, derived from my knowledge of American forms, I am inclined to consider that, if any subfamily divisions are to be adopted, regard for equivalency, or the just coördination of the characters involved, requires a larger number of subfamilies than have usually been recognized—very possibly the full number, eight, admitted by Dr. Gill. The excellent analysis of the family given by this writer (see p. 3, note) is herewith presented:—

XVI.—MUSTELIDÆ.

Sub-families.

I. Skull with the cerebral portion comparatively compressed backwards; and with the rostral portion comparatively produced, attenuated, and transversely convex above; anteorbital foramen small and opening forwards. Feet with little developed or no interdigital membrane [and the species, with few exceptions, not aquatic].
  A. Auditory bulla much inflated, undivided, bulging, and convex forwards; periotic region extending little outwards or backwards. Palate moderately emarginated.

1. Last molar of upper jaw ($M \frac{1}{1}$) transverse, (with the inner ledge inflated at its inner angle;) sectorial tooth with a single inner cusp.
    a. $M \frac{1}{2}$; first true molar (sectorial) of lower jaw followed by a second (tubercular) one. Toes short, regularly arched, and with the last phalanges bent up, withdrawing the claws into sheaths. (Gray.) [Martens and Weasels.] . . . . MUSTELINÆ.
    b. $M \frac{1}{1}$; first true molar (sectorial) of lower jaw only developed. Toes straight, with the last phalanges and claws extended; the latter non-retractile. (Gray.) [Extra-limital.] . . . . MELLIVORINÆ.
2. Last molar of upper jaw ($M \frac{1}{1}$) enlarged and more or less extended longitudinally.—$M \frac{1}{2}$. Toes straight, with the last phalanges and claws extended; the latter non-retractile. (Gray.) [Badgers.] MELINÆ.
B. Auditory bulla elongated and extending backwards close to the paroccipital process. (Flower.) Palate moderately emarginated.
    1. Last molar of upper jaw ($M \frac{1}{1}$) transverse; (with the inner ledge narrowed inwards): sectorial tooth with two inner cusps. [Extra-limital.] . . . . HELICTIDINÆ.
C. Auditory bulla inflated, undivided, with the anterior inferior extremity pointed and commonly united to the prolonged hamular process of the pterygoid. (Flower.) Palate moderately emarginated.
    1. Last molar of upper jaw ($M \frac{1}{1}$) transverse; (with the inner ledge compressed.) [Extra-limital.] . . . . ZORILLINÆ.
D. Auditory bulla little inflated, transversely constricted behind the meatus auditorius externus and thence inwards; in front flattened forwards; periotic region expanded outwards and backwards. Palate deeply emarginated.
    1. Last molar of upper jaw ($M \frac{1}{1}$) quadrangular, wide, but with an extended outer incisorial ledge. [Skunks.] . . . . MEPHITINÆ.
II. Skull with the cerebral portion swollen backwards and outwards; and with the rostral portion abbreviated, high and truncated forwards, and widened and depressed above; anteorbital foramen enlarged and produced downwards and backwards. Feet with well-developed interdigital membrane, and adapted for swimming. [The species highly aquatic, one of them marine.]
A. Teeth normal, 36 ($M \frac{1}{2}$, PM $\frac{4}{3}$, C $\frac{1}{1}$, I $\frac{3}{3} \times 2$): sectorial tooth (PM $\frac{4}{.}$) normal, efficient, with an expanded inner ledge; the other molars submusteline. Posterior feet with normally long digits. [Otters.]
LUTRINÆ.
B. Teeth very aberrant, 32 ($M \frac{1}{2}$, PM $\frac{3}{3}$, C $\frac{1}{1}$, I $\frac{3}{2}$—the lower inner incisors being lost — $\times 2$): sectorial tooth (PM $\frac{4}{.}$) defunctionalized as such, compressed from before backwards; the other molars also with blunted cusps. Posterior feet with elongated digits. [Sea Otter.]
ENHYDRINÆ.

Of the foregoing eight subfamilies, three, namely, the *Mellivorinæ*, *Helictidinæ*, and *Zorillinæ*, each of which consists of a single genus, are confined to the Old World. No one of the subfamilies is peculiar to North America; but the *Mephitinæ*, or Skunks, are not found in the Old World, where they are represented by the African *Zorillinæ;* they occur in South as well as North America. The *Melinæ,* or Badgers, are common to North and Middle America and the Eastern Hemisphere, but do not occur in South America. The Sea Otter, sole representative of the *Enhydrinæ,* inhabits both coasts of the North Pacific. The *Lutrinæ,* or ordinary Otters, are of general distribution in both hemispheres. The *Mustelinæ,* or true Weasels, Martens, &c., are of very general distribution, as already indicated; such is especially the case with the typical genus *Putorius.* The genera *Mustela* and *Gulo* chiefly inhabit the higher latitudes; *Galictis* is peculiar to South America.

The North American forms of the family down to the genera are exhibited in the following synoptical table:—

 Subclass *Monodelphia.*
  Super-order *Educabilia.*
   Order *Feræ.*
    Suborder *Fissipedia.*
     Super-family *Arctoidea.*
      Family *Mustelidæ.*
       Subfamily *Mustelinæ.*
        Genera *Gulo.* (The Wolverene.)
          Mustela. (The Martens.)
          Putorius. (The Weasels.)
       Subfamily *Mephitinæ.* (The Skunks.)
        Genera *Mephitis.*
         Spilogale.
         Conepatus.
       Subfamily *Melinæ.* (The Badgers.)
        Genus *Taxidea.*
       Subfamily *Lutrinæ.* (The Otters.)
        Genus *Lutra.*
       Subfamily *Enhydrinæ.* (The Sea Otter.)
        Genus *Enhydra.*

SCHEDULE OF DIFFERENTIAL CHARACTERS OF THE NORTH AMERICAN GENERA.

Various characters by which the subfamilies and genera are differentiated are exhibited in the following schedule:—

	Mustelinæ			Mephitinæ		Melinæ		Lutrinæ	Enhydrinæ
	Gulo.	Mustela.	Putorius.	Mephitis.	Spilogale.	Conepatus.	Taxidea.	Lutra.	Enhydra.
Dental formula: I. $\frac{3-3}{3-3}$, C. $\frac{1-1}{1-1}$, Pm. $\frac{4-4}{4-4}$, M. $\frac{1-1}{2-2}=\frac{18}{20}=38$	×	×							
I. $\frac{3-3}{3-3}$, C. $\frac{1-1}{1-1}$, Pm. $\frac{3-3}{3-3}$, M. $\frac{1-1}{2-2}=\frac{16}{18}=34$			×	×	×		×		
I. $\frac{3-3}{3-3}$, C. $\frac{1-1}{1-1}$, Pm. normally $\frac{2-2}{3-3}$, sometimes $\frac{3-3}{3-3}$, M. $\frac{1-1}{2-2}=\frac{14}{18}$ or $\frac{16}{18}=32$ or 34							×		
I. $\frac{3-3}{3-3}$, C. $\frac{1-1}{1-1}$, Pm. $\frac{4-4}{3-3}$, M. $\frac{1-1}{2-2}=\frac{18}{18}=36$								×	
I. $\frac{3-3}{2-2}$, C. $\frac{1-1}{1-1}$, Pm. $\frac{3-3}{1-1}$, M. $\frac{1-1}{2-2}=\frac{16}{16}=32$									×
PM. and M. normal—angular, trenchant, or acute	×	×	×	×	×	×	×	×	
abnormal—rounded, blunt, tuberculous									×
Back upper M. quadrate, transverse, much wider than long	×	×	×					×	
about as wide as long				×	×		×		
triangular, the hypothenuse postero-exterior						×			
irregularly oval; all corners rounded off									×
Back upper Pm. strictly sectorial, linear, with small anterior interior cusp	×	×	×						
triangular, owing to size of the inner ledge				×	×	×	×		
resembling the back upper M.									×
Upper Pm. 4-4, the anterior one comparatively well developed	×	×	×						
minute, crowded out of line								×	
3-3, the anterior one comparatively well developed				×	×		×		
or 2-2; when 3-3, the anterior very minute						×			×
Lower I. 3-3, the usual carnivorous formula	×	×	×	×	×	×	×	×	
2-2, the inner pair lacking									×
Lower sectorial without obvious inner tubercle of middle lobe	×	×							
with slight but evident inner tubercle			×						
with strongly developed inner tubercle				×	×	×	×	×	
Rostrum of skull so short that root of zygoma is nearly or quite opposite fore end of nasal bones								×	×
moderately produced; root of zygoma more nearly opposite hind end of nasals	×	×	×	×	×	×	×		
Brain-case comparatively compressed backward, little broader behind than before, with straightish or little convex lateral outline	×	×	×	×	×	×			
widened backward, with quite straight lateral outlines							×		
much widened backwards and swollen outward, with very convex lateral outlines								×	×
Frontal region very short, broad, flat on top								×	×
lengthened, narrowed, very convex transversely	×	×	×	×	×	×	×		
Bony palate ending opposite back upper molars				×	×		×		
produced back of the molars, but not half-way to end of pterygoids						×			
produced far back of the molars—half-way or more to ends of pterygoids	×	×	×					×	×
Postorbital processes moderate, slight or obsolete	×	×	×	×	×	×		×	×
strong, transverse, acute							×		
Antorbital foramen bounded above by slender maxillary process, large, subtriangular, or oval, presenting downward-forward								×	×

NORTH AMERICAN MUSTELIDÆ.

Schedule of differential characters of the North American genera—Continued.

	Mustelinæ			Mephitinæ			Melinæ	Lutrinæ	Enhydrinæ
	Gulo	Mustela	Putorius	Mephitis	Spilogale	Conepatus	Taxidea	Lutra	Enhydra
Antorbital foramen bounded by stout process, presenting more or less vertically	×	×	×	×	×	×	×	..	..
Aperture of nares in two planes, approaching the vertical and horizontal								×	×
one plane, or nearly so, more or less oblique	×	×	×	×	×	×	×	..	..
Auditory bullæ at maximum of inflation, with shortest and least tubular meatus							×	×	..
much inflated, with moderate constriction into the tubular meatus	×	×	×					..	..
little inflated, much constricted across the meatus				×	×	×		×	×
Mastoids little developed, outward or backward	×	×	×					..	..
more developed, outward				×	×	×		..	..
much developed, downward								×	×
Periotic region contracted, bringing paroccipitals close to auditory bullæ	×	×	×				×	..	..
expanded, removing paroccipitals from bullæ, and horizontal				×	×	×		..	..
expanded, removing paroccipitals from bullæ, very oblique								×	×
Glenoid fossæ shallow, open, without anterior ledge, presenting more forward than downward, never locking condyles							×	..	..
moderately deep and close, with anterior ledge presenting downward-forward, never locking condyles	×	×	×	×	×	×		..	..
very deep and close, with strong anterior and posterior ledges, sometimes locking condyles								×	×
Coronoid process of jaw in profile conical, erect, apex forward of condyle	×	×	×	×	×	×	×	..	..
obtusely falcate, sloping, apex overhanging condyle								×	×
Lower border of jaw straightish from symphysis to posterior angle	×	×	×					..	..
usually ascending posteriorly, in straight or concave line				×	×	×	×	×	×
Toes scarcely or not webbed, with ordinary ratio of lengths	×	×	×	×	×	×	×	..	..
fully webbed, with ordinary ratio of lengths								×	..
those of the hind feet elongated, with extraordinary ratio of lengths								..	×
Fore claws long, stout, little curved, highly fossorial	..	..	..	×	×	×	×	..	..
moderate or short, curved and acute, not fossorial	×	×	×					×	×
Body very stout; size very large; tail bushy, short; appearance somewhat bear-like	×							..	..
rather slender or extremely so; size medium and small; tail long, terete		×	×					..	..
stout; size medium and small; tail long, very bushy				×	×		×	..	..
stout, much depressed; size medium; tail short, distichous								×	..
stout, cylindrical; size large; tail long, conical, close-haired						×		..	×
Habits chiefly terrestrial	×							..	..
terrestrial and highly arboreal		×						..	..
strictly terrestrial and more or less fossorial			×			×	×	×	..
aquatic (fluviatile, lacustrine, or maritime)					×			×	..
aquatic (marine)									×

Such a table as this might be indefinitely continued, but the foregoing analysis of leading differential characters suffices for present purposes.

We may finally sum and amplify the differential characters of the foregoing table, with others, in the following expressions, diagnostic of the five subfamilies here adopted :—

DIAGNOSES OF THE FIVE NORTH AMERICAN SUBFAMILIES.

1. MUSTELINÆ.—Teeth of ordinary Carnivorous pattern, 38 or 34 in number, according to varying number of premolars, whether $\frac{4-4}{4-4}$ (*Gulo, Mustela*) or $\frac{3-3}{3-3}$ (*Putorius*); the number unequal in the two jaws, $\frac{18}{20}$ or $\frac{16}{18}$; incisors constantly $\frac{3-3}{3-3}$; canines $\frac{1-1}{1-1}$, as in all *Mustelidæ*; and molars $\frac{1-1}{2-2}$, as in all *Mustelidæ* excepting *Mellivorinæ*. Molar of upper jaw much wider than long; its long axis transverse to the axis of the dental series, longitudinally constricted across the middle. Posterior upper premolar (the large "sectorial" tooth) narrow and linear, with a small distinct spur projecting inward from its antero-interior corner. Rostral part of skull moderately produced, sloping in profile, very obliquely truncated, transversely convex, the hind ends of the nasals more nearly opposite the roots of the zygoma than their fore ends are.* Cerebral portion of skull comparatively compressed backward, little broader behind than before, with moderately convex lateral outlines. Postorbital processes moderately developed. Anteorbital foramen small, oval or subcircular, presenting upward forward (*Gulo*) or more or less downward forward (*Mustela, Putorius*). Posterior nares thrown into one common conduit by absence of bony septum. Bony palate produced far back of molars,—half-way (more or less) to ends of pterygoids; interpterygoid space longer than wide. Auditory bullæ much inflated, with moderate constriction of the tubular meatus.† Little or no expansion of periotic region behind the bullæ, with which the paroccipitals appear in contact. Mastoids little developed, presenting outward or backward. Glenoid fossæ shallow, the anterior ledge slight; condyles never locked. Coronoid process of mandible erect, conical in profile, the posterior outline with forward upward obliquity (*Mustela, Putorius*,—more nearly vertical in *Gulo*), the apex in advance of the condyle. Feet with ordinary development and ratio of

---

* It is curious to observe that an aquatic species of *Putorius* (*P. vison*, the Mink) tends to approach the aquatic Otters (*Lutrinæ* and *Enhydrinæ*) in the relative shortness of rostrum, its less oblique truncation, flatness on top, &c.

† Here again the aquatic *Putorius vison* approaches the other aquatic species of different subfamilies in the comparative flatness of the bullæ.

digits; digits incompletely or not webbed. External appearance and habits variable, according to the genera and species, none strictly fossorial; progression digitigrade and subplantigrade; size from nearly the maximum to the minimum in the family; body never much depressed, nor tail conical or distichous. Perinæal glands moderately developed. No peculiar subcaudal pouch. Nature highly predacious.

2. MEPHITINÆ.—Teeth of ordinary Carnivorous pattern, 34 or 32 in number, according to varying number of premolars, whether $\frac{3-3}{3-3}$ (*Mephitis* and *Spilogale*) or indifferently $\frac{3-3}{3-3}$ or $\frac{2-2}{3-3}$ (*Conepatus*): the number unequal in the two jaws, $\frac{16}{18}$ or $\frac{14}{18}$. Incisors, canines, and molars as in the last subfamily. Molar of upper jaw quadrate, about as wide as long (varying in detail with the genera). Posterior upper premolar with a large inner shelf, giving a triangular shape to the tooth. Rostral part of skull moderately produced, and otherwise much as in the last (aperture of nares very oblique in *Conepatus*); cerebral portion as in *Mustelinæ*. Postorbital processes slight or obsolete. Anteorbital foramen very small, circular, sometimes subdivided into two or more canals. Posterior nares completely separated by a bony septum reaching to the end of the bony palate. Bony palate ending opposite last molars (*Mephitis*, *Spilogale*) or a little back of them, but not half-way to ends of pterygoids, (*Conepatus*). Auditory bullæ little inflated, with much constriction of the tubular meatus. Mastoids well developed, outward. Periotic region flattened and expansive behind the bullæ, the surface nearly horizontal, the paroccipitals remote from the bullæ. Glenoid shallow, presenting much forward as well as downward, without anterior wall, never locking condyle. Coronoid process of jaw conical in profile, erect, wholly in advance of condyle (except in *Conepatus*, which, in this respect, singularly resembles *Enhydra*). Feet with ordinary development and ratio of digits; digits not webbed. Form stout; tail very bushy; pelage long; colors black and white. Habits strictly terrestrial, more or less fossorial; progression plantigrade; movements slow. Size moderate and small. No peculiar subcaudal pouch. Perinæal glands extraordinarily developed, affording a means of offence and defence.

3. MELINÆ.[*]—Teeth of ordinary Carnivorous pattern, 34 in

---

[*] The characters here given are drawn entirely from the American genus *Taxidea*, and will require modification in order to their applicability to the subfamily at large.

number (in the North American genus); Pm. $\frac{3-3}{3-3}$; the number unequal in the two jaws, $\frac{16}{18}$; incisors, canines, and molars as in the last. Molar of upper jaw triangular, the long side postero-exterior. Posterior upper premolar substantially as in *Mephitinæ*. Rostral portion of skull as in the foregoing; cerebral portion conical, rapidly widening backward, with nearly straight lateral outlines. Postorbital processes moderately well developed. Anteorbital foramen large, subtriangular, presenting vertically. Posterior nares as in *Mephitinæ*. Bony palate produced back of the molars, as in *Mustelinæ*. Auditory bullæ very highly inflated, with little constriction across the short tubular portion. Periotic region much as in *Mustelinæ*, the paroccipitals close to the enormous bullæ. Mastoids moderately developed, outward. Glenoid fossa very deep, with prominent anterior as well as posterior walls, at length locking in the condyle. Coronoid process as in the foregoing. Feet with ordinary development and ratio of digits, not webbed. Body stout, extremely depressed; tail short, stout, flattened; size medium; snout somewhat hog-like. Progression plantigrade. Terrestrial and highly fossorial; fore claws highly developed. Perinæal glands moderately developed. A peculiar subcaudal pouch.

4. LUTRINÆ.—Teeth of ordinary Carnivorous pattern, 36 in number; Pm. $\frac{4-4}{3-3}$; the number equal in the two jaws, $\frac{18}{18}$; incisors, canines, and molars as before. Molar of upper jaw quadrate. Back upper premolar substantially as in *Mephitinæ* and *Melinæ*. Rostral part of skull extremely short, bringing the fore ends of the nasals nearly or quite opposite the anterior root of the zygoma, the sides of the rostrum erect, the top flat. Cerebral portion of the skull much swollen backward, with strongly convex lateral outlines. Postorbital processes variable (highly developed in the North American species, slight or wanting in some others). Anteorbital foramen very large, presenting obliquely downward as well as forward, circumscribed above by a very slender maxillary process. Posterior nares as in *Mustelinæ*. Bony palate produced far back of molars. Auditory bullæ very flat. Periotic region expanded, removing the paroccipitals from the bullæ, but the surface not horizontal as in *Mephitinæ*, but very oblique. Mastoids highly developed, downward. Glenoid much as in *Melinæ*, deep, sometimes locking condyle. Coronoid as in the foregoing. Feet with ordinary development and ratio of digits, which are fully webbed. Claws variable, sometimes rudimentary or wanting. Body stout, but elongate and cylin-

drical; tail long, conical, tapering, sometimes dilated, close-haired; muzzle very obtuse. Highly aquatic in habits. Pelage whole-colored.

5. ENHYDRINÆ.—Teeth very aberrant in general pattern, the molars and premolars without trenchant edges or acute angles, but tuberculous, 32 in number, and of equal number in both jaws, brought about by incisors $\frac{3-3}{2-2}$ and premolars $\frac{3-3}{3-3}$, the canines and molars remaining as before. Molar of upper jaw irregularly oval; back upper premolar defunctionalized as a "sectorial" tooth, and substantially similar to the molar. Proportions of rostral and cerebral parts of the skull substantially as in *Lutrinæ*, but rather an exaggeration of that conformation. Postorbital processes moderate. Anteorbital foramen very large, triangular, presenting downward and forward; the bridge over it very slender. Posterior nares as in *Lutrinæ*. Palate produced far back of molars; interpterygoid space very wide, the emargination rather wider than deep. Auditory bullæ, periotic region, mastoids, and glenoids as in *Lutrinæ*. Coronoid sloping backward, obtusely falcate, its apex overtopping condyle. Hind feet with extraordinary development and ratio of digits, being transformed into Seal-like flippers; otherwise general configuration and external appearance substantially as in *Lutrinæ*. Highly aquatic and marine.

ON THE ODORIFEROUS ANAL GLANDS OF THE MUSTELIDÆ.

Throughout this family of Carnivores are found special secretory apparatus in the perinæal region, which furnish a strongly odorous fluid. These glands are so highly developed, and play such a part in the economy of the animals, that special notice is to be taken of them. A classification of the *Mustelidæ* has even been proposed, based chiefly upon their modifications in the different genera. They early attracted attention, and have long been generally known to zoölogists. Quite recently a French anatomist, M. Chatin, has made them a special study, publishing a very important and interesting paper upon the subject.* This paper, so far as it relates to the *Mustelidæ* (for the author has studied the odorous anal glands of various other animals), I have translated for incorporation with the present work; under heads of the several species beyond will be found

---

* Recherches pour servir à l'histoire anatomique des glandes odorantes des mammifères. Par M.-J. Chatin. <Annales des Sciences Naturelles, 5ᵉ sér., tome xix, pp. 1-135, planches i-ix, 1874.

the matter relating to them. Here I introduce M. Chatin's descriptions of the parts as they appear in *Mustela foina*, for the same type of structure obtains throughout the subfamily *Mustelinæ*. I also bring in the author's résumé of the several modifications of structure found in the family at large, with extracts from his proposed classification of the family, as based primarily upon these organs, though I should add that I do not indorse his views without qualification.

1.—*Description of the glands in Mustela foina, as illustrating their structure throughout the subfamily Mustelinæ.**

The anal glandular apparatus being essentially the same throughout the *Mustelinæ*, the following description of the parts as they appear in *Mustela foina* will suffice:—

The anal orifice is found at the bottom of a fossa covered with thin, smooth, whitish integument, with a slightly raised border, the rudiment of a fold which is much more highly developed in the Skunk. At each side of this fossa, in a small special depression, in front of which this fold lies, is found an umbilicated papilla, through the narrow orifice of which the milky-whitish secretion of the anal gland exudes. Within the perinæum are two lateral masses, each as large as a small bean, bound together by one muscular envelope. The anal gland is 11 millimetres long and 6 across the middle. Upon removal of the muscular coat, which is rather delicate, the secretory part comes into view; its exterior is studded with nipple-like eminences; its substance is like that of the anal glands of most Carnivores. The parenchymatous tissue mainly consists of laminated fibres, elastic fibres, nerve tubes, and capillaries; the striped muscular fibres do not penetrate the substance of the organ. The culs-de-sac are of an average diameter of 0.04 millimetre; they are sometimes varicose or moniliform, and inclose a granular substance. In the middle of the gland is a small receptacle for the product of secretion, which is voided through a short duct opening on the edge of the anus, as above said.

It seems improbable that a scanty supply of merely disagreeably musky liquid can effectively answer in any way as a means of defence. The simple fact that it does not appear to be repugnant to the animals which may be supposed inimical

---

* For the modifications of the structure of the organs in Skunks and Badgers, see subfamilies *Mephitinæ* and *Melinæ*.

to Martens and Weasels, is sufficient to invalidate such a hypothesis. It is true that it is emitted when the animals are angered, terrified, or put in pain; but these are merely circumstances of irritation akin in many respects to other forms of excitement. It is more probable that the secretion subserves a purpose in the sexual relation, as it is undeniably a means whereby the sexes may discover and be attracted toward each other.

2.—*Résumé of the several types of structure of the odoriferous glands in Mustelidæ.*

The Ferrets and Martens exhibit one general plan of structure of the anal glands. At each side of the termination of the rectum, there is an oval body consisting of a tunic of muscular striped fibres enveloping a mass of glands, in the midst of which is a receptacle of variable capacity, containing a liquid differing little in its properties, which is poured out through a short duct opening upon a pore at each side of the anus.

In the Badgers, Skunks, and Ratels, there are decided modifications of this plan. In the last two named, the true anal glands alone exist, and these are quite different from those of the *Mustelinæ*. Instead of a thin and simple muscular envelope of the gland, we find a thick fleshy tunic, formed of two layers of interlaced fibres, capable of sudden strong compression of the receptacle. This latter is not a small simple sac with laminar walls, such as is found in the centre of the gland of *Mustelinæ*, but is an enormous reservoir, with a dense resisting fibrous coat, always containing a considerable quantity of the follicular product. The glandular substance is not spread all over this central capsule, but is restricted to a particular portion, and contrasts by its dark color with the white surface of the envelope of the pouch. The contents of the receptacle are sufficiently offensive to justify the profound and universal disgust which these animals excite in consequence of their curious and very efficacious means of defence. The voiding of the liquid must be sudden; and it does not suffice that the receptacle is large and powerfully muscular; the offensive liquid must be directed far backward, so as to flow as little as possible upon the rectal mucous membrane; consequently the opening is large and upon the summit of an umbilicated papilla, around which rests a cutaneous fold, which in a measure directs the discharge.

This general plan is further modified in the Badgers, where not only are there anal glands of a usual type, but also in their neighborhood is found, in both sexes, a racemose cluster of glands, the secretion of which is turned into the subcaudal pouch, which is generally described as appertaining to the anus; but its form is peculiar, and its contents, moreover, are of a different character from those of the anal glands proper. In some respects this pouch resembles the large reservoirs of viverreum of the Civets, and, as in these cases, is sparsely hairy. Thus the Badger is a special case in its own family, where it distantly represents, in this respect, the *Viverridæ*. These last have, in addition to anal glands, a secretory apparatus for special products, though even here species of *Herpestis* have anal glands like those of various *Mustelidæ*.

3.—*Résumé of M. Chatin's views of the classification of the family, as based on the odoriferous glands.*

"This is one of the least homogeneous families of *Carnivora*, if we include in it, after Van der Hoeven and others, such different animals as the Otter, Pole-cat, Badger, Skunk, Marten, and Ratel. It is surprising that types so distinct as these should have been suffered to remain thus far in an association as intimate as it is unphilosophical, and it is easily seen how Milne-Edwards was enabled to form three families out of the components of so miscellaneous an assemblage as that of the *Mustelidæ*. In the configuration of the limbs, as well as in their entirely peculiar habits, the Otters may represent one family (*Lutridæ*); then come the true *Mustelidæ*, embracing *Mustela*, *Putorius*, &c.; and, finally, the family *Melidæ*, consisting of *Mephitis* (with *Conepatus*, &c.), *Meles* (*Taxidea*, &c.), and *Mellivora*.

"Now, these three divisions correspond with as many modifications of the perinæal secretory apparatus: the two former, *Lutridæ* and *Mustelidæ*, offer in a general way a single pair of glands opening on the border of the anus, one on each side, furnished with a receptacle for the product of secretion.

"In the *Melidæ*, the Badgers on the one hand and the Skunks and Ratels on the other form two quite distinct sections. In these latter genera are likewise found a single pair of anal glands, but these are quite different from those of the *Mustelidæ*. The receptacle has a remarkable capacity; the follicular mass, instead of spreading over it, occupies but a small portion of its surface; while the secretion, which is always plentiful, here acquires an unparalleled fetor. In the Badgers, on the contrary, these anal glands are not the only secretory organs; there being in addition a particular subcaudal pouch surrounded by a racemose gland, which produces a peculiar liquid.

"This brief summary of the leading modifications of the perinæal glands of *Mustelidæ* suffices to show that several different types are included in that group . . . . ."*

ON THE EXTINCT MUSTELINÆ OF NORTH AMERICA.

The following fossil species of North American *Mustelinæ* have been described:—

1. Mustela mustelina, (*Cope*).

Aelurodon mustellinus, *Cope*, Palæont. Bull. no. 14, July 25, 1873, 1.
Martes mustellinus, *Cope*, Ann. Rep. U. S. Geol. Surv. for 1873, 1874, 520.
Mustela parviloba, *Cope* (change of name on reference of the species to *Mustela*).

Pliocene. Loup Fork epoch.

"A small, single-rooted second molar of the lower jaw. First molar sectorial, with a rather narrow posterior heel, one-third its length, and a small inner tubercle at the base of the second outer cusp. Last premolar with a short posterior heel, and distinct outer tubercle on the posterior side of the cusp. Margin of jaw strongly everted below masseteric fossa.

"*Measurements.*

	M.
"Length of three last molars	0.018
"Length of sectorial molars	.010
"Width of sectorial molars (greatest)	.005
"Height of posterior cusp (greatest)	.005

"This species was about as large as the domestic cat, and less than one-third that of *Aelurodon ferox*, Leidy." (*Quoted from the second reference above cited.*)

2. Mustela nambiana, (*Cope*).

Martes nambiana, *Cope*, Proc. Acad. Nat. Sci. Phila. 1874, 147.
"? Putorius nambianus", *Cope*.

From the Santa Fé (N. Mex.) Marls. Pliocene.

"Represented by a mandibular ramus which supports three teeth. The anterior blade of the sectorial is rather obtuse. The first premolar is one-rooted; the second and third are without

---

* But M. Chatin, regarding the family in the perspective of his special studies, may be considered not to have given due weight to other points of structure, the sum of which, as I believe, indicates that the *Mustelidæ*, as defined in the present work, are a homogeneous and natural assemblage of genera, of the grade usually held to represent family value.

posterior coronal lobe, but exhibit small basal lobes, both anterior and posterior. The anterior of the second is rather elevated, and the entire crown is directed obliquely forwards. Canine compressed. Mental foramina below the second and third premolars.

"*Measurements.*

	M.
"Length of three premolars	.006
"Elevation of anterior lobe of sectorial	.002
"Depth of ramus at anterior lobe of sectorial	.003

"This species is of smaller size than the *M. mustelinus*, Cope, and the sectorial tooth less elevated and trenchant."—(*Orig. descr.*)

3. Galera macrodon, *Cope.*

Galera macrodon, *Cope,* Proc. Phila. Acad. Nat. Sci. 1869, 155 (see also 132).—*Leidy,* Extinct Mamm. Dak. Nebr. 1869, 369, pl. xxx. f. 1, 2, 3.

Post-pliocene deposits in Charles County, Maryland, associated with remains of *Dicotyles torquatus* and a *Manatus*.

"This species is based on the greater portion of the right ramus of the mandible of an adult, containing three molars in place, the alveolæ [*sc.* alveoli] of the first and of the last, with a considerable portion of that of the canine.

"The alveolus indicates a canine of large size. The basis of the first premolar is turned obliquely outwards, and is two-rooted. The second and third premolars are separated by a space: they have well-marked cingula, but neither posterior nor internal tubercles. The sectorial is elongate, more than twice as long as wide, the inner tubercle well-marked, acute, the posterior lobe flattened, elongate; anterior lobe narrowed. Alveolus of the tubercular molar longitudinal, receiving a flattened fang with a groove on each side. Inferior face of ramus below anterior line of coronoid process, broad rounded, turned outwards. Masseteric ridge only reaching the latter below near the apex of the coronoid process, and not extending anterior to the line of the posterior margin of the tubercular molar. Ramus narrow at first premolar.

	In.	Lin.
"Length of ramus from posterior margin of canine to ditto of tubercular	1	5.5
"Ditto to posterior margin sectorial	1	3.
"Ditto third premolar		3.75
"Ditto sectorial molar		6.
"Width of same (posterior lobe)		2.8
"Depth ramus at posterior margin first premolar		7.5
"Ditto ramus at posterior margin sectorial		8.25
"Width ramus at posterior margin symphysis		4.5

"This species appears to have been perhaps rather larger than the Galera barbata (Gray) of Brazil, and of a rather more slender muzzle. As compared with that species, it exhibits many peculiarities. The third premolar is smaller, and the first, the sectorial, and the tubercular [are] relatively larger. In G. barbata, the first molar has but one root, and the mandibular ramus [is] thicker and deeper. The masseteric ridge advances to opposite the middle of the sectorial molar, and is continued on the inferior margin of the ramus, much anterior to its position in the G. macrodon.

"The discovery of this species adds another link to the evidence in favor of the extension of neotropical types* over the nearctic region during the post-pliocene epoch. Of thirty continental North American species enumerated by Leidy (Ancient Fauna of Nebraska, 9) all but thirteen may be said to be characteristic of that, or closely allied to the species of the present period of North America. Of the thirteen, one (Elephas) is characteristic of the old world, of one (Anomodon) affinities [are] unknown, and eleven are represented by members of the same family or genus now living in South America."—(*Quoted from the original article.*)

4. **Galera perdicida,** *Cope.*

Hemiacis perdicida, *Cope*, Proc. Acad. Nat. Sci. Phila. 1869. 3 (named, not described).
Galera perdicida, *Cope*, Proc. Amer. Philos. Soc. 1869, 177, pl. iii. figs. 1, 1 a.—*Leidy,* Ext. Mam. Dak. and Nebr. 1869, 445.

From limestone breccia, Wythe County, Virginia. Post-pliocene.

"This is a small carnivore of the Lutrine group of the Mustelidæ, apparently allied to Mephitis and Lutra. [The generic name given, however, is that of one of the *Mustelinæ*.] It is only represented by a left ramus of the mandible, with dentition complete. Its characters are as follows: Dentition $\frac{1}{3}, \frac{1}{1}, \frac{1}{3}, \frac{1}{2}$. The tubercular molar is relatively as in the allied genera, but without sharp tubercle; the sectorial characterizes the genus as distinct from the two mentioned [*Lutra* and *Mephitis*]. The posterior lobe is without the marked internal and external acute tubercle seen in *Mephitis*, nor the tubercular crest of Lutra, but is rounded and slightly concave. The median crests,

---

* "The genus Galera, Gray, is here regarded as distinct from *Galictis* Bell (Grisonia Gray), as it possesses an internal tubercle on the inferior sectorial, which is wanting in the latter."—(*Loc. cit.*)

inner and outer, are strongly developed, and with the anterior, quite as in Mephitis.

"The jaw pertained to an adult individual of smaller size than the common skunk, *Mephitis chinga*. The bases of the crowns of the first and second premolars, and to the outer side of the canine are surrounded by a well marked cingulum. The length of the crown of the molar is greater in proportion to the length [?] than in the skunk. The axis of the coronoid process is as in it, at right angles to that of the ramus. The latter is straighter on the inferior border than in the skunk, and exhibits a marked difference in the angle being nearly on the same line, and not raised above it, as in the species of American skunks and others, figured by Baird.

"*Measurements.*

	Lines.
"From angle to outer incisive alveolus	15.6
"Depth at coronoid	8.
"From base condyle to tubercular molar	5.
"Length of sectorial molar	3.6
"Width of sectorial molar	1.2
"Height from basal shoulder	2.
"Depth ramus at tubercular	2.7
"Depth ramus at Pm. 2	3.1
"Length of crown of canine	3.

"There are two mental foramina in the specimen, one below the third, the other below the first premolar. The crown of the canine is contracted and curved; slightly flattened on the inner side." (*Quoted from the original description.*)

I do not know the skull of *Galera*. As figured, the jaw of *G. perdicida* differs from that of *Mephitinæ* and *Lutrinæ*, as usually presented, in the straightness of the inferior border, agreeing in this respect with *Mustelinæ*. It closely resembles, among recent forms, the genus *Putorius*, from which, however, the character of the sectorial lower molar, with its strong acute inner tubercle of the middle lobe, as in *Mephitis* (and *Lutra*), perfectly distinguishes it. I should not be surprised, however, if the relationships of this form proved to be actually with *Mephitis*, especially with *Spilogale*. In a specimen of the latter before me from Georgia, the lower border of the jaw is quite as straight as that figured by Professor Cope; in size, the specimen agrees better with the figure than it does with some other specimens of *Spilogale* before me; the general shape is the same; there are two mental foramina exactly as described and figured; and

I fail to note, in the figure or description, any decided differences in dentition from *Spilogale*. In fine, it may be questioned whether "Galera perdicida" is even specifically distinct from *Spilogale putorius*. The fossil was found, it will be remembered, amongst remains of numerous species not distinguishable from recent ones.*

ON THE DERIVATION AND SIGNIFICATION OF THE NAMES APPLIED TO THE MUSTELIDÆ.

To treat of this interesting topic I cannot, perhaps, do better than give a version of Dr. E. von Martens's article, *Ueber Thiernamen*,† so far as it relates to the animals of the present family. This valuable article, as it seems to me, places the subject in a clear light, and gives, in a sufficiently concise and convenient form, just the information that is required for an understanding of the etymology and philological bearing of the names used in various languages to designate the species of *Mustelidæ*. Study of this subject, which is sadly neglected in ordinary zoölogical writings, is essential to the proper appreciation of the technical or binomial names; the older ones being, as will be seen, not necessarily of Greek or Latin origin, as commonly assumed. Thus, for instance, the generic name *Gulo* comes simply by translation into Latin of the Scandinavian and Russian names, which refer to the voracity of the animal.

DACHS [*Meles vulgaris*].—For this remarkable animal, no Greek name can be determined with certainty, although it is stated by late investigators, as Fiedler and Lindenmeyer, to exist in Greece; for it is at least a hazardous interpretation to identify the species with the τρόχος, "runner", of which Aristotle (Gen. 3, 6) speaks on the authority of Herodorus of Heraklea. The Latin *Meles* of Pliny, 8, 38, 58, is decidedly more certain: *sufflatæ cutis distentu ictus hominum et morsus canum arcent;* the Badger, of course, does not inflate its skin, but, nevertheless, its thick hide enables it to withstand bites and blows. Less pertinent is a passage in Varro *De Re Rust.* 3, 12, 3, where *maelis* is written. Isidor of Sevilla (seventh century

---

* Some time after the foregoing was written, I addressed to Professor Cope a note on the subject, stating my views; and in reply I learned that Professor Cope "had for some time suspected" that the animal was a *Mephitis*.

† "Ueber Thiernamen." Von E. von Martens in Berlin. In: Der Zoologische Garten; the portions relating to the *Mustelidæ*, here translated, being at pp. 251-256 and pp. 275-281 of Jahrg. (or vol.) xi (1870).

after Christ) writes *melo*, genitive *melonis*; and, in the vicinity of Bologna, according to the statement of Diez, the Badger is still called *melogna*. Elsewhere, however, this word is obsolete, being replaced in the living European languages by various others, entirely different.

The German word *dachs* may be traced back to the early period of the Middle Ages: in the quack prescriptions of Marcellus of Bordeaux, in the ninth century, is found *adeps taxoninus*, Badger's fat, and *taxea*, used by the above-mentioned Isidor as the definition of *adeps*, fat, with reference to a still earlier author, is probably the same; the short form *das*, as the word still runs in Dutch, is found in the German vocabulary of the ninth century; the nun Hildegard, in the twelfth, wrote *dahsis*; Albertus Magnus, in the thirteenth, *daxus*. The form *taxus* or *taxo*, as a name for the animal itself, occurs in the Latin vocabulary from the period of the eighth century; it may be that this term is related to the pure Latin name of the yew-tree, *taxus* of Cæsar and Virgil (*Taxus braccata* Linn.), agreeably to which the initial *t* straightway becomes fixed in the Romanic names of the animal, in the Italian, *tasso*; in the Spanish, *tejon* (and *tesajo*, smoked meat); the Portuguese *texugo*; while the Old French had its *taisson*, of which only *tanière* (from *taisniere*), meaning particularly a Badger-burrow, and, generally, the den of a wild beast, remains in modern French. The poet Tasso, and the founder of the German postal system, Taxis, derive their family name from *dachs*, Badger, as the old Roman agitator Sp. Maelius probably also did. The word itself may be originally German, and have become naturalized in France, Spain, and Italy with the migrations of German races. To derive it from the Sanskrit *taksha* (Greek τέκτων), a carpenter, to be taken in the sense of an architect, is rather far-fetched. Another series of names of the Badger in Northern Europe begins with *B*, as the French *blaireau*, the English *badger*, the Danish *brok*,* and the Russian *borsuk*; but it is not certain that these are all etymologically related. *Blaireau*,† in Middle-Age Latin *blerellus*, is interpreted by Diez as the diminutive of the mediæval Latin *bladarius*, a grain-merchant (Romanic *biado*, late French *blé*, grain); and in support of this it is argued that the English name of the animal, *badger*, signifies also a dealer in grain. Such connection requires us to

---

* "Brock" is also found as an English provincialism.—Tr.

† Which is corrupted, in America, into Braro, Brairo, and Prarow.—Tr.

invent the certainly erroneous explanation that the animal lays up a store of provisions in its domicile, as if it drove a trade in grain. Diefenbach's derivation from the Celtic, originally Cymric, word *blawr*, gray, seems to me to be nearer the mark; it would then be "the little gray beast"; and it is corroborative of this that the animal is called, in Picardy, *grisard;* in Sweden and Denmark, *gräving* or *gräfling*, that is to say, *Grauling*, "a gray or grizzly beast". But the proper Celtic name of the animal is *broc;* in the Gaelic, Irish, and Bretonic remarkably like the Danish *brok*, and somewhat similar to *borsuk*, which prevails in Poland, Russia, and Siberia; there this name for the Badger is current among the Bashkirs, Kirghiz, and Buchares, and is rendered *borz* by the Magyars; so we may consider it a primitive Turanian word, the more so since the South Sclavonic uses another term, in Carniola, *jozarec* or *jasbez;* in Bohemia, *gezwec*. The Wallachian, *jezure* or *esure*, which has been incorrectly considered as from the Latin *esor*, eater, is probably related.

VIELFRASS [*Gulo luscus*].—According to the latest investigations, the Glutton inhabited Middle Europe nearly to the Alps, in the period of the Lake-dwellers (*Pfahlbauten*, literally pile-buildings), together with the Reindeer; and of its occurrence in Germany, even in the last century, two cases are given, one at Frauenstein in Saxony, by Klein, 1751, the other at Helmstädt in Brunswick, by Zimmermann, 1777, both, unfortunately, without the particulars. Though both these zoölogists saw the stuffed specimen, neither gives the date of capture, the first only stating that it occurred under Augustus II, who died in 1733. These can only have been stray specimens, since no contemporaneous or previous writer mentions the occurrence of the animal in Germany. The species was entirely unknown in the Middle Ages, making its first appearance in literature through Michow, a physician of Cracow (de Sarmatia Asiana et Europæa, 1532), as Lithuanian and Moscovitic, and through Bishop Olaus Magnus, of Upsala, 1562, as an animal of North Sweden, thus nearly at the limit of its present distribution. What we can gather from the name of the animal accords perfectly with this. In Europe, names are only found in the vernacular proper of Scandinavia and Russia, *järf* or *jerv* of the former, and *rossomaka* of the latter, both of which are given by the above-mentioned historians; all German, French, Latin, and such, are book-names, intended to denote the voracity of the animal, and point back to the well-known account of Olaus, as the German *Vielfrass*, the Latin *Gulo*, the French *Glouton*, the English *Glutton*. It has often been asserted that the German *Vielfrass*, in the sense of *glutton*, is a misunderstanding, it being derived from the Swedish word *fjäll*, Norwegian *fjall*, rock or cliff; but this I cannot credit, first, because the second syllable is not accounted for on such supposition (*fjäll*—järf is remote, and the animal is nowhere so called, but simply *järf*); secondly, because both the Swedish Olaus Magnus and the Norwegian Bishop Pontoppidan give its voracity special prominence, and from this trait derive the name jerf (*gierv*, "gierig", greedy?), translated *Gulo* and *Vielfrass*. Another Norwegian clergyman, H. Ström, gives, indeed, the designation *Fieldfrass*, besides *jerf*, to the animal, which is of rare occurrence in his locality, but with the explicit remark that *Fieldfrass* was, beyond doubt, derived from the German word *Vielfrass*. This is thus exactly contrary to the usual German acceptation; and, in fact, "Felsenfrass" would be a singular appellation.

ZOBEL [*Mustela zibellina*].—The name appears as early as the latter half of the Middle Ages, under many variations, as the modern Latin, *sabelus, zibellina;* German, *zebel* (as early as the ninth century, according to Graff), *zobel;* Provençal, *sebeli;* English and old French, *sable;* Swedish, *sabel;* Russian, *sobol;* Finnish, *soboli*—in every case meaning a northern peltry. In the East, we find another variation, *samur*, in the Crimea and Armenia, and thence to Servia and Wallachia. The name is probably of Turanian origin.

MARDER [*Mustela martes, M. foina*].—This word now occurs in Germanic and Romanic languages, in both either with or without the second R, as the Spanish and Portuguese *marta*, in the former as a feminine noun, and likewise the French *la marte*, though in some dialects *la martre*, the Provençal *mart*, Italian *martora* and *martorella;* the English *martin* [or, oftener, *marten*—TR.] appears to be an easy way of saying *martern*, still in use in some localities; Dutch *marter*, Swedish *mard*, Danish *maar*. Seeking for the earliest form of the word, we first find *martes* in Martial, the Spanish-born Roman poet; but this can scarcely be an old Latin word, as it is not found in Pliny or other classical writers; and Martial often introduced foreign words into his Latin. In Anglo-Saxon, it only appears as *meardh;* whilst, on the other hand, in Germany, we find *martarus* used by Hildegard and Albertus Magnus, in the twelfth and thirteenth centuries. The resemblance to the German verb "martern" [to torment] is obvious; in fact, "martern" might be defined " to act like a marten ", the proper implication being, not the sanguinary murders the marten commits, but the palpable torment which it designedly inflicts. Another derivation comes decidedly nearer—*martyr*, meaning a person tortured, from *martyrium*, torture, whence the verb first arose. The resemblance in sound may have occasioned the second R in those cases in which it appears. We might also seek to establish a connection between " marder ", a marten, and " Mörder ", German for a murderer; but the T, which occurs in a majority of the forms of the word, is against this, as is also the fact that the German name occurs in many languages to which " Mord " and " Mörder " do not belong.

A second Romanic name of the Marten is *faina;* Spanish and Italian the same, Portuguese *fuinha,* French *la fouine;* in some dialects with *a* in place of *u,* as in certain Italian localities *faina,* in Provence *faguino, fahino,* Old French *fayne;* Catalonian *fagina,* Belgic *faweina,* in the Canton of Graubündten further modified into *fierna.* The obsolete German names of certain pelts, *Fehe, Feh-wamme,* are very likely related. The word is not Latin as the name of an animal; but it may be inquired, with respect to the later forms, whether it does not probably signify *marta fagina,* Beech-marten, as one of the two European species of the genus is often named; properly the Tree- or Pine-marten, in distinction from the Stone- or House-marten, since the former lives in the forest, the latter about buildings; though very curiously, the Stone-marten [*Mustela foina*] is the *Martarus* or *Martes fagorum* of Albertus Magnus and afterward of Ray, whilst the Pine-marten [*M. martes*] is distinguished as *M. abietum,* " Marten of the firs". The precise distinction between *fouine, foina* = Stone-marten, and *marte, martes* = Pine-marten, moreover, may have been first set forth by Buffon and Linnæus, and have obtained rather among zoölogists than among the people at large; the more valuable Pine-marten ["Edel-marder", literally "noble marten"] took the commonest name, leaving the less popular one for the other rarer species. From this *fouine,* the French have formed the verb *fouiner*, to pry into or rummage about.*

The Celtic, Sclavonic, and Finnish names are entirely different, as are the Cymric *bela*,† the Russian, Polish, Bohemian, and Crainish *kuna*, Finnish and Laplandish *nätä*; with which the Magyaric *nyest* or *nest* accords.

ILTIS [*Putorius fœtidus*].—The German name is found under many variations, according to localities, particularly in North Germany, as *iltnis*, *eltis*, Danish *ilder*, Swedish *iller;* furthermore, with *k*, *ilk*, *ulk*, according to Bechstein in Thuringia even *Haus-unk*, which is the well known name of a reptile [toad]; and again with *b*, *elb-thier*, *elb-katze*, which has been sought to be derived from *elben* = elves, the nocturnal sprites; but the oldest form of the word known to me, *illibenzus* of Albertus Magnus (thirteenth century), is little unfavorable to this etymology. The Dutch *bunsing* stands entirely alone. The Romanic languages name the species simply from its bad smell, as the Italian *puzzola*, French *putois*, mediæval Latin *putorius*, the *pusnais* of French animal-fable, which is the same as *punaise*, a bed-bug. The second portion of the English name, *pole-cat*, is of obvious meaning; agreeably to which we find in Diefenbach (Celtica, ii, p. 435) that in Wales, in early times, the animal was kept, or, more likely, suffered to remain, about houses, to destroy mice.‡ Another English name, *fitcher*, *fitchet* [or *fitch*—TR.], related to the old French *fissan*, apparently indicates the same capacity in which the animal was employed or regarded. The Sclavonic languages have a particular word, *tschor*, *tschorz*, or *tscher*, in Carniolan *tvcor*, in Roumanian *dihor*.

By Pliny (8, 55, 84), this species is called *viverra*, probably an Iberian word no longer occurring in later languages, and which Linnæus first reapplied in zoölogy to the Civet-cats. Since the Middle Ages, however, two forms of the name of this animal have simultaneously appeared, the first without *t*, *furo* of Isidor of Sevilla (seventh century), whence the present Por-

---

* "Durchsuchen, durchstöbern"; so defined by the writer, but other authority defines *fouiner* to slink off, to sneak away; used only in trivial style. But either meaning is sufficiently characteristic of the animals.—TR.

† Obviously related to the modern French *belette*—see beyond.—TR.

‡ The whole English word, *pole-cat*, is by some simply rendered "Polish-cat", as if the animal were originally from Poland. In America, the word has been very commonly transferred to the Skunks, *Mephitis:* Catesby's polcat is such, and Kalm's fiskatta is translated *pole-cat*.—TR.

rtuguese *furqo*, and the Spanish *huron*, transferred by the Spanish colonists to the South American *Galictis vittata*, and the North American *Mustela huro* Fr. Cuv., and *furetus* of the Emperor Frederick II, considered as French by Albertus Magnus, with which the present French *furet*, English *ferret*, Celtic *fured* and *fearaid*, German *frett*, are all related. The *-et* may be a diminutive form, or be a part of the original word; it is slighted by the etymologist Isidor, who somewhat gratuitously finds in it the Latin *fur*, thief. The word cannot be Arabic, for Isidor died in 636, before the irruption of the Arabs into Africa. But if, as Shaw states, the Weasel is called *fert* in Barbary, the probability is that the word, like others, is common to the North African pre-Arabic and the Iberian pre-Romanic languages, and that it is this very animal which Strabo calls the North African (Libyan) Weasel.*

WIESEL [*Putorius vulgaris*].—This word is found in most of the Germanic languages: Swedish *wessla*; English *weesel* or *weasel*; Dutch *wezel*. It may be traced back to late mediæval German and Anglo-Saxon. The Swabian verb *wuseln*, to skip about ("*sich rasch bewegen*") like any small creature, may readily be derived from *wiesel*, notwithstanding the difference in the vowel. In this case again, as in the instance of *dachs*, the same word recurs in Spanish, but without the diminutive termination, as *veso*. It is found in mediæval Latin of the twelfth century, and was by the Romanic colonists bestowed upon an American Musteline animal (*Putorius vison*, the representative of the European Mink). The ordinary French term for the Weasel, *belette*, is diminutive of the old French *bele*, from the Celtic and the present Welch *bela*, a marten, and also occurs under a different modification in North Italy, which was certainly once inhabited by Celts. It may all the more readily have been preserved in French, since it may be considered related to *belle*, pretty, and be so interpreted. Certainly in many languages the Weasel derives its name from its neat and elegant ways, as the Italian *donnola* and Portuguese *doninha*, little lady; the Spanish *comadreja*, god-mother; the

* According to Rolleston (Journ. Anat. and Phys. i. 1867, p. 47 *seq.*) the Cat and the Marten were both domesticated in Italy nine hundred years before the period of the Crusades, and the latter, *Mustela foina*, was the "cat" or γαλῆ of the ancients, who, furthermore, called *Mustela martes* γαλῆ αγρία, and designated *Viverra genetta* as ταρτησσία γαλῆ.—TR.

andereigerra* of the inhabitants of Biscay, meaning the same as the Portuguese word just given; the late Greek νυμφιτα, νιφιτζα, a bride; the Bavarian *Schönthierlein*, "pretty little creature"; the English *fairy* (Diez). The Sclavonic tongues have an entirely peculiar series of names: *laska, lasika, lastiza,* and the like.

In Greek and Latin proper, we find for the Musteline animals only three names, which are all different from those which are better known in living languages, and of the present existence of which we only find isolated instances; these are ἰκτίς, γαλέη, and *mustela*.

Pliny uses *mustela* in different places for native and exotic *Mustelidæ*, without furnishing the means of nicer discrimination of the species; he indicates their mousing capacity; and Palladius *De Re Rust.* 4, 9, 4, says that they were kept for this purpose. The name appears to be derived from *mus*, and to mean "a mouser"; for I cannot agree with Sundevall in recognizing in the second syllable the Greek θήρα, a hunt; since ϑ does not become *t* in Latin. According to Risso, the Weasel is called *moustelle* to this very day in Nice, and in Lorraine, according to Diez, *moteile*; this is a partial persistence of the name which, among the Romans, not only indicated the Weasel as the species best known to them, but also included the other Musteline animals in general. So it was also with the Greek γαλέη (Batrachomyomachia, 9) or γαλῆ (Arist. Hist. An. 2, 1, and his not very well written book 9, chap. 6), the best-known Greek species of the Marten family, yellowish, white beneath, and a mouser; whilst the fable that it was a transformed maiden (Ovid, Metam. 9, 306–323; Galanthis, with the express statement that the beast still lived about houses) accords well with the complimentary names already mentioned. Thus *mustela* is primarily our Weasel [*Putorius vulgaris*], though occasionally other species receive the same name, as, for example, an African one, in Herodotus, 4, 192. More difficult to explain is the second Greek name, ἡ ἰκτίς, the skin of which, according to Homer (Iliad, 10, 333), made a night-cap for a Trojan hero, and which, according to Pseudo-Aristotle, Hist. An. 9, 6, was of the size of a small Maltese dog ("*Malteser Hündchens*"), like a Weasel, white underneath, and fond of honey. This latter circumstance caused Cetti to sep-

---

* Precisely the same as the Latin *muliercula*.—Tr.

arate his *boccamela* ("honey-mouth"—as we should say, "having a sweet tooth"), which is, however, a species scarcely distinguishable from *P. vulgaris* (*cf.* Zool. Gart. 1867, p. 68). Aubert and Wimmer, on the other hand, argue for *Mustela foina*, as this animal is common in Greece, where it is still called ἰκτίς; the latter position is certainly well taken, and the Marten, as the larger animal, better fulfils the Homeric indication just given; but the expression "white underneath" is only true of the throat of the Martens, for both species of *Mustela* are dark-colored on the belly, and in this respect very different from the Weasel. For the rest, it is much more probable that Aristotle named both the Marten and the Weasel together, than that he distinguished two kinds of Weasels and knew nothing whatever of the Marten.

HERMELIN [*Putorius erminea*].—Though this name sounds like a foreign word, it is nevertheless probably of German origin, since not only are there several provincial variations of less strange accent, like *Heermänchen* and *Härmchen*, but there is also the simple *harmo* of old German manuscripts of the ninth to the eleventh century (Graff, althochdeutscher Sprachschatz). From this came *harmelin*, of the twelfth century, simply the diminutive. The name went with the peltries into foreign lands, becoming the Italian *armellino*, the Spanish *armino*, the French [and English] *ermine*—originally, with Albertus Magnus, who had many French forms of names, *erminium*,—and came back to the German as *Hermelin*, with a foreign accent, on the last syllable. The she-fox Ermeleyn, in the Fable of "Reinecke Fuchs" ["The Beasts at Court"], obviously derived her name from this animal. In Lithuanian, we find *szarmu* or *szarmonys* as the name of the same animal, which is the same as *harmo*, according to the rules for the rendering of the sound, just as the Lithuanian *szirdis* is the German *herz*. The interpretation of *Hermelin* as the "Armenian Mouse" is thus virtually refuted. The Swedes call the animal *ross-kat* and *le-kat*, the latter probably shortened from *Lemmingskatze*, since the creature is destructive to Lemmings. In North France, we find for the Ermine the name *roselet*, obviously indicating its reddish color, and with this corresponds the fabulous name Rüssel, offspring of the Ermeleyn. The South European languages have no special name of their own for the Ermine, since it is there found only in the mountains, as the Southern Alps and the Balkan for example.

NÖRZ [*Putorius lutreola*].—This animal is at once proclaimed to be East European by its name; for the word, first used in Germany by the Saxon mineralogist Agricola, in 1546, is Sclavonic; the Russian is *norka*, the South Russian *nortschik*, the Polish *nurek*, from the verb *nurka*, to dive. The Swedes alone, in whose country the animal also appears, have a particular name for it, *mänk*, which is the source of the *mink* or *minx* applied to the different North American species [*P. vison*].

OTTER [*Lutra vulgaris*].—To the comparative philologist this word offers a field as broad as it is difficult, for the names of the animal in various European languages are enough alike to be compared, yet sufficiently dissimilar to be questioned as the same word; the initial particularly differs in a suspicious manner: *otter*, *lutra*, ἐνυδρίς. In Sanskrit and Zend,* we find for an aquatic animal, of what kind is not known with certainty, but which may easily have been the Fish-otter, the name *udra-s*, derived from the root *ud*, water (Latin *udus*, Greek ὕδωρ). With this agrees perfectly the Lithuanian *udra*, the Curlandic and Livonian *uderis*, and, with slight change of the initial, *wydra*, which obtains throughout the Sclavonic tongues, the Roumanian *vidre*—all of which are actual names of the Otter. In the Germanic languages, the *u* becomes *o; otr* in the old Northern sagas, *ottar* in old mediæval German, *otter* in the present German, Dutch, Danish, and Swedish, though in the latter the early initial *u* sometimes reappears, giving *utter*. The change of *d* into *t* is the rule in the rendering of the sound of Sanskrit, Greek, Lithuanian, and Sclavonic in the Germanic languages, although in pure German this consonant properly changes into sharp *s* (ὕδωρ, water—"*wasser*"), as is not, however, the case with the name of the animal.

In Greek, we find, as the name of the Otter, ἐνυδρίς, Herod. 2, 72, and 4, 109, ἐνυδρὶς, Arist. Hist. An. 1, 1, and 8, 5, or ἔνυδρος, Aelian Hist. An. 11, 37, nearly always mentioned in connection with the Beaver; also the forms, agreeing better with the Sanskrit, ὕδρος, ὕδρα, the former for an actual serpent (Ilias, 2, 723, Arist. Hist. An. 2, 17, 83), the latter for a fabulous serpent like monster (Hesiod, Theogon. 413, &c.).

In Latin, we find only *lutra*, Plin. 8, 30, 47, which differs not only in the initial, but also in the *t*, though the Latin should agree with the Greek and Sanskrit and differ from the Ger-

---

*Zend: the language of the Avesta, or ancient sacred writings of the Persians. The people who used it were a branch of the Asiatic Aryans.—TR.

manic in respect to the consonants. This *lutra* obtains in modern Romanic languages with little variation; French, *la loutre;* Italian and Portuguese, *lontra;* Asturian, *londra;* in some Italian dialects, *lodra, ludria* (preserving the primitive *d?*), and *lonza* (which bears lightly upon the name *unze* [cf. *onza, onça, ounce*] among the cats [*Felidæ*]; Provençal, *luiria* or *loiria*. The *n* in many of these names may simply be a matter of easy pronunciation. Curiously enough, we find in Norway, far removed from Romanic influence, a name of the Otter of similar sound, *slenter*.

The Spaniard says *nutria*. This may be an arbitrary corruption of *lutra;* but when we recall the Greek ἔνυδρις, and consider that many Spanish names of animals are nearer the Greek than the Latin (for example, *golondrina*=χελιδών [a swallow], and *galapago* in the first two syllables = χελώνη [a turtle]), it seems very likely that *nutria* is derived from ἔνυδρις; and it may be seriously questioned whether the latter is actually compounded of ἐν and ἴδωρ, not rather that the ν represents the *l* in *lutra*, and that the ε is simply a prefix, as in ἔλαχυς = the Sanskrit *lag hus*=the Latin *levis*. Initial *l* and *n* are sometimes interchangeable, as for instance in the Greek λίτρον and νίτρον, the Latin *lamella*=the Provençal *namela* (Curtius, Griechische Etym. 395). The primitive Indo-Germanic word from which all the above are conjecturally derived probably did not begin with a pure vowel, since a consonant precedes it in so many of the foregoing forms, as the *v* in Sclavonic, the *l* in Latin, and the rough aspirate in Greek.

The German word *otter*, when it signifies a snake, is feminine; when used for the quadruped it is indifferently masculine or feminine. The former is justifiable, inasmuch as the old Northern *otr* or *otur* is masculine; to make it feminine may be partly on account of its identity with the name of the serpent, partly from its analogy with the Romanic *lutra*. Albertus Magnus furthermore converted *lutra* into the masculine form, *luter*. In the Middle Ages, finally, there arose the Latin word *lutrix*, as the name of a snake, formed from *lutra* by analogy with *natrix*, and apparently furnishing an imitation of the double employ of *otter*.

On account of its similarity in form and its kindred signification, I cannot refrain from mentioning in this connection the word *natter* [viper, a kind of snake]. In spite of the Spanish *nudria*, I believe that it has nothing to do with *otter*, though the two are often confounded by persons not learned in natural history, or considered of similar signification. It is an old word, appearing in the Latin of Cicero as *natrix* (Qu. Acad. 2, 28); in the Gothic of Ulfilas as *nadrs*, masculine moreover, Ev. Luc. 3, 7, where the Greek text has ἔχιδνα, and Luther translated "*otter*", but at that time already feminine in the old Northern *nadhra*. The same word is also found in Celtic. This wide diffusion of the word makes it probable that the Latin *natrix* is not to be interpreted as a swimmer, as if from *nare* =*natare;* in general, people take "*natter*" for a poisonous serpent, not simply as a water-snake, and the specific application of the term to the *Coluber natrix* Linn. is of later origin. Many philologists derive the word from an old root, *na* (German *nähen*, Latin *neo*, Greek νέω), in the sense of coiling ("*umschnüren* "); cf. Latin *necto*.

We may briefly treat of other names of the Otter. The Celtic languages have a particular term, Gaelic *dobran*, Cymric *dyfrgi*. The Tartaric *kama* has probably given name to the largest tributary of the Volga. In many, particularly Asiatic, languages, our animal is called by some equivalent of "water-dog" or "river-dog"; as in the Dekan *pani-cutta;* in the Canaries (and also in the East Indies), *nir-nai;* Malayan, *andjing-ayer;* whilst the κύνες ποτάμιοι of Aelian, 14, 21, appear to have been Otters.

CHAPTER II.

Subfamily MUSTELINÆ: The Wolverene.

The genus Gulo—Generic characters—Gulo luscus, the Wolverene—Synonymy—Habitat—Specific characters—Description of external characters—Measurements—Anal glands—Description of the skull and teeth—Measurements of skulls, European and American—Nomenclature of the species—Relation of the European and American animal—General history, geographical distribution, and habits of the species—Its distribution in the Old World.

HAVING already presented the characters of the subfamily *Mustelinæ* with detail sufficing for present purposes, I may at once proceed to consider the genera composing the group. These are: *Gulo; Galictis; Mustela; Putorius.* The second of these is not represented in North America. *Putorius* is susceptible of division into several subgenera. These genera will be treated in successive chapters, the present being devoted to the genus *Gulo*.

The Genus GULO. (Storr, 1780.)

< **Mustela**, *Linn.* Syst. Nat. i. 10th ed. 1758, and of many authors.
=< **Ursus**, *Linn.* Syst. Nat. i. 10th ed. 1758, and of some authors.
< **Meles**, *Pall.* Spic. Zool. xiv. 1780; also of *Bodœrt*, 1784.
=**Gulo**, *Storr,*[*] Prod. Meth. Mamm. 1780, and of late authors generally. (From Klein.)
< **Taxus**, *Tiedem.* Zoöl. i. 1808.

[*] This extremely rare work has lately been made the subject of a critical essay by Prof. T. Gill, who examined a copy in the library of the Surgeon-General, U. S. Army, at Washington ("On the 'Prodromus Methodi Mammalium' of Storr". By Theodore Gill. Extracted from the Bulletin of the Philosophical Society of Washington, October, 1874. Philadelphia: Collins, printer, 1876. 8vo. pamph., 1 p. l., pp. i–xiii). The full title, as quoted by Gill, is as follows:—

PRODROMVS METHODI MAMMALIVM.— — | Rectore Vniversitatis magnificentissimo | serenissimo atqve potentissimo | dvce ac domino | Carolo | dvce Wvrtembergiæ ac Tecciæ regnante, | rel. rel. | — | Ad institvendam | ex decreto gratiosæ facvltatis medicæ | pro legitime conseqvendo | doctoris medicinæ gradv | inavgvralem dispvtationem | propositvs | præside | GOTTL. CONR. CHRIST. STORR | medicinæ doctore, hvivs, chemiæ et botanices | professore pvblico ordinario | vniversitatis H. T. pro-rectore, | respondente | Friderico Wolffer, | Bohnlandense. | — | *Tvbingæ, d. Jul. MDCCLXXX.* | — | Litteris Reissianis. [4to, 43 pp., 4 tables.]

GENERIC CHARACTERS.—*Dental formula:* i. $\frac{3-3}{3-3}$; c. $\frac{1-1}{1-1}$; pm. $\frac{4-4}{4-4}$; m. $\frac{1-1}{2-2} = \frac{18}{20} = 38$ (as in *Mustela*). Sectorial tooth of lower jaw (anterior true M.) without an internal cusp (usually evident in *Mustela*). Anteorbital foramen presenting obliquely upward as well as forward, canal-like, and opening over interspace between last and penultimate premolars. Skull little constricted at the middle; rostral portion relatively shorter, stouter, and more obliquely truncated anteriorly than in *Mustela*. General upper outline of the skull in profile more arched. Mastoids and auditory tubes more produced, the whole periotic region decidedly more prominent. Zygomatic arch very high behind, at first ascending vertically, then giving off a posterior convexity. Depth of emargination of palate about equal to distance thence to the molars. Skull, as a whole, massive, finally developing strong ridges.

Vertebral formula: c. 7; d. 15; l. 5; s. 3; cd. 15 or 16. (*Gerrard*.)

Size much above the average for this family, and nearly at a maximum (*Galictis* alone, of this subfamily, is said to be larger). Form very stout, and general appearance rather Bear-like than Weasel-like; organization robust. Legs short and stout. Tail short (about as long as the head), bushy, with drooping hairs. Pelage shaggy. Ears low. Soles densely hairy, with six small naked pads. Claws strong, acute, much curved. Coloration peculiar.* Anal glands moderately developed. Progression incompletely plantigrade. Habits chiefly terrestrial.

Notwithstanding the remarkably peculiar outward aspect of *Gulo* in comparison with its allies, it is very closely related to the Martens in structure, forbidding more than generic distinction from *Mustela*. The dental formula is the same. In addition to the cranial characters above given, it may be stated that the skull is relatively as well as absolutely more massive than that of the arboreal Martens, in coördination with the much more robust and sturdy organization of the Wolverene.

Detailed descriptions of the skull and teeth, as well as of the external characters of the genus, are given beyond under the head of the single known species, *G. luscus*.

The generic name is the Latin *gulo*, a glutton, in allusion to the voracity of the animal. The obvious relation of the word is with the Latin *gula*, throat or gullet, also used figuratively for appetite or gluttony; and in various languages the vernacular name of the species is a word of similar signification. "Gulo" was the original specific name in the binomial nomenclature; but its application to the present animal was originally simply by translation into Latin of the Scandinavian and Russian vernacular (cf. *anteà*, p. 22).

---

*In the pattern of coloration, however, we discern the trace of the same character that is fully developed in *Mephitis mephitica*—the light bands, converging over the rump, being similar to the stronger white stripes which mark the Skunk.

The Wolverene.

Gulo luscus.

PLATE I.

(A. Old World references.)

Gulo, *antiquorum.*—"*Gesn.* Quad. Vivip. 1551, 623, fig.—*Ol. Mag.* Hist. Gent. Sept. 1555, 605.—*Aldrov.* Quad. Dig. 1645, 178.—*Scheff.* Lappon. 1673, 339.—*Charlet.* Exercit. 1677, 15.—*Rzacz.* Hist. Nat. Polon. 1721, 212.—*Linn.* S. N. 2d-5th eds. 1740-7, 44.—*Klein,* Quad. 1751, 83, pl. 5.—*Hill,* Hist. An. 1752, 546, pl. 27.—*Jonst.* Theatr. 1755, 131, pl. 57."

Mustela rufo-fusca, medio dorsi nigro, *L.* Fn. Suec. 1st ed. 1746, 2, no. 6; S. N. 6th-7th eds. 1748, 5, no. 1.—*Kram.* Elench. Veg. et Anim. 1756, 311.

Mustela gulo, *L.* Fn. Suec. 2d ed. 1761, 3, no. 14; S. N. i. 10th ed. 1758, 45, no. 3; S. N. i. 12th ed. 1766, 67, no. 5.—*Gunn.* Act. Nidros. i21, pl. 3, f. 5.—*Houtt.* Natura. ii. ——139, pl. 14, f. 4.—*Mull.* Zool. Dan. Prod. 1776, 3, no. 11.—*Erxl.* Syst. An. 1777, 477, no. 15.—*Fab.* Fn. Grœnl. 1780, 21, no. 12.

Ursus gulo, *Schreb.* Säug. iii. 1778, 525, pls. 144 (Act. Holm. 1773) and 144* (Buff.).—*Zimm.* Geog. Gesch. ii. 1780, 276, no. 162.—*Gm.* S. N. i. 1788, 104, no. 8.—*Shaw,* G. Z. i. 1800, 460, pl. 104.—*Turt.* S. N. i. 1806, 64.—*Cuv.* "Tabl. Élém. ——, 112."—"*F. Cuv.* Dict. Sci. Nat. xix. 79, f. —."

Meles gulo, *Pall.* Spic. Zool. xiv, 1780, 25, pl. 2; Z. R. A. i, 1831, 73, no. 20.—*Bodd.* Elench. An. i. 1784, 81, no. 5.

Taxus gulo, *Tiedem.* Zool. i. 1808, 377.

Gulo borealis, "*Nilss.* Illum. Fig. till Skand. Fn."—"*Retz.* Fn. Suec. 1800, 25."—*Cuv.* R. A. i. 1817, —.—*Wagn.* Suppl. Schreb. ii. 1841, 246.—*Keys. & Blas.* Wirb. Eur. 1840, 66.—*Schinz.* Syn. Mamm. 1844, 347.—*Blas.* Wirb. Deutschl. 1857, 209, figs. 119, 120 (skull).—*Brandt.* Bemerk. Wirb. N. E. Russl. 185-, 20.—*Gray,* P. Z. S. 1865, 120.

Gulo sibiricus, *Pall.* "Sp. Zool. xiv. t. 2".—(*Gray.*)

Gulo arcticus, *Desm.* Mamm. i. 1820, 174.—*Less.* Mam. 1827, 142.—*Fisch.* Syn. 1829, 151.—*Gieb.* Säug. 1855, 786.—*Fitzinger,* Naturg. Säug. i. 1861, 341, f. 70.

Gulo vulgaris, *Griff.* An. Kingd. v. 1827, 117, no. 331.—*H. Smith,* Nat. Lib. xv. 1842, 209.

Gulo leucurus, "*Hedenborg*".—(*Gray.*)

Rossomaka, *Russian.*—"*Nieremb.* Hist. Nat. 1635, 188.—*Rossomack, Bell.* Trav. i. 1763, 221.—*Rosomach, Rytsch.* Orenb. Topog. i. 1772, 237.—*Rosomak, Steller,* Beschr. Kamt. 1774, 118."

Veelvraat, "*Isbr.* Reize naar China, 1704, 21.—*Houtt.* Nat. Hist. Dier. ii. 1761, 189, pl. 14, f. 4."—*Dutch.*

Vielfrass, *Klein,* op. et loc. cit.—*J. G. Gm.* Reise Sibir. iii. 1751, 492.—*Müller,* Natura. i. 1773, 265, pl. 14, f. 4 (ex. Houtt.).—*Von Martens,* Zool. Gart. xi. 1870, 253 (philological).—*German.*

Vielfras, *Hallen,* Naturg. Thiere, 1757. 548.

Goulon, *Bomare,* Dict. d'Hist. Nat. ii. 1768, 343.

Glouton, *Bomare,* tom. cit. 333.—*Buff.* Hist. Nat. xiii. 1765, 278; Suppl. iii, 240, pl. 48. - *French.*

Glutton, *Penn.* Syn. Quad. 1771, 196.—*English.*

Jerf, Jærr, Filfras, *Norwegian.*

Järf, Jerv, Filfrass, *Swedish.*—*Genberg,* Act. Stockh. 1773, 222, pl. 7, *.

Gleddk, *Laplanders.*

(B. American references.)

Coati ursulo affinis americanus. *Klein,* Quad. 1751, 74.

Ursus ~~fere~~ hudsonis, *Briss.* Quad. 1756, 260, no. 3.

SYNONYMY OF GULO LUSCUS.   35

Ursus luscus, *L.* S. N. i. 1758, 47, no. 2; 1766, 71, no. 4 (based on *Brisson* and *Edwards*).—
   Erxl. Syst. Anim. 1777, 167, no. 5.—*Schreb.* Säug. iii. 1778, 530.—*Zimm.* Geogr. Gesch.
   ii. 1780, 276, no. 169.—*Gm.* S. N. i. 1788, 103, no. 4.—*Shaw*, G. Z. i. 1800, 462, pl. 103,
   lower fig. (after Edwards).—*Turt.* S. N. i. 1806, 64.
Ursus luscus, *Fabric.* Fn. Grœnl. 1780, 24, No. 14.
Meles luscus, *Bodd.* Elench. An. i. 1784, 80.
Gulo luscus, *J. Sab.* Franklin's Journ. 1823, 650.—*E. Sab.* Suppl. Parry's 1st Voy. 1824,
   p. clxxxiv.—*Rich.* App. Parry's 2d Voy. 1825, 292.—*Rich.* F. B.-A. i. 1829, 41.—*Fisch.*
   Syn. 1829, 154.—*Godm.* Am. N. H. i. 1831, 185, pl.—, lower fig.—*Ross*, Exp. 1835, 8.—
   H. Smith, Nat. Lib. xv. 1842, 209.—*De Kay*, N. Y. Zoöl. i. 1842, 27, pl. 12, f. 2.—*Gray*,
   List Mamm. Br. Mus. 1843, 6c.—*Aud. & Bach.* Quad. N. A. i. 1849, 203, pl. 26.—
   Thomps. N. H. Vermont, 1853, 30.—*Baird*, Stansbury's Report, 1852, 311 (Great Salt
   Lake, Utah); M. N. A. 1857, 181.—*Billings*, Canad. Nat. and Geol. i. 1857, 241.—*Ross*,
   op. cit. vi. 1861, 30, 441.—*Maxim.* Arch. Naturg. 1861,—; Verz. N.-Am. Säug. 1862, 35.—
   Gerr. Cat. Bones Br. Mus. 1862, 96 (includes both).—*Coues*, Am. Nat. i. 1867, 352.—
   Dall, Am. Nat. iv. 1870, 221 (Yukon).—*Allen*, Bull. M. C. Z. i. 1870, 177 (Massachu-
   setts).—*Merr.* U. S. Geol. Surv. Terr. 1872, 662 (Wyoming).—*Allen*, Bull. Essex. Inst.
   vi. 1874, 54 (Montgomery, Colorado).—*Trippe, apud Coues*, Birds N. W. 1874, 224,
   in text (Clear Creek County, Colorado).—*Coues & Yarrow*, Zoöl. Expl. W. 100
   Merid. v. 1875, 61 (Wahsatch Mountains and localities in Utah).
Gulo arcticus, var. A., *Desm.* Mamm. i. 1820, 174, no. 267.—*Harl.* Fn. Amer. 1825, 60.
Gulo wolverene, *Griff.* An. Kingd. v. 1827, 117, no. 332.
Carcajou, *La Hontan*, Voy. 1703, 81.—*Sarrasin*,* Mém. Acad. Sci. Paris, 1713, p. 12.—*Bo-
   mare*, Dict. d'Hist. Nat. i. 1769, 423.—*French Canadians*. (*Not of F. Cuvier*, Suppl.
   Buff.) (Also, *Carkajou, Karkajou.* Compare Cree Indian names.)
Carcajou or Queequehatch, *Dobbs*, Hudson Bay, 1744, 40.
Quickhatch or Wolverene, *Edw.* Birds, ii. pl. 103.—*Ellis*, Hudson's Bay, i. 1750, 40, pl. 4.
   (*Quickehatch* and *Quiquihatch* are also found. Compare Cree Indian names.)
Wolverene, *Penn.* Syn. Quad. 1771, 195, no. 40, pl. 20, f. 2: Hist. Quad. ii. 1781, 8, pl. 8:
   Arct. Zoöl. i. 1784, 66, no. 21.—*Hearne*, Journ. —, 373.—*Church*, Cab. Quad. ii. 1805,
   pl. —. (Also, *Wolverene, Wolveren, Wolverin, Wolverine, Wolvering*.)—*Volverene*,
   Less. Man. 1827, 142 (in text).
Gröste americanische Halbfuchs, *Hall.* Naturg. Thiere, 1757, 518.
Wolfbeer, *Houtt.* Natuur. Hist. Dieren, ii. 227.—*Wolfsbar*, *Müll.* Naturs. i. 1773, 285.
Ours de la baye de Hudson, *Briss.* op. et loc. cit.
Okeecoohawgew, Okeecoohawgees, *Cree Indians.* (Obvious derivation of *Quickhatch*, if
   not also of *Carcajou*.)

HAB.—*Arctogœa*. In America, the whole of the British Provinces and Alaska, south in the United States to New England and New York, and still further in the Rocky Mountains, to at least 39°.

SPECIFIC CHARACTERS.—Sub-plantigrade, thick-set, shaggy, bushy-tailed, with thick legs and low ears; blackish, with a light lateral band meeting its fellow over the root of the tail, thus encircling a dark dorsal area; forehead light; 2-3 feet long; tail-vertebræ 6-9 inches.

Description of external characters.†

The form of this animal indicates great strength, without corresponding activity. The body is heavy and almost clumsy, supported upon thick-set and rather low legs; the walk is incompletely plantigrade. The back is high-arched, the general

---
* Special paper: Histoire d'un animal nommé Carcajou en Amérique, &c.
† Taken from a mounted specimen, from Great Salt Lake, Utah, in the National Museum.

figure drooping both before and behind, both tail and head being carried low. The general appearance is strikingly that of a Bear cub, with the addition of a bushy tail, though there is somewhat of the elongation which characterizes the *Mustelidæ*. The head is broad and much rounded on every side, with rather short and pointed muzzle, wide apart eyes, and low ears, altogether not very dissimilar from that of *Mustela pennanti*. The jaws, however, are rather Canine in appearance. The muffle and septum of the nose are naked, the former for about half an inch from the end of the snout. The eyes are remarkably small. The ears are low, much broader than high, obtusely rounded, well furred on both sides, scarcely overtopping the fur of the parts. The whiskers are few and short; there are other similar bristles about the head. The pelage, as usual, is of two kinds; there is a short under-fur, a kind of coarse kinky wool scarcely an inch long, which is mixed with the longer stiffer and straightish over-hairs, which are about four inches long on the sides, flanks, and hips, giving the animal a shaggy aspect, like a Bear. On the fore parts, and especially the head, however, the coat is much shorter and closer. The tail is clothed with still longer hairs, measuring some six or eight inches, drooping downward and conferring a peculiar shape, as if this member were deficient at the end. The tail-vertebræ are one-fourth, or rather more, of the length of head and body. The legs are very stout and the feet large; the track of the animal resembles that of a small Bear, but it is less completely plantigrade. The palms and soles are densely furry; but the balls of the digits are naked, and among the hairs may be discerned small naked pads at the bases of the digits, as well as a larger one beneath the carpus, the correspondent to which on the heel is apparently wanting. The fourth front digit is longest; then comes the third, fifth, second, and first, which last is very short. On the hind feet, the third is longest, the fourth little shorter; then follow the second, fifth, and first.

In color, the Wolverene is blackish, or deep dusky brown, with a remarkable broad band of chestnut or yellowish-brown, or even fading to a dingy brownish-white, beginning behind the shoulders, running along the sides, and turning up to meet its fellow on the rump and base of the tail, circumscribing a dark dorsal area. There is a light-colored grayish area on the front and sides of the head. On the throat, and between the fore legs, there is a patch, or there are several irregular spots of light color, as in *Mustela*. The legs, feet, most of the tail, and under parts generally, are quite blackish. The claws are whitish, strong, sharp, much curved, and about an inch long.

"The color of the fur varies much according to the season and age. The younger animals are invariably darker in the shadings than the old, which exhibit more of the grey markings. . . . In some specimens the yellowish fringing of the sides and rump is almost entirely white and of larger extent, leaving but a narrow stripe on the centre of the back dark. In such the hoary markings of the head would be of greater extent, and descend, most probably, to the shoulders."—(Ross, *l. c.*)

Measurements of seven specimens of Gulo luscus.

Orig. number.	Locality.	Sex.	From tip of nose to—					Tail to end of—		Length of—			Height of ear.	Nature of specimen.
			Eye.	Ear.	Occiput.	Tail.	Vertebræ.	Hairs.	Forefoot.	Hind foot.				
356	Fort Simpson, H. B. T...	♂	2.60	3.00	6.25	26.50	7.40	12.40					Fresh.
1092	Yukon River, Alaska ...		3.10	5.50	6.70	31.00	7.60	13.60	3.00	7.10			...do.
1093	......do	♀	3.00	5.10	6.75	29.25	7.60	13.10	3.00	6.75			...do.
1637	Peel's River (Dec.)		2.70	5.15	6.50	29.00	9.25	13.00	3.50	6.70			...do.
1664	......do	♀	2.75	4.80	6.50	27.00	8.00	13.00	3.10	6.50			...do.
*	Mackenzie's River.......	♂	2.80	6.10	6.90	34.80	8.00	13.00	4.40		2.00		...do.
†	Montana, U. S...........					36.00	8.90	14.00					Dry.

* From Ross. Longest hairs of body 4.00; of tail 7.50; upper canines 0.90; lower 0.75.
† From Baird.

Anal glands.

The anal glands of this animal are stated to be of about the size of a walnut; the fluid yellowish-brown and of the consistency of honey. The discharge is by the usual lateral papillæ within the verge of the anus. The scent is fœtid in a high degree.

Description of the skull and teeth. (See Plate I.)

The massiveness of the skull of *Gulo*, in comparison with that of *Mustela*, is as striking as its superiority in size. In general form, the prominent peculiarity is the strong convexity of the upper outline in profile. From the highest point, just behind the orbits, the skull slopes rapidly downward behind; and the frontal declivity is also much greater than in *Mustela*. There is much more of a frontal concavity, and the plane of the nasal orifice is extremely oblique. These features of the profile rather suggest a Feline than a Musteline skull, although, of course, the resemblance is still far from complete. There is a strong character in the zygoma: in *Mustela* a simple arch; here a nearly horizontal beam borne posteriorly upon an upright base, with a strongly convex backwardly projecting elbow. The same straightness requires a prominent process for definition of this part of the orbit. The zygoma is laminar and quite deep, much more so than in *Mustela*. Viewed from above, the zygomata are widely divergent from before backward. The anteorbital foramen is comparatively small, and appears over the fore border of the sectorial tooth. Prominent characters are observed in the paroccipital and mastoid, which form great processes of abutment against the bullæ, the same being only moderate in *Mustela*, and merely indicated in the smaller Weasels. The palate is very broad for its length, with straight (not a little concave) sides; measured across its broadest point, it forms very nearly an equilateral triangle with the sides. The posterior emargination is moderate, broadly U-shaped. The bullæ auditoriæ are only inflated on less than the interior half, the rest being greatly contracted and drawn out into a long tubular meatus (one extreme, of which the other is seen in the slender-bodied species of *Gale*—compare descriptions). The basioccipital space is somewhat wedge-shaped, owing to the divergence posteriorly of the bullæ. The pterygoids are very stout at base, but soon become laminar, and terminate in long, slender, hamular processes. Even in young skulls, the lambdoidal crests are as strong and flaring as in the oldest of *Mustela*, and terminate in the very prominent mastoids. The occipital surface is considerably excavated beneath these crests; the median superior protuberance is great. The condition of the sagittal crest varies, as usual, with age. In the youngest specimens, it is single and median for but a little way, then gradually divaricates on either hand to the supraorbital process; in the oldest, the divarication only begins more than half-way forward, a high, thin crest occupying the rest of the median line. The general shape of the brain-box, viewed from above, is, in consequence of the breadth and depression of the skull behind, neither the ovate nor the somewhat cylindrical, as obtains in *Mustela* and *Putorius*, but rather trapezoidal, somewhat as in *Taxidea*. The body of the under jaw is shaped exactly as in

Mustela, though it is more massive, but the coronoid is different. Its back border rises straight and perpendicularly, the anterior border curving strongly backward to meet it in a rounded obtuse apex. In *Mustela*, the borders gradually approximate to each other and meet more acutely. *M. pennanti* alone is much like *Gulo* in this respect.

Reviewing general cranial characters from the small *Gale* to the large *Gulo*, we see with increase of mere size a corresponding increment of massiveness; a graduation in obliquity of the plane of the end of the muzzle; a lengthening and constriction (on the whole) of the rostrum; an increase of the convexity of the upper profile; a depression of the zygomata from regular arches to a shape higher behind and more nearly horizontal in continuity; enlargement of paroccipitals and mastoids; constriction and lateral elongation of the bullæ into auditory tubes; and a flattening and widening behind, and corresponding contraction in front of, the brain-box.

The dentition shares the general massiveness of the cranium. Compared with those of *Mustela*, the teeth, if not relatively larger, are more swollen and stouter, with bulging sides, blunt points, and dull edges. The back upper molar is placed so far inward, out of line with the rest, that its outer border scarcely projects outside the inner border of the next. It has the same general character as in *Mustela*. The median constriction is slight, the inner more strongly regularly convex, with raised brim and crescentic ridge inside this; the outer is double convex (convex with an emargination), higher than the other, with an irregularly tuberculous surface. The antero-internal spur of the last premolar is low and little more than a mere heel, scarcely to be called a cusp. Turgidity aside, this tooth otherwise repeats the same in *Mustela*. The next premolar abuts against the reëntrance between the spur and main body of the last one, rather than lies in continuation of the same axis. The foremost premolar is relatively smaller and more crowded than the same in *Mustela*; it rests directly against the canine, to the inner side of the general axis of dentition. It would seem that but little more crowding would cause this tooth to permanently abort. The great canines are extremely stout at the base, rather blunt, and have a strong forward obliquity. Of the six upper incisors, the lateral pair are, as usual, much larger (wider and deeper, though little, if any, longer) than the rest. They are usually found much abraded by rubbing against the under canines. The other incisors are all alike, smaller and evenly set; all show indication of trilobation, with a large middle and minute lateral lobes, best seen from behind, where, at the point where the teeth flatten toward the tips, ridges divaricate, the termination of these ridges forming the lateral lobes. The inferior incisors are irregularly set, the middle one on each side being crowded back out of the general plane. The outermost pair are broader than the rest, and seem longer, viewed from the front, since more of the tooth is exposed from the alveolus. The next, partially displaced pair, viewed from the front, seem the smallest of all; but this is due to their position. Viewed from behind, their size is seen to be much greater than that of the middle pair. All the incisors are obscurely lobate at end. The under canines are shorter, stouter, and more curved than the upper; most of their surface is striate. The anterior lower premolar, like the same tooth in the upper jaw, is very small, displaced inward, and apposed against the canine. The next premolar is markedly increased in size, and set in the jaw with its longitudinal axis very oblique to the general axis of dentition, as if turned partially around for want of room. The next two premolars are much larger still and massive; they both show a single central pointed conical cusp, whose sides are bevelled down all around to the rimmed base of the tooth, but there is no indication of the secondary cusp halfway up the back edge of the main cusp, as in *Mustela pennanti, martes, americana,* and perhaps all of this genus. Similarly, on the great sectorial lower molar, there is no sign of a secondary cusp on the inner face of the main cusp, as is so plainly seen in *M. martes,* and which also exists in less degree in *M. pennanti, americana,* and *foina.* These differences of the two back under premolars and front under molar are, perhaps, the strongest dental peculiarities of *Gulo* as compared with *Mustela.* Besides this, the two main cusps of the anterior lower premolar are subequal in size and elevation instead of very unequal, as in *Mustela,* where the hinder one is much the highest. The posterior tuberculous portion of this tooth is relatively much smaller. As in allied genera, the back lower molar is small, subcircular, tuberculous, not calling for special remark.

In a large proportion of the skulls which come to hand, the canines and sectorial teeth are found cracked, even split entirely in two or broken off, apparently a result of the desperate exertions the captured animals made to free themselves from iron traps.

I append measurements of a very large and another rather small American skull, with those of a specimen from Lapland.

Measurements of skulls, European and American.

	1041, Lapland.	6234, Pool's River.	7157, Ft. Good Hope.
Total length from apex of intermaxillary to occipital protuberance	5.50	5.75	6.00
Greatest width (zygomatic)	3.35	3.55	3.75
Distance between orbits	1.40	1.45	1.50
Nasal bones, length	.....	1.00	1.00
Upper incisors from front to hinder margin of palate	2.80	2.85	3.10
Upper molars and premolars, length taken together	1.65	1.65	1.75
Lower molars and premolars, length taken together	1.95	1.95	2.10
Lower jaw, length to back of condyle	3.65	3.70	4.00
Lower jaw, height of coronoid above condyle	1.70	1.65	1.85
Greatest width of palate	1.80	1.80	1.95
Least width of skull	1.30	1.35	1.55
Intermastoid width	3.00	3.20	3.50
Interparoccipital width	1.95	2.05	2.35
Foramen magnum, width	0.70	0.75	0.80
Width across supraorbital protuberance	1.75	1.75	1.80

NOMENCLATURE OF THE SPECIES—RELATION OF THE EUROPEAN AND AMERICAN ANIMAL.

This animal has received a great variety of names, both technical and vernacular. Nearly all barbarous tribes of Northern regions in both hemispheres, as well as civilized nations, have each bestowed some appellation; and in some cases at least the latter have adopted an aboriginal name, with more or less modification, while in all cases the book-names of the species appear to be derived from the vernacular. Thus, "*quickhatch*" of the English residents of British America is obviously an Anglicism of the Cree or Knisteneaux word, and I agree with Sir John Richardson that *carcajou* of the French Canadians is probably derived from the same source. I have no idea what the meaning of the more frequent term *wolverene* may be; none of its various spellings furnish a clue, beyond the obvious *wolf*, which is however of wholly uncertain applicability here. *Gulo, glutton, glouton*, are self-explaining, in allusion to the voracity of the animal; this is also the meaning of the Swedish, Russian, and German names above quoted. *Gulo* was adopted by Linnæus as the specific name of the European animal, which he placed in the genus *Mustela*. This author separated the American as *Ursus luscus*—an absurd name indeed.

"Luscus" signifies blind of one eye, mope-eyed; as is said to have been the unfortunate condition of a specimen imported from Hudson's Bay, some time in possession of Sir Hans Sloane, and described by Edwards, upon whose account Linnæus based his *Ursus luscus*. Linnæus was frequently capricious, and sometimes facetious, in bestowing names; while some of those he gave were wholly inappropriate. Thus the *Paradisea apoda* ("footless"), the common bird of Paradise, was so called for no other reason than that the skins imported into Europe used to lack the feet, these having been removed in the preparation of the specimens by the natives. This taxidermal accident not only gave rise to the name, but to the general belief that the bird had no feet, and to various fabulous accounts of its habits as a consequence of such condition. It is deplorable that an accidental deformity of one particular individual should be thus perpetuated as the designation of a species; the more so, as it is the name which, according to strict rules of nomenclature, must prevail. It may, however, be fairly questioned whether it should not be set aside, under the accepted ruling that priority shall not be entitled to precedence when the first name involves a palpable error, or is wholly inept, as in the present instance. The specific term *gulo* being used for the genus, the name *borealis* would come next in order, should *luscus* be ignored on these considerations.

The foregoing synonymatic lists show that this species has not escaped subdivision into nominal ones, and that varieties have been generally recognized. But the close similarity of the animals from the two continents did not escape some of the earlier writers, among them even those of slight scientific acquirements or experience. Thus Shaw, in 1800, states of the Wolverene, of which he reproduces Edwards's figure, that "this appears to be no other than a variety" of *Ursus gulo*. Desmarest allowed varietal distinction from the animal be called *G. arcticus*. Cuvier endorsed the specific validity which earlier writers had generally admitted; this error Griffith perpetuated, and, calling one *Gulo vulgaris*, the other *G. wolverene*, introduced at once two new synonyms. At least, if these names did not originate with him, I have not found them in previous works. A certain "*Gulo leucurus* Hedenborg", quoted by Gray, I have not had an opportunity of verifying. In the foregoing synonymy I separate the American from the Old World quotations merely for the convenience of reference, and must not be understood as implying that any distinction, varietal or specific, subsists between the Glutton and the Wolverene.

In comparing numerous American skulls with one from Lapland, I detect in the former a tendency to less constriction of the cranium behind the postorbital processes. This is an interesting correlation with one of the more pronounced differences in the skulls of *M. martes* and *M. americana*. But this is the only discrepancy I find, and it is not, moreover, uniformly exhibited to any appreciable degree. The identity of the animals of the two continents is to be considered fairly established, whatever range of variation in size and color either may present.

Pallas notes a curious supposed character in urging a critical comparison of the two forms. " Pilos Guloni esse triquetros notavit Baster (Act. Harlemens. vol. xv.) sed hoc an in Americano? nostrati pili teretes", he says, on p. 75 of the " Zoographia".

HISTORY AND HABITS OF THE SPECIES.

The written history of the Glutton or Wolverene dates from an early period in the sixteenth century, when the animal is mentioned with little interval of time by several writers in much the same extravagant terms. The first appearance of the animal in literature is said by von Martens to have been in 1532, at the hands of Michow, a physician of Cracow, in the work *De Sarmatia Asiana et Europæa*. Olaus Magnus (1562), to whom is commonly attributed the earliest mention, though he thus appears to have been anticipated, gives a most extraordinary account, made up of the then current popular traditions and superstitions, and tales of hunters or travellers, unchecked by any proper scientific enquiry; although, to do him justice, he does not entirely credit them himself. We may be sure that such savory morsels of animal biography did not escape the notice of subsequent compilers, and that they lost nothing of their flavor at the hands of the versatile and vivacious Buffon. Endorsed for two centuries by various writers, each more or less authoritative in his own times, and, moreover, appealing strongly to the popular love of the marvellous, the current fables took strong root and grew apace, flourishing like all "ill weeds", and choking sober accounts. Coming down to us through such a long line of illustrious godfathers, they were treated with the respect generally accorded to long years, and furnished the staple of professedly educational text-books. Probably no youth's early conceptions of the Glutton were uncolored with romance; the general picture impressed upon the susceptible mind of that period being that of a ravenous monster of insatiate voracity, matchless strength, and supernatural cunning, a terror to all other beasts, the bloodthirsty master of the forest. We cannot wonder at the quality of the stream, when we turn to the fountain-head of such gross exaggeration. We find it gravely stated that this brute will feast upon the carcase of some large animal until its belly is swollen as tight as a drum, and then get rid of its burden by squeezing itself between two trees, in order that it may return to glut itself anew—an alleged climax of gluttony to which no four-footed beast attains, and for the parallel of which we must refer to some of the most noted gormandizers of the Roman Empire. We have indeed reliable accounts of such gastronomic exploits, but they are not a part of those records which are generally accepted as zoölogical. In one of the old zoölogical works of some celebrity, there is a very droll picture of a Wolverene squeezing itself between two trees, with a most anxious expression of countenance, the fore part of the body being pressed thin, while the hinder is still distended, and the large pile of manure already deposited being rapidly augmented with further supplies. Still in the track of the marvellous, we read how the Glutton, too clumsy and tardy of foot to overtake large Ruminants, betakes itself to the trees beneath which they may pass, and there crouches in wait for its victim; it drops like a shot upon the unsuspecting Elk, Moose, Reindeer, and, fastening with claws and teeth, sucks the blood and destroys them as they run. That nothing may be left undone to ensure success, the animal has the wit to throw down moss or lichens to attract its prey, and to employ the friendly services of Foxes to drive the quarry beneath the fatal spot. I allude to these things, not that such gross exaggerations longer require refutation, but because they are a part, and no inconsiderable one, of the history of the species; and because, as we shall see in the sequel, a perfectly temperate and truthful narration of the creature's actual habits sufficiently attests the possession of really remarkable qualities, which need be but caricatured for transformation into just such fables. We may remember, also, that the history of the Wolverene is mixed in some cases with that of other animals, some of whose habits have been attributed to it. Thus Charlevoix

(Voy. Amer. i, 201) speaks of the "carcajou or quincajou, a kind of cat", evidently, however, having the Cougar (*Felis concolor*) in view, as appears from the rest of his remarks. Such habit of lying in wait for their prey is common to the Cougar, Lynx, and other large Cats. Not to prolong this portion of the subject, I may state briefly, that the animal whose characteristics will be fully exposed in the course of this article is simply an uncommonly large, clumsy, shaggy Marten or Weasel, of great strength, without corresponding agility, highly carnivorous, like the rest of its tribe, and displaying great perseverance and sagacity in procuring food in its northern residence when the supply is limited or precarious, often making long uninterrupted journeys, although so short-legged. It is imperfectly plantigrade, and does not climb trees like most of its allies. It lives in dens or burrows, and does not hibernate. It feeds upon the carcases of large animals which it finds already slain, but does not destroy such creatures itself, its ordinary prey being of a much humbler character. It is a notorious thief; not only of stores of meat and fish laid up by the natives of the countries it inhabits, the baits of their traps, and the animals so caught, but also of articles of no possible service to itself; and avoids with most admirable cunning the various methods devised for its destruction in retaliation.

All the earlier accounts referred to the animal of Europe and Asia. I have not found the terms "Carcajou" and "Wolverene", nor any allusion to the American form, until early in the eighteenth century. La Hontan speaks of it in 1703, likening it to a large fierce Badger; Lawson has been quoted in this connection, he having attributed to the Lynx some of the fabulous accounts of the Glutton; but it is evident that his remarks neither apply, nor were intended to apply, to the Wolverene. Catesby speaks of an animal "like a small bear" which exists in the Arctic portions of America; this reference is among the earlier ones to the Wolverene, those which confound it with other species being excluded.* We have other definite accounts of the Wolverene, near the middle of the eighteenth century,

---

* The Wolverene has been confused not only with the Lynx and Cougar in early times, but also quite recently with the American Badger, *Taxidea americana*. Thus F. Cuvier (Suppl. Buffon, i, 1831, 267) treats at length of "Le carcajou, ou Blaireau Américain", his whole article being based upon the Badger, to which he misconceives the name Carcajou to belong. Paul Gervais also speaks of the "*Carkajou* ou Blaireau d'Amérique" (Proc. Verb. Soc. Philom. Paris, 1842, 30).

as those of Klein, Ellis, Dobbs, Edwards, and Brisson. *Ursus luscus* of Linnæus arose in a way already narrated, and the species may be considered to have been well known from this period, although it was for a long time very generally supposed to be different from the Glutton of Europe and Asia.

The various American biographies of this animal are without exception more or less incomplete and unsatisfactory; even those which are shorn of obvious exaggeration are, in large part, a compilation of earlier statements. They have, however, steadily improved, the latest, that of Audubon and Bachman, being by far the best, although Sir John Richardson's was an excellent contribution. The account which Pennant gave in 1784 (Arct. Zool. pp. 66-68) is purged of some of the fables, yet curiously shows how their effects will linger. He scouts the idea of such excessive gluttony as had been attributed, yet relates the moss-throwing story, and represents the Wolverene as "a beast of uncommon fierceness, the terror of the Wolf and Bear; the former, which will devour any carrion, will not touch the carcase of this animal, which smells more fœtid than that of the Pole-cat". Pennant traces its distribution as far north as Copper River, to the countries on the west and south of Hudson's Bay, Canada, and the tract between Lakes Huron and Superior. He gives a fair description, and adds:—"It hath much the action of a Bear; not only in the form of its back, and the hanging down of its head, but also in resting on the hind part of the first joint of its legs." "The Kamtschatkans", he naïvely continues, "value them so highly as to say, that the heavenly beings wear no other furs." Richardson gives some interesting particulars, among them none, so far as I am aware, that are not accurate. In a passage he quotes from Graham's MSS., we see a probable basis for the fabulous accounts that the Fox is the Wolverene's provider or abettor in the chase—for it is the well-nigh universal rule that fable is founded on facts exaggerated, distorted, or perverted. Alluding to the Wolverene's notorious habit of following Marten roads, Mr. Graham remarks that the animal tears the captured Martens to pieces or buries them at a distance in the snow. "Drifts of snow often conceal the repositories thus made of the martens from the hunter, in which case they furnish a regale to the hungry fox, whose sagacious nostril guides him unerringly to the spot. Two or three foxes are often seen following the Wolverene for this purpose." Richardson discredits the accounts which had come down from Buffon of the destruction of Beavers by the Wolverene. "It must be only in summer," he says, "when those industrious animals are at work on land, that it can surprise them. An attempt to break open their house in winter, even supposing it possible for the claws of a Wolverene to penetrate the thick mud walls when frozen as hard as stone, would only have the effect of driving the beavers into the water to seek for shelter in their vaults on the borders of the dam."*

Hearne gives a much more credible account of the depredations of the Wolverene upon another of the valuable fur-bearing animals of the north—the Fox—during the period of reproduction. Being directed by scent to the burrow of the Fox, which its great strength enables it to enlarge if necessary, it enters and destroys the whole family. In evidence of its amazing strength, of that sort most effective in pulling, pushing, and prying, the same author mentions that a Wolverene had been known to upset the greater part of a pile of wood nearly seventy yards around, in order to get at some provisions which had been deposited in this câche. Audubon's article, although entertaining and accurate, is chiefly a compilation from previous accounts, as he appears to have met with the animal in a state of nature but once, the result of which occurrence is his principal contribution to the subject. This was in Rensselaer County, near the banks of the Hoosac River. He tracked a Wolverene in the snow to its den, which was among rocks, and shot it after prying away some heavy fragments. "There was a large nest of dried leaves in the cavern, which had evidently been a place of resort for the Wolverene during the whole winter, as its tracks from every direction led to the spot. It had laid up no winter store, and evidently depended on its nightly excursions for a supply of food. It had however fared well, for it was very fat."

The fur of the Wolverene is highly valued both by civilized and uncivilized people. A number of skins sewn together makes a very beautiful carriage robe or hearth-rug, and the pelts are in common use for these purposes. The Indians and Esqui-

---

* An anonymous writer, doubtless General D. S. Stanley ("D. S. S., Fort Sully, Dakota"; American Naturalist, ii, 1868, p. 215), notes the depredations committed by the Wolverene upon Beaver, in the following terms:—"The wolverene follows the Beaver and preys upon them; in northern latitudes, the wolverene is almost always present where the beaver is abundant. The beaver has a beaten path on the bank of the stream near his lodge. There the wolverene lies in wait for him, and often cuts short his career."

manx use the fur as they do that of the Wolf, for fringing their garments, the skin being cut in strips for this purpose. I have already given (p. 2) some statistics of the trade in this kind of pelt, which indicate the comparative standing of the animal among the fur-bearing species of this country. The following methods of its capture are taken from Gibson:—*

"The wolverine is a dangerous foe to many animals larger than itself, and by the professional hunter it is looked upon as an ugly and dangerous customer. There are several methods of trapping this horrid creature, and in many localities successful trapping of other animals will be impossible without first ridding the neighborhood of the wolverines. Dead-falls of large size will be found to work successfully, baiting with the body of some small animal, such as a rat or squirrel. A piece of cat, beaver or muskrat flesh is also excellent, and by slightly scenting with castoreum success will be made sure. Several of these traps may be set at intervals, and a trail made by dragging a piece of smoked beaver meat between them. The gun-trap, . . . will also do good service in exterminating this useless and troublesome animal. Steel traps of size No. 3 or 4 are commonly used to good purpose. . . . In all cases the trap should be covered with leaves, moss or the like, and the bait slightly scented with castoreum. Like all voracious animals, the perpetual greed of the wolverine completely overbalances its caution, and thus renders its capture an easy task."†

The Wolverene is an animal of circumpolar distribution in both hemispheres. In North America, it exists in all suitable country north of the United States to the Arctic Coast, and even on some of the islands of the Polar Sea, traces of its presence having been discovered on Melville Island, about latitude 75°. Our notes upon its distribution in this country may relate chiefly to its southern limits. Of an erroneous quotation, by which it has been supposed to occur as far south as Carolina, I have already spoken. Its southern limit has been fixed more properly between 42° and 43°; this is probably nearly correct for the eastern portions of the continent, aside from what recession of the species northward may have recently occurred, although, as we shall see, the species reaches

---

* Complete American Trapper, [etc.] p. 200. New York. 16mo. 1876.
† A statement at variance with the experience of others, as detailed on a following page, from which it would appear that the wary creature is particularly difficult to entrap.

farther south in the West. In Massachusetts, according to Mr. Allen, it still lingered a few years since, in that portion where the Canadian, as distinguished from the Alleghanian, fauna is represented. But the Massachusetts reports are all probably traceable to a Hoosac Mountain record some years prior. Dr. Hitchcock and Dr. De Kay both quote Dr. Emmons for this, although the species is not given in the latter's report. In New York, it was rare in the time of Audubon and De Kay: the former notes specimens from Reusselaer (1810) and Jefferson (1827) Counties. Dr. Z. Thompson, writing in 1853, states that it was then extremely rare in Vermont, none having been met with to his knowledge for several years. Though occasionally found when the country was new in all parts of the State, it was never very plentiful, and for years had been known only in the most wooded and unsettled parts. I have met with but one record of its presence in the United States from west of New York to the Rocky Mountains, though it is to be presumed that it inhabits, or has lately done so, the wooded portions of our northern frontier. Maximilian speaks (*l. c.*) of the occurrence of the species on the western border of Canada and near the mouth of the Red River of the North, and surmises that the species may extend to the Missouri River, especially as he saw a skin, but without indication of locality, at one of the trading posts. I never saw the Wolverene in Dakota or Montana, where most of the country is altogether too open. Baird, however, speaks of its occasional occurrence in the Black Hills, and registers a specimen from "northwest of Fort Union"* (probably Montana, toward the base of the Rocky Mountains); and Mr. C. H. Merriam (as recorded *l. s. c.*) procured a specimen on the Yellowstone River, Wyoming, in August, 1872. In the Rocky Mountains, as was to have been expected, its extension southward has been traced to the farthest known point, between 40° and 39°. Professor Baird notes a specimen obtained by Captain Stansbury from the Great Salt Lake, Utah, which lies wholly south of 42°. This individual is still (1877) preserved, mounted, in the National Museum. It is probable that its extreme limit is even somewhat farther than this, reaching in the mountains to the borders of Arizona and New Mexico and

---

*This locality (Fort Union), frequently mentioned in the works of Audubon, Baird, and others, no longer exists as such, being now a heap of rubbish. It is replaced by Fort Buford, commanding the mouth of the Yellowstone, at the extreme southwest corner of Dakota, adjoining the southeast corner of Montana.

corresponding latitudes in California. Of this, I was assured by hunters whose statements I had no reason to doubt, and who were evidently acquainted with the species. But I could not confirm their statements by actual observation, and, for all that is positively known, the Salt Lake record remains the southernmost, excepting that very recently furnished by Mr. Allen (*op. cit.*). He saw the skin of an individual taken in the vicinity of Montgomery, Colo., near the limit of timber, and the animal was stated to be not uncommon. This locality is somewhat south of $40°$, and the occurrence is strongly corroborative of the accounts I received, as just mentioned. I have myself lately seen a mounted specimen among a collection of animals made by Mrs. M. A. Maxwell, in the vicinity of Boulder, Colo. I have no record from the region west of the main chain of the Rockies in Oregon or Washington Territory, although it is not to be presumed, upon this negative evidence, that the species does not occur there.

The Wolverene ranges, as we have seen, in greater or less abundance, all over the northern portions of this country. It appears to be particularly numerous in the Mackenzie River region, and it fairly infests the whole country bordering the lower portions of this river and the west side of the mountains. From this country, many accounts have reached me, from variour officers of the Hudson's Bay Company, through the liberality of the Smithsonian Institution, which placed in my hands all the matter represented in its archives upon the mammals of the far north. These manuscripts witness the wonderful cunning and sagacity of the beast, as well as its ferocity, and represent it to be the greatest enemy with which the hunters and trappers have to contend in the pursuit of fur-bearing animals. Messrs. Kennicott, Macfarlane, Ross, and Lockhart have each recorded their experiences, which together afford the material for a complete biography.

The hunter, says Mr. Lockhart, may safely leave an animal he has killed for one night, but never for a second time, without placing it in a strong câche of logs. The first night the Wolverene is pretty sure to visit the place, but will touch nothing. The next night he is certain to return, and, if he can possibly get at the meat, he will gorge himself, and then make away with the rest, which he cunningly hides, piece by piece, under the snow, in different directions. At every câche he makes he voids his urine or drops his dirt, probably to prevent

Foxes, Martens, or other animals from smelling the hidden meat and digging it up. Câches must be made of green wood, and be exceedingly strong, or the animal will certainly break into them. He has been known to gnaw through a log nearly a foot in diameter, and also to dig a hole several feet deep in frozen ground, to gain access to the coveted supply. Should he succeed in gaining entrance for himself, and yet be unable to displace the logs sufficiently to permit of removal of the meat, the brute will make water and dirt all over it, rendering it wholly unfit to be used; even a dog will then scarcely touch it.

To the trapper, the Wolverenes are equally annoying. When they have discovered a line of Marten traps, they will never abandon the road, and must be killed before the trapping can be successfully carried on. Beginning at one end, they proceed from trap to trap along the whole line, pulling them successively to pieces, and taking out the baits from behind. When they can eat no more, they continue to steal the baits and câche them. If hungry, they may devour two or three of the Martens they find captured, the remainder being carried off and hidden in the snow at a considerable distance. The work of demolition goes on as fast as the traps can be renewed.

The propensity to steal and hide things is one of the strongest traits of the Wolverene. To such an extent is it developed that the animal will often secrete articles of no possible use to itself. Besides the wanton destruction of Marten traps, it will carry off the sticks and hide them at a distance, apparently in sheer malice. Mr. Ross, in the article above quoted, has given an amusing instance of the extreme of this propensity:—"The desire for accumulating property seems so deeply implanted in this animal, that like tame ravens, it does not appear to care much what it steals so that it can exercise its favorite propensity to commit mischief. An instance occurred within my own knowledge in which a hunter and his family having left their lodge unguarded during their absence, on their return found it completely gutted—the walls were there but nothing else. Blankets, guns, kettles, axes, cans, knives and all the other paraphernalia of a trapper's tent had vanished, and the tracks left by the beast showed who had been the thief. The family set to work, and by carefully following up all his paths recovered, with some trifling exceptions, the whole of the lost property."

Though very clumsy animals, the Wolverenes manage to capture, at times, such prey as Hares or Grouse, and they successfully attack disabled Deer. We have already seen how they destroy Foxes in their burrows; and they are usually found in excellent condition. They also feed on offal or carrion; in fact, anything that they can catch or steal. Their own flesh is only eatable in the extreme of starvation. They bring forth in burrows under ground, probably old Bear washes, and have four or five young at a birth. It is very rarely that they are discovered at this period or whilst suckling their young. One reason, however, may be that they reproduce late in June and early in July, when the mosquitoes are so numerous that no one who can avoid it goes abroad in the woods. The rutting season is in the latter part of March. The female is ferocious in the defense of her young, and if disturbed at this time will not hesitate to attack a man. Indeed, Indians have been heard to aver that they would sooner encounter a she-bear with her cubs than a Carcajou under the same circumstances. In October, when the rivers set fast, the Wolverenes reappear in families, the young still following their dam, though now not much her inferior in size. They are full grown when about a year old. In early infancy, the cubs are said to be of a pale cream color.

The Wolverene may be captured in wooden traps similar to those used for Martens, but of course made on a much larger scale, as the animal's strength is enormous, even for its size. The traps are sometimes built with two doors. But so great is the cunning and sagacity of the beast, that the contrivance for its destruction must be very perfect. The traps are covered up with pine-brush, and made to resemble a câche as much as possible; the Wolverene is then likely to break in and get caught. The bait, ordinarily the conspicuous feature of a trap, must in this instance be concealed, or the animal will either break in from behind, or, failing in this, will pass on his way. It is sometimes also taken in steel traps, or by means of a set gun; but both these methods are uncertain, great "medicine" being required to outwit the knowing and suspicious beast.

The eyesight of the Wolverene is not very bright, but his sense of smell is extremely acute.

"The winter I passed at Fort Simpson", writes Mr. Lockhart, "I had a line of Marten and Fox traps, and Lynx snares, extending as far as Lac de Brochet. Visiting them on one occasion I found a Lynx alive in one of my snares; and being indisposed to carry it so far home, determined to kill and skin it before it should freeze. But how to câche the skin till my return? This was a serious question, for Carcajou tracks were numerous. Placing the carcase as a decoy in a clump of willows at one side of the path, I went some distance on the opposite side, dug a hole with my snow-shoe about three feet deep in the snow, packed the skin in the smallest possible compass, and put it in the bottom of the hole, which I filled up again very carefully, packing the snow down hard, and then strewing loose snow over the surface till the spot looked as if it had never been disturbed. I also strewed blood and entrails in the path and around the willows. Returning next morning, I found that the carcase was gone, as I expected it would be, but that the place where the skin was câched was apparently undisturbed. 'Ah! you rascal,' said I, addressing aloud the absent Carcajou, 'I have outwitted you for once.' I lighted my pipe, and proceeded leisurely to dig up the skin to place in my muskimoot. I went clear down to the ground, on this side and on that, but no Lynx skin was there. The Carcajou had been before me, and had carried it off along with the carcase; but he had taken the pains to fill up the hole again and make everything as smooth as before!

"At Peel's River, on one occasion, a very old Carcajou discovered my Marten road, on which I had nearly a hundred and fifty traps. I was in the habit of visiting the line about once a fortnight; but the beast fell into the way of coming oftener than I did, to my great annoyance and vexation. I determined to put a stop to his thieving and his life together, cost what it might. So I made six strong traps at as many different points, and also set three steel traps. For three weeks I tried my best to catch the beast without success; and my worst enemy would allow that I am no green hand in these matters. The animal carefully avoided the traps set for his own benefit, and seemed to be taking more delight than ever in demolishing my Marten traps and eating the Martens, scattering the poles in every direction, and câching what baits or Martens he did not devour on the spot. As we had no poison in those days, I next set a gun on the bank of a little lake. The gun was concealed in some low bushes, but the bait was so placed that the Carcajou must see it on his way up the bank. I blockaded my path to the gun with a small pine tree which completely hid it. On my first visit afterward I found that the beast had gone up to the bait and smelled it, but had left it untouched. He had next pulled up the pine tree that blocked the path, and gone around the gun and cut the line which connected the bait with the trigger, just behind the muzzle. Then he had gone back and pulled the bait away, and carried it out on the lake, where he laid down and devoured it at his leisure. There I found my string. I could scarcely believe that all this had been done designedly, for it seemed that faculties fully on a par with human reason would be required for such an exploit, if done intentionally. I therefore rearranged things, tying the string where it had been bitten. But the result was exactly the same for three successive occasions, as I could plainly see by the footprints; and what is most singular of all, each time the brute was careful to cut the line a little back of where it had been tied before, as if actually reasoning with himself that even the knots might be some new device of mine, and therefore a source of hidden danger he would prudently avoid. I came to the conclusion that *that* Carcajou ought to live, as he must be something at least human, if not worse. I gave it up, and abandoned the road for a period.

"On another occasion a Carcajou amused himself, much as usual, by tracking my line from one end to the other and demolishing my traps, as fast as I could set them. I put a large steel trap in the middle of a path that branched off among some willows, spreading no bait, but risking the chance that the animal would 'put his foot in it' on his way to break a trap at the end of the path. On my next visit I found that the trap was gone, but I noticed the blood and entrails of a hare that had evidently been caught in the trap and devoured by the Carcajou on the spot. Examining his footprints I was satisfied that he had not been caught, and I took up his trail. Proceeding about a mile through the woods I came to a small lake, on the banks of which I recognized traces of the trap, which the beast had laid down in order to go a few steps to one side to make water on a stump. He had then returned and picked up the trap, which he had carried across the lake, with many a twist and turn on the hard crust of snow to mislead his expected pursuer, and then again entered the woods. I followed for about half a mile farther and then came to a large hole dug in the snow. This place, however, seemed not to have suited him, for there was nothing there. A few yards farther on, however, I found a neatly built mound of snow on which the animal had made water and left his dirt; this I knew was his cache. Using one of my snowshoes for a spade I dug into the hillock and down to the ground, the snow being about four feet deep; and there I found my trap, with the toes of a rabbit still in the jaws. Could it have been the animal's instinctive impulse to hide prey that made him carry my trap so far merely for the morsel of meat still held in it? Or did his cunning nature prompt him to hide the trap for fear that on some future unlucky occasion he might put his own toes in it and share the rabbit's fate?"

This propensity of the Wolverene to carry off traps receives confirmation from other sources. In Captain Cartwright's Journal (ii, 407), a similar instance is recorded in the following terms:—"In coming to the foot of Table Hill I crossed the track of a Wolvering with one of Mr. Callingham's traps on his foot: the foxes had followed his bleeding track. As this beast went through the thick of the woods, under the north side of the hill, where the snow was so deep and light that it was with the greatest difficulty I could follow him even on Indian rackets, I was quite puzzled to know how he had contrived to prevent the trap from catching hold of the branches of trees or sinking in the snow. But on coming up with him I discovered how he had managed: for after making an attempt to fly at me, he took the trap in his mouth and ran upon three legs. These creatures are surprisingly strong in proportion to their size; this one weighed only twenty-six pounds and the trap eight; yet including all the turns he had taken he had carried it six miles."

The ferocity of the Wolverene, no less than its cunning, is illustrated in some of the endless occasions on which it matches its powers against those of its worst enemy. A man had set a gun for a Carcajou which had been on his usual round of demolition of Marten traps. The animal seized the bait unwarily, and set off the gun; but owing to careless or improper setting, the charge missed or only wounded it. The Carcajou rushed upon the weapon, tore it from its fastenings, and chewed the stock to pieces. It is added to the account of this exploit that the beast finished by planting the barrel muzzle downward upright in the snow; but this may not be fully credited. The stories that pass current among trappers in the North would alone fill a volume, and they are quite a match for those that Olaus Magnus set down in his book centuries ago. How much wiser are we in our generation? Is there anything new under the sun? But we need not go beyond the strict fact to be impressed with the extraordinary wit of the beast, whom all concur in conceding to be "as cunning as the very devil".

With so much for the tricks and the manners of the beast behind our backs, roaming at will in his vast solitudes, what of his actions in the presence of man? It is said that if one only stands still, even in full view of an approaching Carcajou, he will come within fifty or sixty yards, provided he be to windward, before he takes the alarm. Even then, if he be not warned by sense of smell, he seems in doubt, and will gaze earnestly several times before he finally concludes to take himself off. On these and similar occasions he has a singular habit—one not shared, so far as I am aware, by any other beast whatever. He sits on his haunches and shades his eyes with one of his fore paws, just as a human being would do in scrutinizing a dim or distant object. The Carcajou then, in addition to his other and varied accomplishments, is a perfect skeptic— to use this word in its original signification. A skeptic, with the Greeks, was simply one who would shade his eyes to see more clearly. To this day, in sign-language among some of the North American Indians, placing the hand to the forehead signifies "white man"—either in allusion to this habit, or to the shade given the eyes by the straight vizor of the military cap, which the Indians see oftener than they desire. Mr. Lockhart writes that he has twice been eye-witness of this curious habit of the Wolverene. Once, as he was drifting down stream in a small canoe, he came within a short distance of one of the animals on the bank; it stopped on perceiving him, squatted on its haunches, and peered earnestly at the advancing boat, holding one fore paw over its eyes in the manner described. Not seeming to take alarm, it proceeded on a few paces, and then stopped to repeat the performance, when Mr. Lockhart, now sufficiently near, fired and killed the beast. On another occasion, when the same gentleman was crossing the Rocky Mountains, a Wolverene, which had become alarmed and was making off, stopped frequently and put up his paw in the same manner, in order to see more clearly the nature of that which had disturbed him.

On other occasions, the Wolverene displays more boldness than this in the presence of man. It has been known to seize upon the carcase of a deer, and suffer itself to be shot rather than relinquish possession, though the hunter had approached within twenty yards of his game. When pressed by the pangs of hunger, still bolder exploits are sometimes performed, as in the instance narrated by Capt. J. C. Ross. In the dead of an Arctic winter, his ship's company were surprised by a visit from a Wolverene, which clambered over the snow wall surrounding the vessel, and came boldly on deck among the men. Forgetful of its safety in the extremity of its need for food, the animal seized a canister of meat, and suffered himself to be noosed while eating.

DISTRIBUTION OF THE SPECIES IN THE OLD WORLD.

This portion of the subject is translated from J. F. Brandt's elaborate article.*

According to Georgi (*l. s. c.* [*i. e.* Geogr. Phys. Beschr. 1786] p. 1547), the Glutton is found in the temperate, and particularly in the cold regions of Russia and Siberia; that is to say, from Lithuania and Curland, where, however, it is rare, to Finland, Kola, Archangel, Wologda, Perm; and in Siberia, from the mountains which bound this country (the Altaï, the Sajan, and Daurian Alps, the Stannovoi, &c.), to the Arctic Tundras. Brincken (Mem. sur la Forêt de Bidlowicza, p. 45) speaks of Gluttons in the forest of Bidlowicza. Eichwald, however, two years later (1830), states that formerly they were only found in some few forests of Podolia and Pinsk (Skizze, p. 237). In 1791, Fischer says (Naturgesch. von Livland, Livonia, 2d edition, p. 141) that the Glutton was already rare in Livonia, though still common in Russia, Poland, Lithuania, and Lapland, as well as in Curland; though in Derschau and von Keyserling's description of the Province of Curland, published as early as 1805, the Glutton is not mentioned among the animals of Curland, and it is likewise wanting in Lichtenstein's Catalogue of the Mammals of Curland, published in 1829 (Bull. Nat. Hist. Moscou). Kessler only mentions the Glutton incidentally, stating that there were reports of its casual appearance, and that a specimen was once captured, though giving no particulars. According to Rezaczynski (Auctuar. Hist. Nat. p. 311), two Gluttons were killed in Podolia at the beginning of the last century. It is, therefore, unquestionable that the Glutton

---

* Bemerkungen über die Wirbelthiere des nördlichen europäischen Russlands, besonders des nördlichen Ural's. Ein Beitrag zur näheren zoologisch-geographischen Kenntniss Nordost-Europa's.

was occasionally found in Curland, Lithuania, and Podolia during the last century, but that it no longer extends so far westward and southward, so that we may rely, concerning its appearance in Russia. upon the statement of Pallas (Zoög. R.-A. i. p. 74) that the animal was seldom found in European Russia, except in the northerly forests, though common in Siberia. In East Siberia, Sarytschew (Reise, i. p. 77) discovered it on the middle portions of the Indigirka. Wrangel (Reise, ii. pp. 274, 238) indicates the occurrence of the Glutton in Werchojansk and the country of the Tschukts. Gebler (Uebersicht d. Katunischen Geb. p. 84) calls Gulo borealis a solitary inhabitant of the Altaï forests, and we once received from him a specimen from the Altaï region. According to von Middendorff, the Glutton is also found on the Boganida River, whence it makes excursions to the Tundra, to plunder the traps set for the Vulpes lagopus. It was lately observed by Wosnessenski in Kamtschatka, where it was more numerous in northern than in southern portions. There, particularly in the Anadyr regions, it is said to inhabit the Tundras rather than the forests. . . . . Georgi (*l. c.*) designates the Ural in general, Lehmann (Brandt in Lehmann's Reise Zoolog. Append. p. 301) and Eversmann, probably more rightly, only the middle and northerly Ural as its habitat. According to Hoffmann's verbal communications, the animal is to be found in the northerly Ural, at least as far as forests exist, as before indicated by Georgi, and seems to be not rare there, for a skin costs but three silver roubles, and the Samojeds are in the habit of trimming their garments with the fur. Ermann (Reise, i. 1, p. 562) states that the Glutton occurs on the Obi River. Schrenck (Reise, i. pp. 10, 66, 97) reports that it is found in the forests of the District of Mesen, particularly on the Pinega River, and sometimes on Onega Lake. The government of Wologda annually delivers 300 to 500 Glutton skins (von Baer and Helmersen, Beiträge, vii. p. 251). I do not recall, after more than twenty years' experience in the government of St. Petersburg, a single instance of a Glutton's having been captured there. Wallenius (Fauna Fenn. p. 11, and Förteckning öfver Sällsk. Samlingar, p. 7) cites the Glutton as inhabiting the Finnish provinces of Tawastland and Osterbotten. We may safely fix its present distribution in the Russian possessions from Finland and Russian Lapland (?) to Kamtschatka, and from the middle Ural and Altaï to the northerly Tundra.

CHAPTER III.

MUSTELINÆ—Continued: THE MARTENS.

The genus *Mustela*—Generic characters, &c.—Analysis of North American species—*Mustela pennanti*, the Pekan or Pennant's Marten—Synonymy—Habitat—Specific characters—Description of external characters—Dimensions—Skull and vertebræ—General history, habits, and geographical distribution—Interpolated matter relating to exotic species of *Mustela*—*M. martes*—Synonymy—Description of its skull and teeth—*M. foina*—Synonymy—Notes on its characters—*M. zibellina*—Synonymy—Measurements of skulls of the three species—Comparative diagnoses of *M. martes*, *americana*, and *foina*—*Mustela americana*, the American Sable or Marten—Synonymy—Description and discussion of the species—Table of measurements—Geographical variation in the skull—General history and habits of the species.

IN this chapter are treated the genus *Mustela* and the two species by which it is represented in North America. Several closely allied species of the Old World are also introduced, as seemed to be required for the adequate discussion of their intimate relationships.

The Genus MUSTELA. (LINN., 1758, *emend.*)

< Mustela, *Linn.* Syst. Nat. i. 10th ed. 1758, and of many authors.
< Viverra, *Shaw,* Gen. Zoöl. i. 1800; not of authors.
< Gulo, *H. Smith,* (fide *Gray*); not of *Storr.*
= Martes, *Authors,* after *Ray.*
> Pekania, J. E. *Gray,* Proc. Zoöl. Soc. Lond. 1865, 107. (Type *M. pennanti*.)
> Foina, J. E. *Gray,* Proc. Zoöl. Soc. Lond. 1865, 107. (Type *M. martes* var. *fagorum*.)
> Charronia, J. E. *Gray,* Proc. Zoöl. Soc. 1865, 108. (Type *M. flavigula* Bodd.)

GENERIC CHARS.—*Dental formula:* i. $\frac{3-3}{3-3}$; c. $\frac{1-1}{1-1}$; pm. $\frac{4-4}{4-4}$; m. $\frac{1-1}{2-2}$ = $\frac{18}{20}$ = 38 (as in *Gulo*; one more premolar,* above and below, on each side, than in *Putorius*). Sectorial tooth of lower jaw usually with an internal cusp. Anteorbital foramen presenting vertically or somewhat downward as well as forward (as in *Putorius*; reverse of *Gulo*); canal-like, opening over interspace between last and penultimate premolars (as in *Gulo*; the opening is over the last premolar in *Putorius*). Skull much constricted at the middle, the rostral portion relatively longer, more tapering than in *Gulo* or

---

* As a not infrequent anomaly, the small anterior premolar which constitutes the increment in the dental formula as compared with that of *Putorius* fails to develop. Thus it is wanting on the right side above in a skull, No. 7159, from Fort Good Hope, though present on the left side and on both sides below. Similarly, an incisor occasionally aborts.

Putorius, and much more obliquely truncated than in *Putorius*, less so than in *Gulo*; frontal profile more or less concave. Nasal bones narrower in the middle than at either end. General upper outline of skull moderately arched. Production of mastoids and auditory bullæ and general prominence of periotic region intermediate between *Gulo* and *Putorius*. Zygomatic arch high behind (usually higher than in front); nowhere vertical, nor developing a posterior convexity. Depth of emargination of palate greater than distance thence to the molars. Skull as a whole less massive than in either *Gulo* or *Putorius*.

Vertebral formula.—According to Gerrard, the vertebræ of *M. martes*, type of the genus, is c. 7, d. 16, l. 6, s. 3, cd. 1^2 or 19; other species of the genus differ in the number of caudals.

Size medium and large for this subfamily. Form moderately stout; claws strong, curved, acute. Tail longer than the head, bushy, cylindrical or tapering. Soles densely furry, with naked pads. Pelage long and soft, but not shaggy; whole-colored, or nearly so, never whitening in winter. Progression digitigrade. Habits highly arboreal as well as terrestrial; not aquatic.

This genus forms the link between *Gulo* and *Putorius*, as will be evident upon comparison of the diagnoses of the three. The skull, however, is lighter than in either of the two other genera of *Mustelinæ*, with more produced and tapering rostrum; in height, relative to length or breadth, it is intermediate. The species have a somewhat Fox-like or Cat-like superficial aspect, rather than that appearance we usually associate with the name of "Weasel", being much stouter-bodied, more copiously haired, and bushier-tailed; one species, indeed, is commonly called black "fox" or black "cat". They appear to be more decidedly arboreal than the Weasels, spending much of their time in trees, and are not aquatic, like several of the Weasels proper. They are agile and graceful in their movements; and, if not really more active than the Weasels, their actions seem to possess a quality of lightness and elasticity different from the writhing and insinuative motions of the very slender-bodied, short-legged Weasels. Although strictly carnivorous, predacious, and destructive to many kinds of small Mammals and Birds, they appear less ferocious and bloodthirsty than the Weasels, whose sanguinary impulses seem insatiable; and at times they exhibit a playful and comparatively amiable disposition.

The name of the genus is the Latin *mustela* or *mustella*, a kind of Weasel; the word being apparently derived from, or related to, the more general term *mus*.* Its adjectival deriva-

---

* This seems to have included, besides Mice, various kinds of small destructive animals, such as now might be collectively referred to as "vermin". Thus, *mus ponticus* is supposed to have probably been an Ermine. The word may be simply a long form of *mus*, like *maxilla* or *axilla* from *mala* and *ala*.

tive, *mustelinus*, refers primarily to general Weasel-like qualities; secondarily, to the peculiar tawny color of most species of Weasels in summer, and is transferable to other animals, just as "foxy" signifies the peculiar red of the Common Fox. For an example, familiar to ornithologists, the "tawny" thrush of Wilson, *Turdus "mustelinus"*, may be cited.

This genus comprises the Martens and Sables, of which there are several species, inhabiting the northern portions of both Old and New Worlds, and particularly abundant in the higher latitudes. Aside from the very distinct Fisher, or Pekan, *Mustela pennanti*, peculiar to North America, the species are so closely related that some authors have contended for their identity. There appear, however, to be good grounds upon which at least three if not four species may be established; one confined to America, the rest belonging to the Old World. The high commercial value of the pelts of these animals, and their corresponding economic importance, has sharpened the eyes of those pecuniarily interested to such degree that numerous kinds of "sable" and "marten" are recognized by the furriers; and the caprices of imperious fashion set a wholly fictitious value upon slight shades of color or other variable conditions of pelage, which have no zoölogical significance whatever. The Sable *par excellence* is the Russian and Asiatic animal, *Mustela zibellina*, a variety of the common European Marten, *M. martes*, or a closely allied species; but, as all Sables are Martens, it is perfectly proper to speak of our species, *M. americana*, as the American Sable or Marten. Some of its fortuitous conditions of pelage—the darker shades—represent the "sable" of furriers, while in the ordinary coloration it may be called by another name. The meanings of the various terms employed to designate animals of this genus are more fully discussed elsewhere, under heads of the several species.

The two North American species of *Mustela* may be readily distinguished by the following characters:—

Analysis of the North American species of Mustela.

Larger: length two feet or more; tail a foot or more, the vertebræ about half the length of head and body, tapering from base to point. Ears low, wide, semicircular. Blackish; lighter on fore upper parts and head; darkest below; no light throat-patch. . . . . *M. pennanti.*

Smaller: length under two feet; tail less than a foot long, the vertebræ less than half the length of the body, uniformly bushy. Ears high, subtriangular. Brown, &c.; not darker below than above; usually a large yellowish or tawny throat-patch. . . . . *M. americana.*

The Pekan, or Pennant's Marten.
Mustela pennanti.
PLATE II.

Mustela pennantii, *Erxl.* Syst. An. 1777, 470, no. 10 (based on the *Fisher of Pennant*: for discussion of name, in question of priority over *canadensis* Schreber, cf. Bd. *op. infrà cit.* p. 151).—*Zimm.* Geogr. Gesch. ii. 1780, 310, no. 208.—*J. Sab.* Frank. Journ. 1823, 651.—*Griff.* Cuv. R. A. v. 1827, 125, no. 354.—*Less.* Man. 1827, 150, no. 405.—*Fisch.* Syn. Mam. 1829, 217.—*Godm.* Am. N. H. i. 1831, 203.—*Bd.* M. N. A. 1857, 149, pl. 36, f. 1.—*Newb.* P. R. R. Rep. vi. 1857, 41.—*Coop. & Suckl.* N. H. W. T. 1860, 92, 114.—*Ross*, Canad. Nat. vi. 1861, 24.—*Gilpin*, Tr. Nov. Scot. Inst. ii. 1870, 9, 59.—*All.* Bull. M. C. Z. i. 1869, 161 (Mass.); Bull. Ess. Inst. vi. 1874, 54 (Colorado).—*Ames*, Bull. Minn. Acad. Nat. Sci. 1874, 69.

Martes pennantii, *Gray*, P. Z. S. 1865, 107; Cat. Carniv. Br. Mus. 1869, 85.

Mustela canadensis, *Schreb.* Säug. iii. 1778, 492, pl. 134 (based on the *Pekan* of Buffon; not *M. canadensis* of Erxl., which is *Putorius vison*).—*Zimm.* Geog. Gesch. ii. 1780, 309, no. 207.—*Bodd.* Elench. An. i. 1784, 86.—*Gm.* S. N. i. 1788, 95.—*Turt.* S. N. i. 1806, 59 (not same name on p. 57, which is the American Otter).—*Kuhl*, Beitr. 1820, 74.—*Desm.* Mamm. i. 1820, 183, no. 284; Ency. Méth. pl. 80, f. 4; Nouv. Dict. xix, 379.—*Harl.* Fn. Amer. 1825, 65.—*Less.* Man. 1827, 149.—*Griff.* Cuv. R. A. v. 1827, 124, no. 353.—*Fisch.* Syn. 1829, 216.—*Rich.* F. B. A. i. 1829, 52, no. 19.—*Rich.* Zoöl. Voy. Blossom, 1839, 10*.—*Fr. Cuv.* Dict. Sci. Nat. xix. 256.—*Martin*, P. Z. S. 1833, 97 (anat.).—*Emmons*, Rep. Quad. Mass. 1840, 38.—*Wagn.* Suppl. Schreb. ii. 1841, 223.—*De Kay*, N. Y. Zoöl. i. 1842, 31, pl. 13, f. 1 (skull).—*Aud. & Bach.* Q. N. A. i. 1849, 307, pl. 41.—*De Kay*, Fifth Ann. Rep. Reg. Univ. N. Y. 1852, 33, pl. (orig. fig).—*Thomps.* N. H. Vermont, 1853, 32.—*Kenn.* Trans. Ill. State Agric. Soc. 1853-4, 578 (Illinois).—*Kneel.* Proc. Bost. Soc. vi. 1858, 418.—*Giebel*, Odontog. 36, pl. 12, f. 1; Säug. 1855, 773.—*Maxim.* Arch. Naturg. 1861, 229.—*Billings*, Canad. Geol. and Nat. ii. 1857, 116.—*Kneeland*, Proc. Bost. Soc. vi. 1859, 418 (skeleton).—*Hall*, Canad. Nat. vi. 1861, 296.—*Maxim.* Verz. N. A. Säug. 1862, 43.

Mustela canadensis var. **alba**, *Rich. op. cit.* 54 ("White Pekan"; albinism).

Mustela (Martes) canadensis, *Schinz*, Syn. Mamm. i. 1844, 334.

Martes canadensis, *Gray*, List Mamm. Br. Mus. 1843, 63.—*Gerrard*, Cat. Bones Br. Mus. 1862, 91.

Viverra canadensis, *Shaw*, Gen. Zool. i. 1800, 429.

Mustela melanorhyncha, *Bodd.* Elench. An. 1784, 88, no. 13 (based on *Fisher* of Pennant).—*Zimm.* in Penn. Arkt. Zoöl. 1787, 85.

Viverra piscator, *Shaw*, Gen. Zoöl. i. 1800, 414 (based on *Fisher* of Pennant).

Mustela nigra, *Turt.* ed. L. S. N. i. 1806, 60 (= *Fisher Weasel* of Pennant).

Mustela godmani, *Fisch.* Syn. Mamm. 1829, 217.—*Less.* Mamm. 1842, 150.

Mustela piscatoria, *Less.* Man. 1827, 150, no. 403 (quotes *pennanti* Erxl. with query).

"**Gulo castaneus et ferrugineus**, *H. Smith.*"—(*J. E. Gray.*)

Pekan, *Charlev.* Nouv. France, iii. 1744, 134.—*Buff.* Hist. Nat. xiii, 1765, 304, pl. 42 (basis of *M. canadensis* Schreb.).—*Bomm*, Dict. d'Hist. Nat. iii. 1768, 401.—*Pekano*, *Scataglia*, An. Quad. iv. 1773, pl. 155, f. 1 (ex Buffon).—*Pekan*, Penn. Syn. Quad. 1771, 234, no. 159; Hist. Quad. 1781, no. 204; Arct. Zoöl. i. 1784, 78, no. 28 (apparently same as the animal of Brisson and Buffon).—*Pekan* of French Canadian. ("Pecan" is also found.)

Pekan-marder, *Maxim. l. c.*

Fisher, *Penn.* Syn. Quad. 1771, 223, no. 157; Hist. Quad. 1781, 322, no. 202; Arct. Zoöl. i. 1784, 82, no. 31 (basis of *M. pennantii*, *Erxl.*).—*Fisher*, American, *Vulgo*.

Marte-pêcheur, *Desm. op. cit.* 184.

Pennant's Marten, *Godm. op. et loc. cit.*

Wejack, *Hearne*, Journ. —, 378. (Also written *Weejack*.)

Otchock, *Cree Indians* (*Richardson*) ‧ *O'tschilik*, *Ojibways* (*Maxim.*) = *Wejack*, *Fur Traders* = *Woodshock*, *Anglicè*.

Thâ-chô, *Chippewayans* (*Ross*).

Black Fox, Black Cat, *Vulgo*.

HAB.—North America, approximately between 35° and 65°, in wooded portions of the country.

Specific Characters.—Black or blackish, lightening by mixture of brown or gray on the upper fore part and head; no conspicuous light throat-patch; generally darker below than above; very large and stout; length 2 or 3 feet; tail over a foot long.

Description of external characters.*

Form.—With its large size, this animal combines a stoutness of form not seen in other species of the genus. The general aspect is rather that of a Fox than of a Weasel, but, in place of the acute muzzle and pointed ears of the former, we have a fuller face, somewhat canine in physiognomy.† The muzzle is thick and short; the prominent nasal pad has the ordinary T-shape, and is definitely naked; it is black. The whiskers are stiff, scant, and short, hardly reaching to the ears; there are other stoutish bristles over the eye, at the corner of the mouth, on the cheeks and chin; they are all black. The eye is rather large and full for this group. The ears are low, but remarkably broad, being about twice as wide at base as high; they are rounded in contour, and well furred, both sides, to the entrance of the meatus. The feet are broad and flat, furred both sides, and armed with very stout, compressed, much curved, acute claws, not hidden by the bristles at their base; they are light-colored. On the palm may usually be seen the following pads (though they are sometimes hidden by the overgrowing fur): one at the end of each digit; a V-shaped area of four nearly coalesced pads, indicated by mere sulci between them, situated opposite the first digit, and indicating the bases of, respectively, the first, the second, the third and fourth combined, and the fifth digit. There is a tenth pad, isolated from the rest, far back, on the wrist, near the outer border. On the hind feet, the arrangement of the naked bulbs is essentially the same, excepting that the hindermost (tenth) one is wanting. The tail-vertebræ are about half as long as the head

---

*From various specimens in the Smithsonian Institution.

†"The physiognomy of the Pekan is very different from that of the Marten. When the latter is threatened, its features resemble those of an enraged cat, but the expression of the Pekan's countenance approaches to that of a dog, though the apparent obliquity of its eyes gives it a sinister look. The head has a strong, roundish, compact appearance, and contracts suddenly to form the nose, which terminates rather acutely. The ears, low and semicircular, are far apart, so as to leave a broad and slightly rounded forehead. They are smaller in proportion than the ears of the Pine-martin. The eyes, situated where the head curves in to form the nose, appear more oblique than they really are."—(Richardson, *l. c.*)

and body. The tail is cylindric-conic, rapidly tapering to a sharp point from the enlarged and bushy base. The general pelage is much coarser than that of the true Martens, and looser, if not longer; it consists of the usual under fur, with long, glossy, bristly hairs intermixed. The pelage is very short on the head.

Color.—Color is very variable, according to age, season, or other fortuitous circumstances; in general, however, a particular pattern, if not also tone, is preserved. The animal is darker below than above, at least on parts of the belly, contrary to the usual rule in this group. The belly, legs, and tail, in most mature examples, are black or blackish-brown, and the hinder part of the body above is much the same. On the rest of the upper parts, however, there is a progressive lightening toward the head, from increasing admixture of light brown and gray shades, which colors, occupying but little, if any, of the length of the hairs on the dark parts of the body, on the lighter parts so increase in extent that they give the prevailing tone, overpowering both the smoky-brown bases and the blackish tips of the hairs. The ear has usually a light bordering. On the under parts, even of the blackest individuals, are usually found irregular white (not tawny or buffy) blotches on the chest, in the arm-pits, and on the lower belly between the thighs. The throat may also show a few white hairs, though I have never observed anything like the conspicuous light gular area commonly displayed by the Marten.

Smaller specimens before me lack much of the general blackishness above indicated; still the feet, tail, and at least a median abdominal area are darker than the upper parts in general, though the darkness is rather brown than black. The light upper parts are pale or hoary gray, overlaid with the blackish tips of the hairs. Both Richardson and Audubon note nearly white specimens.

Dimensions.—Of the full-grown animal, about 30 inches from nose to root of tail (many specimens are only about 2 feet long, while others a third larger than this are noted). Tail-vertebræ about 14 inches (12 to 16), the terminal hairs 2 to 4 inches longer. Nose to eye 2 inches; to occiput, over curve of head, 6½; ear 1 inch high, about twice as broad; distance between tips of ears 7 inches; hind foot 4½; fore leg, from elbow, 6 or 7 inches; hind leg, from hip, nearly 12. Individuals are said to range in weight from 8½ to 18 pounds.

Skull and vertebræ.

Cranium.—The skull of this species is instantly distinguished from that of *M. martes* by its obviously superior size. The largest of six examples before me measures 4.40 in extreme length by 2.40 in greatest zygomatic width. The under jaw is 3.00 in length. There are other points. The zygomatic arch is notably lower. The skull is more contracted behind the orbits. The lambdoidal (occipital) crest is stronger and more flaring; its termination as a broad flange back of the meatus auditorius is conspicuous when the skull is viewed from above, whereas in the skull of *M. martes*, held in the same position, the terminations of this crest are almost hidden by the bulge of the brain-box. The bony palate is more narrowly and deeply emarginate behind. The bullæ auditoriæ are relatively smaller and flatter; the meatus is absolutely smaller. Some other minor points might be established. I observe no noteworthy dental peculiarities, aside from superior size of the teeth. This skull exceeds in length the large fossil one mentioned by Prof. Baird from the Bone Cave of Pennsylvania, which is little over 4 inches long. Several New York skulls are less than 4 inches in length by little over 2 in greatest breadth. One skull, of a very old animal, in which the sutures are all obliterated, is remarkably massive, and broad for its length, measuring only just 4 inches long by full 2.40 in breadth. This series of skulls, like others in this group, shows that the character of the sagittal crest, or elevation, is wholly transitory; in old specimens, the crest is a thin laminar ridge, while in others there is a median longitudinal elevation half an inch or more in width. The lambdoidal crest is subject to the same modifications. The constriction of the skull back of the supraorbital processes also increases with age.

Vertebræ:—c. 7, d. 14, l. 6, s. 3, cd. 20 or 21 (Gerrard). Kneeland (*loc. suprà cit.*) gives the caudals as 20; the rest of his formula agrees with Gerrard's. Of the 14 ribs, he gives 10 as apparently "true" (sternal).

The Pekan is much the largest of the genus, and indeed of the whole Weasel kind (subfamily *Mustelinæ*), excepting only the Wolverene and Grison. In size, as in some other points of form, vigor, and ferocity, it approaches the Wolverene, and is obviously the connecting link between *Mustela* and *Gulo*. It has no immediate representative in the Old World.

GENERAL HISTORY, HABITS, AND GEOGRAPHICAL DISTRIBUTION OF THE SPECIES.

As this species is confined to North America, and as it presents marked zoölogical characters, its written history is less extensive and less involved than that of animals which have a circumpolar distribution in both the Old and New World. In tracing up this matter, we go back to the works of Buffon, Brisson, and Pennant, all of whom appear to have described the animal from the same specimen—one in the cabinet of M. Aubry at Paris. It is the *Pekan* of Buffon, 1765, and the *Fisher* of Pennant, Syn. Quad. 1771. Pennant's account of his *Fisher* is unmistakable; but he describes, in addition, the *Pekan* of Buffon, not recognizing in it the same species. These two accounts furnished for many years the bases of all the scientific binomial names imposed by various authors. The *Mustela canadensis* of Schreber, 1777, is the *Pekan* of Buffon; the *M. pennantii* of Erxleben, 1777, and *M. melanorhyncha* of Boddaert, 1784, are the *Fisher* of Pennant. This is perfectly plain; but a question of priority arises between the names *pennantii* Erxl. and *canadensis* Schreb., owing to some uncertainty of actual date of publication of the works of Erxleben and Schreber, since the supposed earlier author quotes the other in various places. Judging, however, by the printed dates of publication, as the proper means of arbitration, *pennantii* of Erxleben takes precedence. The question is, however, further complicated by the fact that Erxleben has also a *Mustela canadensis* (p. 455), which included both the *Vison* and *Pekan* of Buffon—the Mink and the Fisher; and many authors have adopted the name for the latter. But, as Prof. Baird has clearly shown, Erxleben's description of *M. canadensis* applies solely to the Mink, and, indeed, will take precedence over *M. vison*, if Brisson be not quotable as an authority in binominal nomenclature. As a summary of the subject, therefore, it may be said that *M. canadensis* Erxl. goes to the Mink, while *M. canadensis* Schreb. and authors sinks to a synonym of *M. pennantii* Erxl.

In later years, various nominal species have been established upon the Pekan, none of which, however, require special discussion.

The name *Fisher*, very generally applied to this species by others as well as authors, is of uncertain origin, but probably arose from some misconception of its habits, or from confounding them with those of the Mink. The name is entirely inapplicable, as the animal is not aquatic, does not fish, nor habitually live upon fish, and it should be discarded, as likely to perpetuate the confusion and misunderstanding of which it has always to a greater or less extent been the cause. *Pekan* is a word of unknown,* or at least of no obvious, application, but is less objectionable, inasmuch as it does not mislead. As to the supposed piscatorial exploits of the Pekan, we find refutation in some of the very earliest accounts of those who, unlike certain compilers of books, had actual knowledge of the animals they recounted. Thus Bartram, who is quoted by Pennant, states that "though they are not amphibious, and live on all kinds of lesser quadrupeds, they are called *Fishers*". Hearne states that they dislike water as much as cats do. In fact, the universal testimony of those who are best informed is that the economy of the Pekan is as nearly as possible like that of the Pine Marten, as indeed one would expect, judging by analogy. Godman, a naturalist who has perhaps not always been fully appreciated, states the case correctly in criticising the same points:—"That it will eat fish when thrown on shore there is little doubt, as almost all the carnivorous animals are delighted with such food: but we have no proof that this Marten is in the habit of *fishing* for itself." Sir John Richardson has a paragraph which may be quoted in continuation of this point, as well as for its affording further insight into the character of the species:—

"The Pekan is a larger and stronger animal than any variety of the Pine Marten, but it has similar manners: climbing trees with facility, and preying principally upon mice. It lives in the woods, preferring damp places in the vicinity of water, in which respect it differs from the Martin, which is generally found in the dryest spots of the pine forests. The Fisher is said to prey much upon frogs in the summer season; but I have been informed that its favorite food is the Canada porcupine, which it kills by biting on the belly. It does not seek its food in the water, although, like the Pine-martin, it will feed upon the hoards of frozen fish laid up by the residents. . . . . It brings forth, once a year, from two to four young."

Doubt has been cast by Audubon upon Richardson's state-

* Compare *Ptan* or *Petan*, the Assiniboine name of the Otter, which may possibly have become transferred with modification to the present species.

ment that the Pekan kills the Porcupine; but its accuracy is attested by Mr. Gilpin in the article above quoted, who states that Porcupine quills have been found in its stomach.

A modified derivation of the name *Fisher* is given by De Kay:—"We are informed by a person who resided many years near Lake Oneida, where the Fisher was then common, that the name was derived from its singular fondness for the fish used to bait traps. The hunters were in the practice of soaking their fish over night, and it was frequently carried off by the fisher, whose well known tracks were seen in the vicinity. In Hamilton County it is still [1842] numerous and troublesome. The hunters there have assured me that they have known a fisher to destroy twelve out of thirteen traps in a line not more than fourteen miles long." The same author continues:—"The hunting season for the fisher, in the northern part of the State, commences about the tenth of October, and lasts to the middle of May, when the furs are not so valuable. The ordinary price is $1.50 per skin; but it is not so fine, nor so highly valued as that of the sable." According to all accounts, the animals were formerly very abundant in the State of New York, where, however, they have latterly become restricted to northern mountainous and thinly settled portions.

The bone caves of Pennsylvania, according to Baird, have furnished numerous remains of Pennant's Marten, among them one skull larger than some recent ones examined (but compare p. 65). The animal may be still found occasionally in the mountains north of Carlisle, in Perry County, where the living animal figured by Audubon was procured.

The distribution of the Pekan is general in wooded districts throughout the greater part of North America. As indicating approximately the southern limit of its distribution (for, like the Marten and Ermine, it is essentially a northern animal), we may refer to its occurrences in North Carolina and Tennessee, as attested by Audubon and Bachman. The parallel of 35° may be near its limit. Mr. Allen recently ascertained its presence in Colorado. West of the Rocky Mountains it was long ago noted by Lewis and Clarke, whose accounts of the "Black Fox" are checked by numerous later observers, as Newberry, Cooper, and Suckley, who found it in Washington and Oregon Territories. From California, however, I have no advices, though the animal probably inhabits at least a part of that State. Dr. Newberry says it is rare in Oregon, but less so in

Washington Territory. According to Dr. Suckley, it is found quite plentifully in the thickly wooded districts along the eastern, and probably also the western, slopes of the Cascade Range, especially in the neighborhood of streams; it also inhabits the Blue Mountains of the same region. In the eastern United States, it must not be presumed that it actually occurs now throughout its ascribed range; for the settlement of the country practically restricts it to the more inaccessible or at least unfrequented wooded districts. Many years ago, as we have already seen, it had become greatly thinned out in the Middle States, and this process has been steadily progressing, until, at the present day, the Pekan is almost unknown in most of the United States east of the Mississippi. Writing in 1853, Mr. Kennicot states it "used frequently to be seen" in Illinois in the heavy timber along Lake Michigan. In New England, according to Mr. Allen, it probably still occurs, though rarely, in the Hoosac ranges. In 1840, Dr. Emmons reported it as occasionally found in the vicinity of Williamstown, Mass., especially in the mountainous ranges which extend through Stamford, Vt. It is stated to be rare in Canada, and not found at all in the populous districts. In Nova Scotia, according to Dr. Gilpin, it was never very plenty, and is being rapidly exterminated, only two hundred at most being taken yearly, chiefly in the high wild region of the Cobequid Hills in Cumberland. In British America, Sir John Richardson states that it is found as far north as Great Slave Lake, latitude 63°; and the specimens I have examined confirm this dispersion, extending it to include Alaska also.

The Pekan is stated to breed but once a year: it brings forth its young in the hollow of a tree, usually 30 or 40 feet from the ground. Two, three, and four young, but not more, so far as I have learned, are produced in a litter. It has been known to offer desperate resistance in defence of its young, as on the occasion when the individual figured by Audubon was procured. This animal, a young one, was kept in confinement for several days. "It was voracious, and very spiteful, growling, snarling and spitting when approached, but it did not appear to suffer much uneasiness from being held in captivity, as, like many other predacious quadrupeds, it grew fat, being better supplied with food than when it had been obliged to cater for itself in the woods." Another mentioned by the same author as having been exhibited in a menagerie in

Charleston, S. C., some months after its capture, continued sullen and spiteful, hastily swallowing its food nearly whole, and then retiring in growling humor to a dark corner of its cage. Hearne, however, has remarked that the animal is easily tamed, and shows some affection at times. When taken very young, it may become perfectly tame, and as playful as a kitten; such was the case with a pair mentioned by Mr. B. R. Ross.

The Pekan is sometimes forced, by failure of other sources of supply, to a vegetarian diet, when it feeds freely upon beech-nuts.

In continuation of the history of this animal, which I have had no opportunity of studying in the living state, the following paragraphs are quoted from the authors just mentioned, as illustrative of its habits and manners:—

"Pennant's Marten appears to prefer low swampy ground; we traced one which had followed a trout stream for some distance, and ascertained that it had not gone into the water. Marks were quite visible in different places where it had scratched up the snow by the side of logs and piles of timber, to seek for mice or other small quadrupeds, and we have no doubt it preys upon the Northern hare, gray rabbit, and ruffed grouse, as we observed a great many tracks of those species in the vicinity. It further appears that this animal makes an occasional meal on species which are much more closely allied to it than those just mentioned. In a letter we received from Mr. Fothergill, in which he furnishes us with notes on the habits of some of the animals existing near Lake Ontario, he informs us that 'a Fisher was shot by a hunter named Marsh, near Port Hope, who said it was up a tree, in close pursuit of a pine marten, which he also brought with it.' . . .

"Whilst residing in the northern part of our native State (New York), thirty-five years ago, the hunters were in the habit of bringing us two or three specimens of this Marten in the course of a winter. They obtained them by following their tracks in the snow, when the animals had been out in quest of their prey the previous night, thus tracing them to the hollow trees in which they were concealed, which they chopped down. They informed us that as a tree was falling, the Fisher would dart from the hollow, which was often fifty feet from the ground, and leap into the snow, when the dogs usually seized and killed him, although not without a hard struggle, as the Fisher was infinitely more dangerous to their hounds than either the gray or red fox. They usually called this species the Black Fox.

"A servant, on one occasion, came to us before daylight, asking us to shoot a raccoon for him, which, after having been chased by his dogs the previous night, had taken to so large a tree that he neither felt disposed to climb it nor to cut it down. On our arrival at the place, it was already light, and the dogs were barking furiously at the foot of the tree. We soon perceived that instead of being a raccoon, the animal was a far more rare and interesting species, a Fisher. As we were anxious to study its habits we did not immediately shoot, but teased it by shaking some grape vines that had crept up nearly to the top of the tree. The animal not only became thoroughly frightened, but seemed furious; he leaped from branch to branch, showing his teeth and growling at the same time; now and then he ran half way down the trunk of the tree, elevating his back in the manner of an angry cat, and we every moment expected to see him leap off and fall among the dogs. He was brought down after several discharges of the gun. He seemed extremely tenacious of life, and was game to the last, holding on to the nose of a dog with a dying grasp. This animal proved to be a male; the body measured twenty-five inches, and the tail, including the fur, fifteen. The servant who had traced him, informed us that he appeared to have far less speed than a fox, that he ran for ten minutes through a swamp in a straight direction, and then took to a tree. . . .

"Species that are decidedly nocturnal in their habits, frequently may be seen moving about by day during the period when they are engaged in providing for their young. Thus the raccoon, the opossum, and all our hares, are constantly met with in spring, and early summer, in the morning and afternoon, whilst in autumn and winter they only move about by night. In the many fox hunts, in which our neighbours were from time to time engaged, not far from our residence at the north, . . . we never heard of their having encountered a single Fisher in the daytime; but when they traversed the same grounds at night, in search of raccoons, it was not unusual for them to discover and capture this species. We were informed by trappers that they caught the Fisher in their traps only by night.

"On several occasions we have seen the tracks of the Fisher in the snow; they resemble those of the pine marten, but are double their size. To judge by them, the animal advances by short leaps in the manner of a mink."

I will supplement this account with the interesting experiences of Mr. B. R. Ross (as recounted *l. s. c.*) with this species in the Mackenzie River region :—" In this district it is not found except in the vicinity of Fort Resolution, which may be considered as its northern limit. In the numerous deltas of the mouth of Slave River it is abundant, frequenting the large grassy marshes or prairies, for the purpose of catching mice, its principal food. In appearance it bears a strong family likeness to both the martin and the wolverene. Its general shape assimilates more to the former, but the head and ears have a greater similitude to those of the latter. It is named by the Chippewayan Indians 'Thá chô,' or great martin. Its neck, legs and feet are stouter in proportion than those of the martin, and its claws much stronger. In color and size it varies greatly. Young full-furred specimens, or those born the previous spring, can scarcely be distinguished from a large martin except by a darker pelage and a less full, more pointed tail. As it advances towards old age, the color of the fur grows lighter, the long hairs become coarser, and the grayish markings are of greater extent and more conspicuous.

"The largest fisher which I have seen was killed by myself on the Rivière de Argent, one of the channels of the mouth of Slave River, about 15 miles from Fort Resolution. It was fully as long as a Fulvus fox, much more muscular, and weighed 18 pounds. In the color of its fur the greyish tints preponderated, extending from half way down the back to the nose. The fur was comparatively coarse; though thick and full. The tail was long and pointed, and the whole shade of the pelage was very light and had rather a faded look. Its claws were very strong and of brown color; and as if to mark its extreme old age the teeth were a good deal worn and very much decayed. I caught it with difficulty. For about two weeks it had been infesting my martin road, tearing down the traps and devouring the baits. So resolved to destroy it, I made a strong wooden trap. It climbed up this, entered from above, and ate the meat. A gun was next set but with no better success, it cut the line and ran off with the bone that was tied to the end of it. As a 'dernier resort' I put a steel trap in the middle of the road, covered it carefully, and set a bait at some distance on each side. Into this it tumbled. From the size of its footprints my impression all along was that it was a small wolverene that was annoying me, and I was surprised to find it to be a fisher. It shewed good fight, hissed at me much like an enraged cat, biting at the iron trap, and snapping at my legs. A blow on the nose turned it over, when I completed its death by compressing the heart with my foot until it ceased to beat. The skin when stretched for drying was fully as large as a middle sized otter, and very strong, in this respect resembling that of a wolverene.

"In their habits the fishers resemble the martins. Their food is much the same, but they do not seem to keep so generally in the woods. They are not so nocturnal in their wanderings as the foxes. An old fisher is nearly as great an infliction to a martin trapper as a wolverene. It is an exceedingly powerful animal for its size, and will tear down the wooden traps with ease. Its regularity in visiting them is exemplary. In one quality it is however superior to the wolverene, which is that it leaves the sticks of the traps where they were planted: while the other beast if it can discover nothing better to hide, will cache them some distance off. It prefers meat to fish, is not very cunning, and is caught without difficulty in the steel-trap. Fishers are caught by methods similar to those employed in fox-trapping."

It may not be generally known that the Pekan successfully assaults an animal as large as the Raccoon; indeed, that the abundance of the latter in some districts depends in a measure upon the rarity of the former. The following letter, addressed to Prof. Baird, in 1857, by Mr. Peter Reid, of Washington County, New York, sufficiently attests these facts:—"Raccoons are more numerous here now than they were at the first settlement of the country, or for some time subsequent. Thirty years ago they were so seldom found, that many boys 15 or 18 years old had scarcely seen one. Before the increase in their numbers I once witnessed a circumstance that satisfied my mind on this score. Whilst hunting, early one winter I found the carcase of a freshly killed sheep, and by the tracks around it in the light snow perceived that a Fisher had surprised a Raccoon at the feast. A hard chase had ensued, the Raccoon tacking at full speed to avoid his pursuer, the Fisher outrunning and continually confronting his intended victim. I saw where at length the Fisher had made an assault, and where a bloody contest had evidently ensued. The Raccoon, worsted in the encounter, had again broken away, and the chase was resumed, but with diminished energy on the part of the Raccoon; the animal had been soon overtaken again, and a still more desperate encounter had taken place. The Coon had failed fast, and it had at length become merely a running fight, when both animals had entered a swamp where it was impossible for me to trace them further; but I have no doubt the Coon was killed. I have witnessed similar engagements between the Mink and Muskrat, the Weasel and House Rat, always ending in the death of the assaulted. The Fisher has been nearly extinct in these parts for about twenty-five years, and this to my mind accounts for the great increase in numbers of the Raccoon."

INTERPOLATED MATTER RELATING TO THE EXOTIC MUSTELÆ, MM. MARTES, FOINA, AND ZIBELLINA.

Before entering upon the discussion of the intimate relationships of the American Sable or Pine Marten with its extralimital allies, some notice of the latter seems to be required in order to a better understanding of the intricate questions concerned. I accordingly present the three exotic species, with such remarks as seem called for and as I am able to offer. The material before me indicates, with little hazard of error, that the American form is specifically distinct from both the Beech Marten and the Pine Marten of Europe. Its relationships with the Asiatic Sable seem to be closer, but these I am unable to discuss satisfactorily, owing to lack of specimens of the Asiatic animal.

Note.—Much of the synonymy relating to these exotic species has been rather summarily compiled at second hand, and should be taken with the allowance for " probable error " which usually obtains in such cases.

1. The European Pine Marten.

Mustela martes.

PLATE III.

Martes, *Antiquorum.*—*Aldrov.* Quad. Digit. 1645. 331.—*Charlet.* Exercit. 1677, 20.—*Wagn.* Helvet. 16c0, 1z1.—*Sibb.* Scots. Illust. 16s4, ii. 11.—*Rzacz.* Polon. 1721, 222; 1736, 314.—*Linn.* S. N. i. 2d ed. 1740, 44.—*Jonst.* Theatr. Quad. 1755, pl. 64.
Martes sylvestris, *Gesn.* Quad. 1551, 863, fig.—*Jonst.* Theatr. Quad. 1755, 156.
Martes arborea, *Schwenckf.* Theriotroph. 1603, 110.
Martes in arboribus, *Agric.* Anim. Subter. 1614, 38.
Martes abietum, *Ray*, Syn. Quad. 1693, 200.—*Klein*, Quad. 1751, 64—*Fleming*, Br. An. 1828, 14.—*Bell*, Brit. Quad. 1837, 174 ; 2d ed. 1874, 217.—*Gerr.* Cat. Bones. Br. Mus. 1862, 90.—*Gray*, List Mamm. Br. Mus. 1843, 63 ; P. Z. S. 1865, 104 ; Cat. Carn. Br. Mus. 1869, 81.—*Fitz.* Naturg. Säug. i. 1861, 325, f. 67.
Martes abietum *vars.* **martes, vulgaris, altaica**, *Gray*, P. Z. S. 1865, 104 ; Cat. Carn. Br. Mus. 1869, 82 (but obviously not *Mustela altaica* Pall., which is a *Putorius*).
Mustela fulvo nigricans, gula pallida, *Linn.* Fn. Suec. 1st ed. 1746, 3, no. 7 ; Syst. Nat. ed. 6th, 1748, 5, no. 2.—*Hill*, Hist. An. 1752, 546, pl. 27.—*Kram.* Elench. An. 1756, 311.

Mustela martes, *Briss.* Quad. 1756, 247, no. 8.—*L.* Fn. Suec. 2d ed. 1761, 6, no. 15; S. N., 10th ed. 1758, 46, no. 5; i. 1766, 67, no. 6.—*Müll.* Zool. Dan. Prod. 1776, 3, no. 12.—*Erxl.* Syst. 1777, 455, no. 4.—*Schreb.* Säug. iii. 1778, 475, pl. 130.—*Zimm.* Geol. Gesch. ii, 1780, 303, no. 197.—*Herm.* Obs. Zool. 45.—*Wildung.* Tasch. 1800, 24, pl. 3.—*Gm.* S. N. i. 1782, 95, no. 8.—*Bechst.* Naturg. Deutschl. i. ——. 769.—*Vselol.* Mém. Soc. Nat. Mosc. i. 1806, 249 (hybridity with cat).—*Turt.* S. N. i. 1806, 59.—*Pall.* Zoög. i. 1811, 85.—*Desm.* Mamm. i. 1820, 181, no. 280; Ency. Méth. pl. 81, f. 4.—*Fr. Cuv.* Mamm. iii. livr. 62; Dict. Sci. Nat. xxix. 255, fig. 1.—*Geoff.* Dict. Class. x. 209.—*Less.* Mam. 1827, 143.—*Fisch.* Syn. 1829, 214.—*Jenyns,* Brit. Vert. 1835, 11.—*Siem.·Piet.* Arch. Naturg. 1839, 251.—*Keys. & Blas.* Wirb. Eur. i. 1840, 67.—*Selys-L.* Fn. Belg. 1842, 8.—*Blainv.* Compt. Rend. xiv. 1842, p. 210 seq. pls.; Ostéogr. 1842, —.—*Burk.* N. Act. Leop. xxv. 1843, 660.—*Schinz,* Syn. 1844, 35.—*Bp.* Fn. Ital. iv. f. —.—*Gleb.* Fn. Vorw. Säug. ——, 56; Odout. ——, 33; Säug. 1855, 774.—*Hensel,* Arch. Naturg. xix. 1853, 17.—*Brandt,* Bem. Wirb. Nord. Eur. Russl. 23; Beit. Kennt. Säug. Russl. i. 1855, pl. iv. —*Midd.* Sib. Säug. 69, pl. 2, f. —.—*Schrenck.,* Reise Amurl. ——, 36.—*Blas.* Wirb. Deutschl. 1857, 213, f. 121. 122.—*Jäckel,* Zool. Gart. xiv. 1873, 457 (albino).

Mustela martes *var.* **abietum,** *L.* S. N. i. 1766, 67.

Viverra martes, *Shaw,* G. Z. i. 1800, 410.

Martes sylvatica, *Nilss.* Skand. Fn. (*Martes sylvestris* Gesn.).

Mustela vulgaris, *Griff.* Cuv. R. A. v. 1827, 123, no. 349.

Marder, *Riding.* Abbild. Thiere. 1740, pl. 19.—*Müll.* Samml. iii. ——, 315; Naturs. 1773, 267.—*Martens,* Zool. Gart. xi, 1870, p. 254 (philological).—*German.*

Martre, *Charlev.* Nouv. France, iii. 1744, 134.—*French.*

Baummarder, *Haller,* Naturg. Vierf. Tb. 1757, 457.—*German.*

Marte, *Briss. l. c.* Buff. Hist. Nat. vii. 186, pl. 22.—*Bom.* Dict. iii. 1768, 37.—*French.*

Marte commune, *Cuv.* R. A. i. 149.

Marter, *Houtt.* Nat. Hist. Dieren, ii. 1761, 193.—*Belgic.*

Maar, *Pontopp.* Dan. i. 1763, 610.—*Danish.*

Martora, *Scatag.* An. Quad. ii, pl. 69. (from Buffon).—*Italian.*

Feldmarder, *Mart.* Buff. Vierf. Tb. iv. 159.

Edelmarder, *Germ.*

Marta, *Spanish.*

Märd, *Swedish.*

Martin,‡*Penn.* Syn. Quad. 1771, 215, no. 154; Brit. Zoöl. 3³, fig.

Pine marten, Sweet Marten, Yellow-breasted Marten.—*English.*

Description of the skull and teeth of M. martes.

(See Plate III.)

The skull and teeth of *M. martes* may be described in general terms to illustrate this part of the structure of the genus, and to serve as a standard of comparison for the other closely related species. The points in which they specially differ from that of *Putorius* are elsewhere summed. The skull indicates considerable strength, particularly in the rostral portions, where it is massive (still it is not so strong relatively as in either *Gulo* or *Putorius*); the cranial part is thinner, and usually gives indication of the cerebral folds within. Most of the sutures are early obliterated; those of the nasals, bullæ auditoriæ, and zygomatic processes of squamosal and malar are the last to disappear. The nasals persist separate from each other long after they fuse with the maxillaries.

The zygomatic width of the skull is more than half its length; these arches are upright, but are borne well away from the skull by the outward obliquity of their roots, both fore and aft. From an egg-shape cerebral part, the skull tapers to a decided postorbital constriction; this is approximately of the same (more or less) width as the rostral part. The cerebral part is rather broader than high. The upper profile of the skull is slightly convex, sloping more rapidly down behind, with a frontal concavity and oblique nasal orifice. The roof of the brain-box is convex in every direction; a temporal "fossa" being only indicated by the ridges (sagittal and lambdoidal), which indicate the extent of the temporal muscle. The sagittal crest divaricates anteriorly to run out to each supraorbital process: in old animals, it is a thin high ridge; in the young, a tablet of greater or less width. The occipital crest rises and flares with age, but is always a thin edge. The occipital depression below this is well marked; the condyles are notably projecting, and connected by a sharp ridge below the foramen magnum. The mastoids are not conspicuous. The bullæ are large, elongate, oblique, convex forward; a slight constriction across them, and some outward prolongation, develops a tubular meatus. Excepting the bullæ, the general floor of the skull is quite flat. The palate is completely ossified some distance back of the molars, and nearly plane. A broad, deep emargination lies between the pterygoids; these are simply laminar, vertical, and terminate in a well marked hamular process. The palatal plates of the intermaxillaries, when not fused, are seen to be of very slight extent; the small incisive foramina do not reach as far back as the hinder border of the canines. The orbits are pretty well defined by the curve of the zygoma and presence of supraorbital processes, but are not otherwise distinguished from the general temporal cavity. The anteorbital foramen is large, high up over the fore edge of the last premolar. The nasal orifice has a well-marked and little irregular bony parietes.

The jaw has a lightly and somewhat irregularly convex inferior profile. The coronoid plate is large, erect, its apex reaching or slightly overlapping the zygomatic arch. The angle of the jaw is a slight sharp process. The condyle is low, about on the level of the teeth, broad from side to side, but very narrow in the opposite direction. Its reception in the glenoid fossa is close, but the articulation does not lock as in *Meles* or *Taxidea*.

The single upper molar is completely tubercular, low, flat, with irregular minor elevations and depressions, much broader transversely than lengthwise, subquadrate in general contour, partly divided by a slight median constriction (both vertical and horizontal), with an inner and outer moiety, whereof the former more or less considerably exceeds the outer in length. The inner border of this inner moiety is always strongly convex, with a raised brim. In typical *M. martes*, the inner moiety is twice (to speak roundly) as large as the outer. In *M. americana*, much as in *foina*, the disproportion is obviously less. The outer border of the outer moiety in *martes* is simply convex; in the other forms just mentioned it is more or less emarginate. The inner moiety shows one tubercle within the brim; the outer has two such.

The next tooth—last premolar—is the largest of all, and sectorial in character, but with a prominent fang projecting inward from the anterior end. In profile, it shows a large, pointed, central cusp, flanked before and behind with a small one. There is quite an excavation between the large central and small posterior cusps. The next two molars, of nearly equal size, are much smaller than the last, but repeat its characters in diminishing degree, *minus* the antero-internal fang. The remaining anterior premolar is very small. It is a simple conical cusp, with a slight heel behind, but none before; it occasionally aborts. The large canines are not peculiar. The six incisors are closely crowded; the outer pair are much larger than the rest; these are all alike. The outer are regularly curved, with an enlarged cingulum around the base; the others start obliquely forward from the jaw, then turn vertically downward with an appreciable angle.

In the lower jaw, of the two molars the hindermost is small, circular, and completely tuberculous. The next is the largest of the under teeth, chiefly sectorial in character, but with a depressed, rimmed, tubercular, posterior moiety. This rim at each of its ends rises into a slight cusp, but the inner one is merely a slight heel to the central cusp, instead of a prominent point as in *M. foina*. The two main cusps of the tooth are much higher, the hinder one highest, compressed, with cutting edge, forming with each other the usual V-shape reëntrance, continued further down as a closed slit. The last premolar is a conical cusp augmented posteriorly by a secondary cusp half as high, and with a heel both before and behind at the base. The next premolar is like the last, but smaller, with a mere trace of the secondary cusp, though it is well heeled fore and aft. On the next premolar, the secondary cusp entirely subsides in a general gentle slope from the summit of the tooth to its base behind, and the front heel is not developed. The first premolar is simply a minute knob. It looks like a tooth hardly yet established, or else about to disappear. The lower canines are shorter, stouter, and more curved than the upper. The six incisors are greatly crowded between the canines, so much so that, through lack of room, one at least sometimes fails to develop, leaving only *five*, as in more than one specimen before me. They are smaller than the upper ones, and not so regular, for one or a pair—most frequently the middle one—on each side is crowded back out of the plane of the rest. As in the upper jaw, the outer pair of under incisors are the largest, and have slightly clubbed and bilobate tips.

2. The European Beech Marten.

Mustela foina.

PLATE IV.

Martes domestica, *Gesn.* Quad. 1551, 865, fig.—*Aldrov.* Quad. Digit. 1645, 332.—*Jonst.* Theatr. Quad. 1755, 156.
Martes saxatilis, *Schwenckfeld*, Theriotroph. 1603, 110.
Martes in saxis, *Agric.* Anim. Subter. 1614, 38.
Martes fagorum, *Ray*, Syn. Quad. 1693, 200.—*Flem.* Br. An. 1828, 14.
Martes saxorum, *Klein*, Quad. 1751, 64.
Mustela foyna, *Briss.* Quad. 1756, 246, no. 7.—*Pall.* Zoögr. R. A. i. 1811, 86.
Mustela foina, *White*, Phil. Trans. lxiv. 1774, 196.—*Erxl.* Syst. An. 1777, 458, no. 5.—*Schreb.* Säug. iii. 1778, 494, pl. 129.—*Zimm.* Geogr. Gesch. ii. 1780, 302, no. 196.—*Gm.* S. N. i. 1788, 95, no. 14.—*Herm.* Obs. Zoöl. 42.—*Wildung.* Tasch. für 1800, —.—*Bechst.* Naturg. i. —, 755.—*Desm.* Mamm. i. 1820, 182; Nouv. Dict. xix. 380; Ency. Méth. pl. 81, f. 1.—*Fr. Cuv.* Dict. Sci. Nat. xxix. 254.—*Is. Geoff.* Dict. Class. x. 209.—*Griff.* An. Kingd. v. 1827, 123, no. 350.—*Jen.* Br. Vert. 1835, 11.—*Selys-L.* Fn. Belg. i. 1842, 9.—*Keys. & Blas.* Wirb. Eur. 1840, 67.—*Schinz*, Syn. Mamm. i. 1844, 336.—*Blainv.* Compt. Rend. xiv. 1842, 210 seq. pls.—*Girb.* Odont. 33, pl. 12, f. 3; Säug. 1855, 775.—*Hensel*, Arch. Naturg. xix. 1853, 17.—*Power*, Ann. Mag. N. H. 2d ser. xx. 1857, 416.—*Brandt*, Bemerk. Wirb. Eur. N. E. Russl. —, 24.—*Blas.* Wirb. Deutschl. 1857, 217, f. 123.—*Jäckel*, Zool. Gart. xiv. 1873, 437 (albino).
Viverra foina, *Shaw*, Gen. Zoöl. i. 1800, 409.

Martes foina, *Bell*, Brit. Quad. 1837, 167; 2d ed. 1874, 208.—*Gray*, List Mamm. Br. Mus. 1843, 63; P. Z. S. 1865, 102; Cat. Carn. Br. Mus. 1869, 86.—*Gerr.* Cat. Bones Br. Mus. 1862, 91.

Mustela martes *var.* **foina**, *L. S. N.* i. 1766, 67.

Mustela foisna, *Chatin*. Ann. Sci. Nat. 5th ser. xix. 1874, 97 (anat.).

Fouine, *Briss. op. loc. cit.*—*Bomare*, Dict. d'H. N. ii. 1769, 242.—*Buff.* Hist. Nat. vii. 161.—*Cuv.* R. A. i. 149.—*French.*

Foina, *Scatag.* Quad. ii. pl. 68.—*Italian.*

Steinmarder, *Hall.* Naturg. Vierf. Th. 1757, 459.—*German.*

Steinmarder *oder* **Buchmarder**, *Meyer*, Vorstell. Thiere. pl. 4.—*German.*

Hausmarder, *Mart.* Buff. Vierf. Th, 147, pl. 61 *a*.—*Schr.* Fn. Boic. i. no. 9.—*German.*

Martin, *Penn.* Syn. Quad. 1771, 215. no. 154; Br. Zoöl. 33. (*House, Stone, and Beech Marten, Martern, Marteron, Martlett.*)

Fuina, *Spanish*.

The Beech or Stone Marten, which seems to be well established as a species, may usually be distinguished from the Pine Marten by the pure white throat and some other external features, as well as by some difference in habits. But stronger characters are found in the skull and teeth. Some differences in the proportions of the skull are obvious, and sufficient to confer a recognizably different physiognomy; the rostral part of the skull is much shorter. The frontal profile above is more sloping; the zygomatic width is relatively greater. The zygoma is regularly arched throughout, instead of rising abruptly behind and then sloping down gradually forward. The anterior root of the zygoma, owing to the shortness of the muzzle, is nearly half-way from the supraorbital process to the end of the skull; it is much further back in *M. martes*. The palate is much shorter and broader for its length. The back upper molar is very notably less massive; its inner moiety is but little larger than the outer; the latter is nicked on the outer border, whereas in *M. martes* the inner moiety of the same is nearly twice as large as the outer, and the border of the latter is strongly convex. In *M. foina*, the inner anterior fang of the last premolar is very small and oblique; in *M. martes*, it is much larger and projects inward at a right angle. The next premolar is appreciably smaller than the same tooth in *M. martes*. These dental peculiarities, taken from specimens before me, are confirmatory of Blasius' diagnosis. The skulls are 3.25 or less in total length by about 1.90 in greatest width; those of *M. martes* are 3.50 or more in length, with a width scarcely greater than in *M. martes*. It seems a slight difference in the figures, but the resulting modification in shape is decided. Similarly, the palate of *M. foina* is about 1.40 in length by 0.90 in greatest width inside the teeth; that of *M. martes* is 1.70 in length, with no greater width. As a practical means of appreciating these differences, let one take the jaw of *M. martes*, and try to fit it to a skull of *M. foina*, or conversely. Cautious and accurate observers, like Daubenton and Bell, have recorded their doubts of the specific distinctness of the two forms; but Bell, at least, has found reason to change his opinion, while the views of many equally good judges are concurrent with those here adopted.

3. The Asiatic Sable.

Mustela zibellina.

Mustela sobella, *Gesn.* Quad. 1551, 869.—*Rzacz.* Anct. Polon. 1736, 317.
Mustela zobela, *Forer.* Allg. Thierb. Gesner, 1669, 347.
Mustela zibellina, *Aldrov.* Quad. Digit. 1645, 335.—*Charlet.* Exercit. 1677, 20.—*Ray*, Syn. Quad. 1693, 201.—*Linn.* S. N. 2d ed. 1740, 44; 6th ed. 1748, 5, no. 7.—*Klein*, Quad. 1751, 64.—*Jonst.* Theatr. Quad. 1755, 156.—*Linn.* S. N. 10th ed. 1758, 46, no. 8; 12th ed. 1766, 68, no. 9.—*J. G. Gm.* N. C. Petrop. v. 338, pl. 6.—*Erxl.* Syst. An. 1777, 467, no. 9.—*Schreb.* Säug. iii, 1778, 478, pl. 136.—*Zimm.* Geogr. Gesch. ii. 1780, 302, no. 196.—*Pall.* Spic. Zool. xiv. 1780, 54, pl. 3, f. 2; Zoög. R. A. i. 180, 83, pl. 6.—*Turt.* S. N. i. 1806, 59.—*Gm.* S. N. i. 1788, 96, no. 9.—*Müll.* Russ. Gesch. iii. 495.—*Desm.* Mamm. i. 1820, 182, no. 282; Nouv. Dict. xix. 382; Ency. Méth. pl. 82.—*Fr. Cuv.* Dict. Sci. Nat. xxix. 1823, 255.—*Is. Geoff.* Dict. Class. x. 210.—*Griff.* An. Kingd. v. 1827, 124, no. 351.—*Less.* Man. 1827, 148.—*Fisch.* Syn. 1829, 216.—*Blaind.* Compt. Rend. xiv. 1842, 210 seq. pls.—*Schinz*, Syn. Mam. i. 1844, 336.—*Gieb.* Säug. 1855, 776.—*Brandt*, Bemerk. Wirb. N. E. Russl. 21.—*Midd.* Sibir. Säug. 68, pl. 2.—*Schrenck*, Reise Amurl. ——, 27.
Mustela martes zibellina, *Briss.* Quad. 1756. 248, no. 9.
Viverra zibellina, *Shaw*, Gen. Zoöl. i. 1800, 411.
Martes zibellina, *Gray*, P. Z. S. 1865, 105 (" *Mustela* " *lapsu*); Cat. Carn. Br. Mus. 1869, 83.
Martes zibellina *var.* **asiatica,** *Brandt*, Beit. Kennt. Säug. Russl. 1855, 6, pll. i, ii, and pl. iii, f. 7, 8, 9 (many " subvarieties " named).
Zobela, *Agric.* Anim. Subter. 1614, 39.
Sebellina, *Scheff*, Lappon. 1673, 343.
Zobel, *Strahlenb.* Eur. u. Asia, 1730, 430.—*J. G. Gm.* Reise, i. 1751, 391.—*Hall.* Naturg. Vierf. Thiere, 1757, 459.—*Müll.* Naturs. 1773, 272.—*Stell.* Kamtscht. 1774, 119.—*Martens*, Zool. Gart. xi, 1870, 254 (philological).
Sabeldier, *Houtt.* Nat. Hist. Dieren, ii. 1761, 204.—*Dutch.*
Zibeline, Buff. Hist. Nat. xiii. 1765, 309.—*Bomare*, Dict. iv. 1768, 630.—*French.*
Cebellina, Cevellina, *Spanish.*—**Zibellino,** *Italian.*—**Sabbel,** *Swedish.*—**Sobol,** *Polish, Russian.*
Sable, *Penn.* Syn. Quad. 1771, 217, no. 156; Hist. Quad. 322, no. 201; Arct. Zoöl. i. 1784, 79, no. 30. (*Saphilinas Pelles*, sable skins, is found in Jornandes; *Zombolines* occurs in Marco Polo.— *Webster*.)

Lack of specimens of this form unfortunately prevents me from bringing it into the discussion upon any original investigations; the views of authors are discussed beyond. I have, however, carefully examined both skins and skulls of *M. martes*, *foina*, and *americana*. Such is the variability of the pelage, that probably no decisive indications can be gathered from comparisons of the skins, however widely these may differ in extreme cases. The skulls and teeth afford the readiest means of separating these three closely-allied forms.

The following measurements of three skulls, selected as fairly expressing averages of *M. martes*, *foina*, and *americana* respectively, will show in what the cranial differences consist. The skull of *M. foina* differs more from those of both *M. martes* and *M. americana* than these latter do from each other; but these latter are readily distinguished by their dental characters.

Measurements of skulls of MM. FOINA, MARTES, and AMERICANA.

	M. foina (Germany).	M. martes (Sweden).	M. americana (Alaska).
	Inches.	*Inches.*	*Inches.*
Total length	3.25	3.50	3.60
Greatest width	1.85	1.95	1.85
Least width (exclusive of muzzle)	0.80	0.80	0.70
Distance between orbits	0.85	0.85	0.80
Upper incisors from front hinder margin of palate	1.55	1.75	1.75
Upper molars and premolars, length taken together	0.95	1.10	1.10
Lower jaw, length, from apex of symphysis to back of condyle	2.05	2.40	2.40
Lower jaw, height angle to top of coronoid	0.90	1.00	1.10
Front border of orbit, end of intermaxillary	0.95	1.10	1.10
Width of muzzle behind canines	0.68	0.70	0.70
Greatest length of zygoma	1.50	1.65	1.70
Greatest width of palate inside teeth	0.85	0.90	0.80
Width across supraorbital processes	1.00	0.95	0.90
Greatest length (longitudinal) of back upper molar	0.23	0.28	0.20
Width of cranium proper	1.40	1.40	1.40

The indications afforded by the foregoing measurements, together with some other cranial and dental characters, may be summed in the following diagnostic paragraphs. It will be seen that most of the cranial points brought out by Prof. Baird (*op. cit.* p. 155) are substantiated, but it must be borne in mind that they are matters of *degree*, which may not always hold, except of averages. The remarkable difference in the back upper molar, as insisted upon by Gray, is the principal character upon which to rely between *martes* and *americana*.

Comparative diagnoses.

M. martes.—Inner moiety of back upper molar one-third longer than outer moiety, and altogether about twice as large (coincidently with which the entire dentelure of *martes* is stronger than in the other two forms, though differences in particular teeth are not readily expressed); outer border of outer moiety regularly strongly convex. Fang of last upper premolar large, transverse. Penultimate under molar with a cusp well developed at the postero-internal base of the main cusp. Sides of muzzle nearly parallel. Supraorbital processes midway between greatest constriction of cranium and anterior root of zygoma; the constriction moderate. Zygomatic width more than half total length of skull.

M. americana.—Inner moiety of back upper molar scarcely longer or larger than the outer [in 25 skulls examined]; outer border of outer moiety double-convex, *i. e.*, with an emargination. Fang of last premolar small, oblique. Penultimate under molar with merely a slight heel at base inside of the main cusp. Sides of muzzle sensibly tapering. Supraorbital processes nearer point of greatest constriction than anterior root of zygoma; constriction great. Zygomatic width about half total length of skull.

M. foina.—Molar and last upper premolar as in *americana*; penultimate lower molar with prominent supplementary cusp as in *martes*. Sides of muzzle sensibly tapering. Supraorbital processes much nearer point of greatest constriction than anterior root of zygoma; constriction slight. Zygomatic width much more than half total length of skull.

The American Sable or Marten.

Mustela americana.

PLATE V.

Mustela martes, *Forst.* Phil. Trans. lxii. 1772, 372.—*J. Sab.* Frankl. Journ. 1823, 651.—*Harl.* Fn. Amer. 1825, 67 (quotes a "*Mustela vison* var.")— *Warden,* Hist. U. S. v. 1819, 613.—*Rich.* F. B. A. i. 1829, 51, no. 17.—*Gapp.* Zoöl. Journ. v, 1830, 203.—*Godm.* Am. N. H. i. 1831, 200.—*Emmons,* Rep. Quad. Mass. 1840, 40.—*De K.* N. Y. Zoöl. i. 1842, 32, pl. 11, f. 2, pl. 19, f. 2 (skull).—*Aud. & Bach.* Q. N. A. iii. 1853, 176, pl. 138.—*Thomps.* N. H. Vermont, 1853, 32.—*Billings,* Canad. Nat. and Geol. ii. 1857, 463.—*Allen,* Bull. M. C. Z. i. 1870, 161 (critical).—*Kenn.* Tr. Ill. State Agric. Soc. for 1853-54, 1855, 578.—*All.* Bull. Ess. Inst. vi. 1874, 54, 59 (Colorado and Wyoming).—*Hall,* Canad. Nat. and Geol. vi. 1861, 295.

Mustela americana, *Turton,* ed. L. S. N. i. 1806, 60.—*Bd.* M. N. A. 1857, 152, pl. 36, f. 2 (skull), pl. 37, f. 1 (skull).—*Newb.* P. R. R. Rep. vi. 1857, 41.—*Kneel.* Proc. Bost. Soc. N. H. vi. 1858, 418.—*Coop. & Suckl.* N. H. W. T. 1860, 92.—*Ross,* Canad. Nat. vi. 1861, 25.—*Gilpin,* Tr. Nov. Scot. Inst. ii. 1870, 10, 59.—*Ames,* Bull. Minn. Acad. Nat. Sci. 1874, 69.—*Coues & Yarrow,* Zoöl. Expl. W. 100 Merid. v. 1875, 61 (Taos, N. M.).—*Allen,* Bull. U. S. Geol. Surv. vol. ii. no. 4, 1876, 328 (skull).

Martes americana, *Gray,* P. Z. S. 1865, 106; Cat. Carn. Br. Mus. 1869, 84.

Martes americana vars. **abietinoides, huro,** et **leucopus,** *Gray, ll. cc.*

Mustela zibellina var. **americana,** *Brandt.* Beit. Säug. Russl. 1855, 16, pl. 3, f. 10 (critical).

Mustela zibellina, *Godm.* Am. Nat. Hist. i. 1831, 208 (refers to true Sable, but the American species described).

Mustela vulpina, *Raf.* Am. J. Sc. i. 1819, 82; Phil. Mag. 1819, 411; Isis, 1834. 452.(Upper Missouri River) (tail white at end).—*Fisch.* Syn. 1829, 215.

Mustela (Martes) vulpina, *Schinz,* Syn. Mamm. i. 1844, 337.

Mustela leucopus, *Kuhl,* Beit. 1820, 74.—*Fisch.* Syn. 1829, 216.

Mustela (Martes) leucopus, *Schinz,* Syn. Mamm. i. 1844, 337.

Martes leucopus, *Gray,* List Mamm. Br. Mus. 1843, 63.—*Gerr.* Cat. Bones Br. Mus. 1862, 91.

Mustela leucopus, *Griff.* Cuv. R. A. v. 1827, 126, no. 357.

Mustela huro, *F. Cuv.* Dict. Sci. Nat. xxix. 1823, 256; Suppl. Bull. i. 1831, 221.—*Is. Geoff.* Dict. Class. x. 211.—*Fisch.* Syn. 1829, 217.

Mustela (Martes) huro, *Schinz,* Syn. Mamm. i. 1844, 337.

Mustela martinus, *Ames,* Bull. Minn. Acad. Nat. Sci. 1874, 69.

Martin or **Marten, Pine Marten, American Sable,** of American writers.

Wawpeestau, Wawbeechius, Wappanow, *Indian* (Richardson).

Description and discussion of the species.*

This animal is about the size of a large House Cat, though standing much lower on account of the shortness of the legs. The length of the head and body is about a foot and a half, more or less; the tail with the hairs is a foot long or less; the tail-vertebræ are less than half as long as the head and body. The tail is very full and bushy, particularly toward the end, the reverse of the tapering pointed shape which obtains in *M. pennanti.* The longer hairs of the tail at and near the end measure about 3 inches. The head is quite broadly triangular, or rather conical, with the contraction of the muzzle beginning

---

* Prepared from numerous specimens in the Smithsonian Institution.

at the site of the eyes. These are oblique, and situated about over the angle of the mouth, midway between the snout and the ears. The latter are quite high, somewhat pointed, though obtusely so, but not regularly orbicular as in *M. pennanti;* their height above the notch is rather greater than their width at base; they are closely hairy on both sides. The longest whiskers reach to the back of the ears; there are other bristles over the eyes, on the cheeks, and chin. The end of the snout is definitely naked in T-shaped area, as usual in this genus. The limbs are short and stout; the feet appear small in comparison with the calibre of the legs. The outstretched hind legs reach more than half-way to the end of the tail. The soles are ordinarily densely furred, only the ends of the pale-colored claws appearing. But in the frequent specimens observed with scant-haired soles, the tubercles may be distinctly seen, without parting the fur; they have the ordinary disposition.

The pelage is long and extremely soft and full. It consists of three kinds of fur. The first is very short, soft, and wool-like, immediately investing the skin, as may be seen upon plucking away both kinds of the longer hairs. The second is soft and kinky, like the first, but very much longer, coming to the general surface of the pelt. The third is the fewer, still longer, glossy hairs, bristly to the roots.

It is almost impossible to describe the colors of the Pine Marten, except in general terms, without going into the details of the endless diversities occasioned by age, sex, season, or other incidents. The animal is "brown", of a shade from orange or tawny to quite blackish; the tail and feet are ordinarily the darkest; the head lightest, often quite whitish; the ears are usually rimmed with whitish; on the throat, there is usually a large tawny-yellowish or orange-brown patch, from the chin to the fore legs, sometimes entire, sometimes broken into a number of smaller, irregular blotches, sometimes wanting, sometimes prolonged on the whole under surface, when the animal is bicolor, like a Stoat in summer. The general "brown" has a grayish cast, as far as the under fur is concerned, and is overlaid with rich lustrous blackish-brown in places where the long bristly hairs prevail. The claws are whitish; the naked nose-pad and whiskers are black. The tail occasionally shows interspersed white hairs, or a white tip.

Upon this subject, I cannot do better than quote again from

the article of Mr. B. R. Ross, who describes the Marten from long experience of its variations:—

"The winter fur of this species is full and soft, about an inch and a half deep, with a number of coarse black hairs interspersed. The tail is densely covered with two kinds of hair, similar to those of the back but coarser. The hairs on the top are longest, measuring 2¼ inches, and giving the end a very bushy appearance. The fur is in full coat from about the end of October until the beginning of May, according to locality. When in such condition the cuticle [*sic*, meaning skin viewed from inside] is white, clean, and very thin. From the latter of these dates the skin acquires a darker hue, which increases until the hair is renewed, and then gradually lightens until the approach of winter, the fur remaining good for some time before and after these changes. When casting its hair the animal has far from a pleasing appearance, as the under fur falls off leaving a shabby covering of the long coarser hairs, which have then assumed a rusty tint. The tail changes later than any other part, and is still bushy in some miserable looking summer specimens now lying before me. After the fall of these long hairs, and towards the end of summer, a fine short fur pushes up. When in this state the pelage is very pretty and bears a strong resemblance to a dark mink in its winter coat. It gradually lengthens and thickens as winter approaches, and may be considered prime after the first fall of snow.

"It is difficult to describe the color of the martin fur accurately. In a large heap of skins (upwards of fifty) which I have just examined minutely there exists a great variety of shades darkening from the rarer of yellowish-white and bright orange, into various shades, of orange brown, some of which are very dark. However, the general tint may with propriety be termed an orange brown, considerably clouded with black on the back and belly, and exhibiting on the flanks and throat more of an orange tint. The legs and paws as well as the top of the tail are nearly pure black. The claws are white and sharp. The ears are invariably edged with a yellowish white, and the cheeks are generally of the same hue. The forehead is of a light brownish gray, darkening towards the nose, but in some specimens it is nearly as dark as the body. The yellowish marking under the throat, (considered as a specific distinction of the pine martins) is in some well defined and of an orange tint, while in others it is almost perfectly white. It also varies much in extent, reaching to the fore legs on some occasions. At other times it consists merely of a few spots, while in a third of the specimens under consideration it is *entirely wanting.*

"After minutely comparing these skins with Prof. Baird's and Dr. Brandt's description of the martins, and the latter gentleman's paper on the sables, I find that the *M. Americana* of this district agrees in general more closely with the latter, and am therefore disposed to coincide with that gentleman in his opinion that they are only varieties. The martins of this district bear a greater resemblance to the sables of Eastern Siberia than to the martins of Europe, holding, as it may be with propriety said, an intermediate position. I am also inclined to believe that the various colors found in these regions are simply varieties of the same species, and that the differences if any, seen in the Zib. [*sic, lege zibellina*] are merely continental. In summer, when the long hairs have fallen off, the pelage of this animal is darker than in winter. The forehead changes greatly, becoming as deeply colored as any other part of the body, which is of an exceedingly dark brown tint on the back, belly and legs. The yellow throat-markings are much more distinct at this season, but vary much both in color and extent, though in only one summer skin are they absolutely wanting. The white edging on and around the ears still remains, but the cheeks assume a grayer tint. The tail is not so full, but from the high North latitude (the Arctic coast) from which these skins were procured it is still rather bushy. One of the specimens has the dark hairs laid on in thin longitudinal stripes, causing a curious appearance."

The last paragraph brings us directly to the consideration of the position which the American Marten holds among its congeners. Upon this vexed question it is incumbent upon me to review the testimony for and against the specific distinction of this animal from the Old World Pine Marten and Sable, and to state clearly the grounds upon which my own conclusions rest. Passing over some earlier accounts, which, owing to inadequacy or lack of point, are entirely superseded by later and better investigations, we may examine four authors who have made the subject a matter of special examination, namely, Gray, Brandt, Baird, and Allen.

In the first place, *M. foina* may be thrown entirely out of the question. It is now almost universally admitted to be a distinct species, even by the most cautious and conservative writers, some among whom, like Bell, were formerly inclined to the contrary opinion. Some external characters, more or less obvious and constant, like the white gular patch, are correlated with perfectly definite and satisfactory cranial and dental peculiarities, as elsewhere detailed in this paper.

In discussing the European and American Pine Martens, to which I will now direct attention, Gray, Brandt, and Baird were agreed upon specific distinction. Allen dissented from such view, reviving the case as presented by Richardson, Audubon, and others. Dr. Gray made the separation entirely upon the character of the posterior upper molar. Dr. Brandt elaborately detailed external characters of size, proportion, color, and character of pelage. Baird adduced certain cranial and dental as well as external features. Allen confined himself to external points. Finding that the accounts of authors are unsatisfactory or conflicting in these respects (as may be truly said to be the case), observing the great admitted range of variation, and not examining the skulls and teeth, he disallowed specific validity. I myself, with ample material before me, do not find sufficient grounds derived from examination of the skins alone for admitting the specific distinction of *M. americana* and *martes* (but it is otherwise when the skull and teeth are considered). Some of the alleged distinctions obviously fail. Thus, there is no difference in the furring of the soles (cf. BAIRD, *op. cit.* p. 154); in the animal from either country, the pads may be exposed or concealed according to season or locality. Many of the minute points of coloration adduced by Brandt cannot be verified, and, indeed, are negatived in the examination of sufficient series of specimens. Prof. Baird has, I think, most pertinently summed the case in the following terms (*l. c.*):—"The Swedish specimens are much larger, although the skulls appear to indicate the same age. The fur is harsher and coarser, and the prevailing tints paler; the tail and feet are not very dark brown, instead of being almost black. The color of the fur at base is lighter. The throat-patch does not touch the fore legs. The tails of the European specimens appear longer in proportion to the body . . . . " This greater length of the tail is also attested by Brandt, who says that the tail-vertebræ in *M. martes* equal one half or more of the length of head and body, and extend nearly one-third beyond the outstretched hind legs. This distinction is confirmed, as an average character, by the specimens before me, though, like other matters of mere degree, it is subject to some uncertainty of determination. I similarly endorse, on the whole, a lighter, grayer, more uniform coloration of *M. martes*, although in the interminable variations of *M. americana* probably no infallible distinctions can be substantiated. But all these points have a certain value when correlated, as they should be, with the cranial and dental peculiarities. These are decided, and, I think, not open to reasonable question as affording good specific characters. Baird has tabulated most of them, and the specimens I have examined confirm nearly all the distinctions he has sought to establish. While he has not, as asserted by Gray, overlooked certain dental peculiarities, he has perhaps not laid the stress upon them which is warranted. Gray rests secure, I think, in basing the primary distinction upon the remarkable features presented by the back upper molar. We may bring the points to mind by saying that in *M. martes* we find an hourglass-shaped tooth with one bulb (the inner) very much larger than the other; while in *M. americana* there is less median constriction, nearly an equality in size of the two bulbs, and an emarginate instead of simply convex exterior contour of the outer bulb. There are coördinated dental characters: the last upper premolar in *M. martes* has a strong, directly transverse, inner fang; the same in *M. americana* is smaller and oblique. The penultimate lower molar in *M. martes* develops a comparatively strong supplementary cusp at the base on the inner side of the main cusp, represented in *M. americana* merely by a slight heel. It is to these dental characters that I primarily refer in predicating, as I do, specific validity of *M. americana*. I coördinate them with the cranial characters elsewhere detailed, and supplement them with the less essential external features already noted, in coming to the conclusion that the American is not the Pine Marten of Europe.

The question then narrows to the characters of *M. americana* in comparison with those of *M. zibellina*, the true "Russian" Sable. Gray separates the two upon dental peculiarities; the Sable having, according to his determination, the same dental characters as *M. martes*. I regret that I have not been able to verify this. If it indeed holds, it would be sufficient to settle the issue between *M. zibellina* and *M. americana*, whatever might then become of the ascribed and supposed differences between the former of these and *M. martes*. Viewing the unquestionably close relations between the American and Asiatic Sables, it becomes very desirable to clear up this point. Without reference to dental or cranial characters, Baird says that "the true Sable is readily distinguishable by the short tail, which does not extend as far as the end of the outstretched hind feet, and by the balls of the toes covered entirely with woolly fur". The latter distinction does not hold, as we have seen; the former is disallowed by Brandt, who finds that in both the Asiatic and American Sable the tail has much the same length, being, without the hairs, about one-third the body, and not reaching as far as the outstretched hind feet. Certain supposed color distinctions which Brandt found in the American specimens he examined are clearly negatived by the more extensive series before me. He, however, finds in the American animal a pelage less dense and lighter-colored, with a less bushy tail, and, upon such considerations, is induced to regard it rather as a variety of the *zibellina* than as a distinct species or as the Pine Marten of Europe. The very close relationships of the American and Asiatic Sables are unquestionable. Brandt properly alludes to intermediate specimens he had seen; Mr. Ross reaffirms such a state of the case; in fine, external characters, when thoroughly sifted, are seen to be inadequate as a means of specific diagnosis. The case really hinges upon the validity of the dental characters ascribed by Gray, of which it is seen that Brandt makes no note. If these characters hold, there is no doubt of the propriety of separating *M. americana* specifically; otherwise, it must be referred to *M. zibellina* as a continental race, as Brandt has done.

In the present state of the case, this may be considered the proper reply to the often-asked question, have we the true Sable in America? The animal is, to all external appearance, indistinguishable except in some of those slight points of pelage which, through the whims of fashion, affect its commercial value, but there may be a technical zoölogical character of importance in the teeth.

I will only add that I see nothing tending to give weight to a supposition that there might be more than one species or variety of Marten on this continent. All the endless diversity in minor points which inspection of large series reveals comes clearly within the range of individual variability as a result of climate, season, age, sex, or other incidents.

Measurements of thirty-four fresh specimens of MUSTELA AMERICANA.

Original number.	Locality.	Sex.	From tip of nose to—				Tail to end of—		Length of—		Height of ear.	Nature of specimen.
			Eye.	Ear.	Occiput.	Tail.	Vertebræ.	Hairs.	Fore foot.	Hind foot.		
1017	Yukon (November)	♂	1.60	3.00	4.40	18.50	7.60	10.00	3.25	4.45	1.40	Fresh.
1018	....do....do		1.70	3.00	4.30	18.75	8.30	12.00	3.40	4.30	.....	...do.
1019	....do..(October)		1.40	2.60	3.80	15.60	6.25	9.85	2.70	3.55	1.30	...do.
1020	....do..(December)		1.45	2.60	3.45	16.00	6.30	9.80	2.90	3.65	1.25	...do.
1021	....do....do		1.55	2.60	3.60	16.20	7.30	10.60	2.80	3.80	1.20	...do.
1022	....do....do		1.65	2.90	3.95	17.25	8.25	11.85	3.10	4.30	.....	...do.
1023	....do....do		1.60	2.90	3.95	17.50	8.00	11.00	3.00	4.30	.....	...do.
1035	....do....do		1.70	2.75	3.90	19.30	8.50	12.00	3.10	4.40	1.55	...do.
1045	....do..(March)		1.50	2.65	3.60	16.60	7.10	10.40	2.50	3.65	1.20	...do.
1046	....do....do		1.75	3.10	4.05	18.75	7.60	11.40	3.10	4.10	1.60	...do.
1649	Peel's River (December)		1.45	2.60	3.50	16.00	6.50	9.80	2.65	3.80	1.30	...do.
1650	....do..(November)		1.45	2.80	3.50	16.70	6.60	10.50	2.80	3.80	.....	...do.
1651	....do....do		1.60	2.90	3.85	17.75	8.00	11.60	3.20	4.30	.....	...do.
1654	....do....do		1.40	2.70	3.45	15.50	6.50	10.10	2.60	3.60	1.35	...do.
1655	....do....do		1.70	2.85	4.10	18.20	7.85	10.60	3.25	4.45	1.65	...do.
1658	....do....(December)		1.45	2.50	3.55	16.30	7.15	10.15	2.60	3.70	1.40	...do.
1662	....do....do		1.65	3.00	3.95	17.60	7.20	10.20	3.10	4.20	1.45	...do.
1663	....do....do		1.60	2.90	3.85	15.30	8.00	12.00	3.00	4.20	1.50	...do.
1706	....do....do		1.60	2.90	3.90	15.40	7.80	11.30	3.10	4.25	1.50	...do.
1634	....do..(November)		1.60	3.00	3.85	17.40	7.90	10.70	3.05	4.10	1.35	...do.
1635	....do....do		1.60	2.85	3.75	17.40	7.30	10.30	3.10	4.20	1.40	...do.
1636	....do....do		1.40	2.50	3.50	16.00	7.00	9.50	2.70	3.70	1.25	...do.
1637	....do....do		1.45	2.80	3.60	16.50	7.25	10.00	2.75	3.75	1.30	...do.
1640	....do....do		1.60	2.80	3.80	17.90	7.90	11.10	3.05	4.10	1.50	...do.
1648	....do..(December)		1.65	3.10	4.10	18.30	7.80	10.80	3.15	4.10	1.55	...do.
1624	....do..(October)		1.55	2.50	3.80	17.00	7.25	10.25	3.00	4.10	1.60	...do.
1625	....do....do		1.60	2.90	3.85	17.10	7.40	10.70	3.00	4.00	1.50	...do.
1626	....do....do		1.60	3.20	4.00	17.00	7.15	10.35	3.00	4.00	1.55	...do.
1627	....do....do		1.60	2.95	3.95	17.00	7.10	10.30	2.80	4.00	1.55	...do.
1628	....do....do		1.60	2.75	3.85	17.60	7.75	11.05	3.00	4.10	1.45	...do.
1629	....do....do		1.60	3.10	4.00	17.50	7.50	11.10	3.05	4.20	1.60	...do.
1630	....do....do		1.60	2.90	3.80	17.50	7.60	10.70	3.00	4.25	1.55	...do.
1631	....do....do		1.50	2.80	3.85	17.70	7.35	9.85	3.00	4.10	1.50	...do.
1632	....do....do		1.40	2.70	3.50	15.80	6.90	9.80	2.65	3.70	1.40	...do.

*As recorded by the collectors on the labels of the specimens.

The foregoing table of careful fresh measurements satisfactorily indicates the average dimensions and range of variation of this species in the higher latitudes. The female is seen to be considerably smaller than the male on an average, though the dimensions of the sexes inosculate. The range is from 15½ to over 19 in length of head and body, with an average near 17½. The tail-vertebræ range from little over 6 to 8½, averaging near 7½. With the hairs, this member ranges from 9½ to 12 inches, being generally about 11 inches. Ear from about 1¼ to 1¾, generally about 1½. Fore foot 2⅔ to 3¾, settling near 3. Hind foot 3⅔ to nearly 4½, generally a little over 4. These extremes, it will be remembered, are those between the largest males and smallest females; neither sex has so wide a range.

GEOGRAPHICAL VARIATION IN THE SKULLS OF M. AMERICANA.

Mr. J. A. Allen has recently* given a table of measurement of length and breadth of forty-six skulls of this species, prepared to show the range of geographical variation. His results are here reproduced, together with his critical commentary on the specific validity of *M. americana*. It will be seen that he abandons his former† position, and endorses the distinctive characters of the dentition of *MM. martes, foina*, and *americana*.

The forty-six male skulls of this species, of which measurements are given below, are mainly from four or five localities differing widely in latitude. A comparison of the average size of a considerable number from each shows a well-marked decrease in size southward. Four skulls from Peel River, the largest, and also from the most northerly locality, have an average length of 3.39, and an average width of 2.07, the extremes being 3.50 and 3.35 in length and 2.12 and 2.02 in width. Nine skulls from the Yukon (probably mostly from near Fort Yukon) give an average length of 3.34 and an average width of 1.93, the extremes being 3.55 and 3.00 in length and 2.15 and 1.73 in width. Five skulls from Fort Good Hope give an average length of 3.24 and an average width of 1.95, the extremes in length being 3.37 and 3.15 and in width 2.05 and 1.73. Ten skulls from the northern shore of Lake Superior average 3.14 in length and 1.76 in width, the extremes in length being 2.23 and 3.02 and in width 1.89 and 1.65. Eight skulls from the vicinity of Umbagog Lake, Maine (Coll. Mus. Comp. Zoöl.), average 2.96 in length and 1.72 in width, the extremes in length being 3.10 and 2.73 and in width 1.85 and 1.50. Five skulls from Northeastern New York average 3.02 in length and 1.61 in width, the extremes being in length 3.10 and 2.92 and in width 1.63 and 1.50. There is thus a gradual descent in the average length from 3.39 to 3.02, and in width from 2.07 to 1.61. The largest and the smallest of the series are respectively 3.55 and 2.92 in length. Several fall as low as 3.00, and an equal number attain 3.50. The difference between the largest and the smallest, excluding the most extreme examples, is one-sixth of the dimensions of the smaller and one-seventh of the size of the larger.

The sexes differ considerably in size, relatively about the same as in *Putorius vison;* but the above generalizations are based wholly on males, and in each case on those of practically the same age, only specimens indicating mature or advanced age being used.

The series of fully one hundred skulls of this species contained in the National Museum presents a considerable range of variation in details of structure, involving the general form of the skull, the relative size of different parts, and the dentition, especially the form and relative size of the last molar. In a former paper,‡ I had occasion to notice somewhat in detail the variations in color our American Martens present, and the difficulty of find-

---

* Bull. U. S. Geological and Geographical Survey of the Territories, vol. ii, no. 4, pp. 328-330 (July, 1876).

† Bull. Mus. Comp. Zoöl. Cambridge, i, pp. 161-167 (Oct. 1869).

‡ "Mammals of Massachusetts", Bull. Mus. Comp. Zoöl. vol. i, pp. 161-167 (Oct., 1869).

ing any features of coloration that seemed to indicate more than a single American species, or that would serve to distinguish this even from the Martens of the Old World. Dr. J. E. Gray, it is true, had already called attention to the small size of the last molar in the American Martens as compared with the size of the same tooth in the Old World Martens; but, as his observation was apparently based on a single American skull, and as I was at the time strongly impressed with the wide range of individual variation I had found in allied groups, even in dental characters, and also with the great frequency of Dr. Gray's characters failing to be distinctive, I was misled into supposing all the Martens might belong to a single circumpolar species, with several more or less strongly-marked geographical races. My friend Dr. Coues some months since kindly called my attention to the validity of Dr. Gray's alleged difference in respect to the size and form of the last molar, which I have since had opportunity of testing. This character alone, however, fails to distinguish *Mustela foina* from *Mustela americana*, in which the last molar is alike, or so nearly so that it fails to furnish distinctive differences. The size and general form of the skull in the two are also the same, the shape of the skull and the form of the last upper molar failing to be diagnostic. The second lower true molar, however, in *Mustela foina* presents a character (shared by all the Old World Martens) which serves to distinguish it from *Mustela americana*, namely, the presence of an inner cusp not found in the latter. In *Mustela flavigula*, the last molar is relatively *smaller* than even in *Mustela americana*, and of the same form. *Mustela martes* differs in its more massive dentition and in the heavier structure of the skull, but especially in the large size of the last molar and the very great development of its inner portion. Hence, while the size and shape of the last upper molar serves to distinguish *Mustela martes* from *Mustela americana*, it fails as a valid distinction between *Mustela americana* and *Mustela flavigula* and *Mustela foina*. As already remarked, however, *Mustela americana* lacks the inner cusp of the second lower molar, which is present in the Old World Martens, or at least possesses it only in a very rudimentary condition.

Measurements of forty-six skulls of Mustela americana.

Catalogue number.	Locality.	Sex.	Length.	Width.	Remarks.
6643	Yukon River	♂	3.55	2.15	
6049	......do......	♂	3.50	1.85	
6085	......do......	♂	3.45	1.83	
6047	......do......	♂	3.37	1.82	
6044	......do......	♂	3.30	1.85	
6051	......do......	♂	3.00	1.73	
6042	......do......	♂	3.27	......	Imperfect.
6046	......do......	♀	3.28	1.82	
9099	Kenai, Alaska	♀	3.30	2.03	
7159	Fort Good Hope	♀	3.37	2.05	
7167	......do......	♀	3.25	1.98	
7168	......do......	♀	3.25	1.93	
7164	......do......	♀	3.25	1.76	
7163	......do......	♀	3.15	1.73	
6081	Peel River	♀	3.50	2.02	
6060	......do......	♂	3.37	2.12	
6063	......do......	♀	3.35	......	Imperfect.
6059	......do......	♀	3.35	......	......do.
3285	Red River	♀	3.40	1.94	

Measurements of forty-six skulls of MUSTELA AMERICANA—Continued.

Catalogue number.	Locality.	Sex.	Length.	Width.	Remarks.
4670	Lake Superior (north shore)	♂	3.23	1.75	
4068	......do......	♂	3.18	1.65	
4664	......do......	♂	3.15	1.65	
4668	......do......	♂	3.16	1.65	
4666	......do......	♂	3.15	1.87	
1675	......do......	♂	3.15	1.83	
4674	......do......	♂	3.15	1.85	
4667	......do......	♂	3.10	1.89	
4672	......do......	♂	3.12	1.65	
4681	......do......	♂	3.02	1.83	
	Washington Territory	♂	3.23	1.90	
	......do......		3.15	1.72	Rather young.
	......do......		3.03	1.55	......do.
	......do......		3.00	......	......do.
1668	Essex County, New York	♂	3.10	1.57	
1163	......do......	♂	3.03	1.63	
3819	......do......	♂	3.00	1.68	
3818	......do......	♂	2.92	1.50	
2245	Saranac Lake, New York	♂	3.03	1.68	
541	Umbagog Lake, Maine	♂	3.10	1.85	
550	......do......	♂	3.00	1.70	
542	......do......	♂	3.00	1.72	
552	......do......	♂	3.00	1.72	
553	......do......	♂	3.00	1.78	
543	......do......	♂	2.90	1.78	
545	......do......	♂	2.92	1.68	
544	......do......	♂	2.73	1.50	

GENERAL HISTORY AND HABITS OF THE SPECIES.

According to the foregoing considerations, the history of this interesting animal, one highly valuable in an economic point of view, is to be disentangled from that of the European and Asiatic species, with which it has always been to a greater or less degree intermixed. The first specific name, so far as I have become aware, is that bestowed in 1806 by Turton, in an edition of the *Systema Naturæ;* if there be an earlier one, it has escaped me. This name, however, appears to have been generally overlooked, or at least unemployed, until of late years revived by Professor Baird. His usage of the term, however, has received but partial support, some of the later writers agreeing with the custom of earlier ones in referring our animal to the European Marten, from which, as I have shown, it is well distinguished. Previous to the appearance of Dr. Brandt's elaborate memoir, only one author, it seems, among those who denied its specific validity, came so near the mark as to refer it to the Asiatic Sable. This was Dr. Godman, but even he used the name under the impression that the true Sable existed in America, as well as the Pine Marten, which he refers to as *M. martes.* As will be seen by reference to the list of synonyms, several nominal species have been established at the expense of the American Sable, upon slight individual peculiarities. The earliest of these is the *M. vulpina* of M. Rafinesque, which represents the occasional anomaly of the tail white-tipped, as alluded to by Mr. Ross in the article already quoted. A similar condition of the feet constitutes Kuhl's *M. leucopus;* while the *M. huro* of F. Cuvier is apparently only light-colored individuals. Dr. Gray seeks to establish these last two varieties, and adds another, *M. abietinoides*, based upon dark-colored examples, with the "throat-spot large or broken up into small spots". But these pretended species are not such, nor even as varieties are they entitled to more than passing allusion, as indicating to what extent some individuals may depart from the usual style of coloration.

Although the American animal was known in very early times, long before it received a distinctive name, having been referred alternately to the European Pine Marten and Asiatic Sable, or to both of these species, very little definite information upon its range and habits was recorded for many years. Pennant, our principal early authority on the animals of the North American fur countries, and the source of much subsequent inspiration on these species, considered it the same as *M. martes,* and drew its range accordingly. He states that it inhabits, in great abundance, the northern parts of America, in forests, particularly of pine and fir, nesting in the trees, bringing forth once a year from two to four young; that its food is principally mice, but also includes such birds as it can catch; that it is taken in dead-falls, and sometimes eaten by the natives. As an article of commerce in comparatively early times, we notice the sale of some 15,000 skins in one year (1743) by the Hudson's Bay Company, and the importation from Canada by the French into Rochelle of over 30,000. "Once in two or three years," he adds, they "come out in great multitudes, as if their retreats were overstocked: this the hunters look on as a forerunner of great snows, and a season favorable to the chase." Such periodicity in numbers thus early noted is confirmed by later observations.

Sir John Richardson has the following observations upon the distribution of the Sable in British America: "The Pine-martin inhabits the woody districts in the northern parts of America, from the Atlantic to the Pacific, in great numbers, and have been observed to be particularly abundant where the trees have been killed by fire but are still standing. It is very rare as Hearne has remarked, in the district lying north of Churchill River, and east of Great Slave Lake, known by the name of Chepewyan or Barren Lands. A similar district, on the Asiatic side of Behring's Straits, twenty-five degrees of longitude in breadth, and inhabited by the Tchutski, is described by Pennant as equally unfrequented by the Martin, and for the same reason,—the want of trees. The limit of its northern range in America is like that of the woods, about the sixty-eighth degree of latitude, and it is said to be found as far south as New England. Particular races of Martins, distinguished by the fineness and dark colours of their fur, appear to inhabit certain rocky districts. The rocky and mountainous but woody district of the Nipigon, on the north side of Lake Superior, has long been noted for its black and valuable Martin-skins. . . . Upwards of one hundred thousand skins have long been collected annually in the fur countries."

But the range of the American Sable is now known to be more extended in both directions than appears from the foregoing. In some longitudes, at least, if not in all, it reaches the Arctic coast, as mentioned by Mr. B. R. Ross, and as attested by specimens I have examined. Mr. Ross states that it is found throughout the Mackenzie River District, except in the Barren lands, to which it does not resort, being an arboreal animal. It occurs abundantly in Alaska, apparently throughout that vast country; and, in short, we cannot deny it a less highly Arctic extension than that of the Asiatic Sable. Along the Pacific side of the continent, west of the Rocky Mountains, the Sable has been traced to the Yuba River of California by Dr. J. S. Newberry, who represents it as not uncommon in Oregon; and Dr. George Suckley procured specimens in Washington Territory. Mr. J. A. Allen found the animal in Wyoming and Colorado, and considers it as common in the last-mentioned Territory in Park County. But however far south it may extend in such longitudes, there is apparently a great stretch of treeless country in which it is not found at all. I obtained no indications of its presence in any of the unwooded portions of Dakota and Montana, which I have explored with special reference to the distribution of the Mammals and Birds. It is represented as common in Canada, Nova Scotia, Newfoundland, and Labrador. In New England, according to Dr. Emmons, writing in 1840, it was not infrequent in the pine and beech forests of Massachusetts, and Mr. Allen states that it is still occasionally seen in the mountains of Berkshire County. It inhabits the mountainous regions of New York and some parts of Pennsylvania; but in tracing its extreme southern limit in the Atlantic States, we see that it has not been found so far south as the Pekan has. I find no indication of its occurrence in Maryland or Virginia. The southern limit, which has been set at about 40° north, is probably correct for this longitude, though in the mountainous regions of the West it may require to be somewhat extended. General considerations aside, its local distribution is determined primarily by the presence or absence of trees, and further affected by the settlement of the country. Being of a shy and suspicious nature, it is one of the first to disappear, among the smaller animals, with the advance of civilization into its woody resorts. In unpeopled districts, even the vast numbers that are annually destroyed for the pelts seems to affect their abundance less materially than the settlement of the country does. Notwithstanding such destruction, they abound in the northern wilds. Even in Nova Scotia, a thousand skins are said to have been exported annually within a few years, and they may justly be regarded as among the most important of the land fur-bearing animals. Respecting their comparative scarcity at times, Mr. Ross has recorded a remarkable fact of periodical disappearance. "It occurs in decades," he says, "or thereabouts, with wonderful regularity, and it is quite unknown what becomes of them. They are not found dead. The failure extends throughout the Hudson's Bay Territory at the same time. And there is no tract, or region to which they can migrate where we have not posts, or into which our hunters have not penetrated. . . . When they are at their lowest ebb in point of numbers, they will scarcely bite at all [at the bait of the traps]. Providence appears thus to have implanted some instinct in them by which the total destruction of their race is prevented."

The Sable is ordinarily captured in wooden traps of very simple construction, made on the spot. The traps are a little enclosure of stakes or brush in which the bait is placed upon a trigger, with a short upright stick supporting a log of wood; the animal is shut off from the bait in any but the desired direction, and the log falls upon its victim with the slightest disturbance. A line of such traps, several to the mile, often extends many miles. The bait is any kind of meat, a mouse, squirrel, piece of fish, or bird's head. One of the greatest obstacles that the Sable hunter has to contend with in many localities is the persistent destruction of his traps by the Wolverene and Pekan, both of which display great cunning and perseverance in following up his line to eat the bait, and even the Sables themselves which may be captured. The exploits of these animals in this respect may be seen from the accounts elsewhere given. I have accounts from Hudson's Bay trappers of a Sable road fifty miles long, containing 150 traps, every one of which was destroyed throughout the whole line twice—once by a Wolf, once by the Wolverene. When thirty miles of this same road was given up, the remaining 40 traps were broken five or six times in succession by the latter animal. The Sable is principally trapped during the colder months, from October to April, when the fur is in good condition; it is nearly valueless during the shedding in summer. Sometimes, however, bait is refused in March, and even early that month, probably with the coming on of the rutting season. The period of full furring varies both in spring and autumn, according to latitude, by about a month as an extreme.

Notwithstanding the persistent and uninterrupted destruction to which the Sable is subjected, it does not appear to diminish materially in numbers in unsettled parts of the country. The periodical disappearances noted by Mr. Ross and the animal's early retreat before the inroads of population are other matters. It holds its own partly in consequence of its shyness, which keeps it away from the abodes of men, and partly because it is so prolific; it brings forth six or eight young at a litter. Its home is sometimes a den under ground or beneath rocks, but oftener the hollow of a tree; it is said to frequently take forcible possession of a Squirrel's nest, driving off or devouring the rightful proprietor. Though frequently called Pine Marten, like its European relative, it does not appear to be particularly attached to coniferous woods, though these are its abode in perhaps most cases, simply because such forests prevail to a great extent in the geographical areas inhabited by the Marten.

The Sable is no partner in guilt with the Mink and Stoat in invasion of the farm-yard, nor will it, indeed, designedly take up its abode in the clearing of a settler, preferring always to take its chances of food supply in the recesses of the forest. Active, industrious, cunning, and predaceous withal, it finds ample subsistence in the weaker Rodents, Insectivora, and birds and their eggs. It hunts on the ground for Mice, which constitute a large share of its sustenance, as well as for Shrews, Moles, certain reptiles, and insects. An expert climber, quite at home in the leafy intricacies of tree tops, it pursues Squirrels, and goes birds'-nesting with success. It is said to also secure toads, frogs, lizards, and even fish. Like the Wolverene and Pekan, it sometimes makes an entrance upon the hoards of meat and fish which are cached by the natives in the higher latitudes. It is said not to reject carrion at times. It has been stated to eat various nuts and berries, as well as to be fond of honey; but we may receive such accounts with caution, viewing the very highly carnivorous character of the whole group to which the species belongs.

The Sable has some of the musky odor characteristic of its family, but in very mild degree compared with the fetor of the Mink or Polecat. Hence the name "Sweet Marten", by which its nearest European ally is known, in contradistinction from Foulimart, or "Foul Marten", a name of the Polecat. With a general presence more pleasing than that of the species of *Putorius*, it combines a nature, if not less truly predaceous, at least less sanguinary and insatiable. It does not kill after its hunger is appeased, nor does a blind ferocity lead it to attack animals as much larger than itself as those that the Stoat assaults with success. Animals like the Rabbit and Squirrel form less of its prey than the smaller Rodents and Insectivores. In confinement, the Marten becomes in time rather gentle, however untamable it may appear at first; it is sprightly, active, with little unpleasant odor, and altogether rather agreeable.

CHAPTER IV.

MUSTELINÆ—Continued: The Weasels.

The genus *Putorius*—Generic characters and remarks—Division of the genus into subgenera—Analysis of the North American species—The subgenus *Gale*—*Putorius vulgaris*, the Common Weasel—Synonymy—Habitat—Specific characters—General characters and relationships of the species—Geographical distribution—Habits—*Putorius erminea*, the Stoat or Ermine—Synonymy—Habitat—Specific characters—Discussion of specific characters and relationships—Table of measurements—Note on the skull and teeth—Description of external characters—Conditions of the change of color—General history and habits of the species—Its distribution in the Old World—*Putorius longicauda*, the Long-tailed Weasel—Synonymy—Habitat—Specific characters—Description—Measurements—General account of the species—*Putorius brasiliensis frenatus*, the Bridled Weasel—Synonymy—Habitat—Specific characters—General account of the species.

CONTINUING with the subfamily *Mustelinæ*, but passing from the genus *Mustela*, we reach the next genus, *Putorius*, which contains the true Weasels or Stoats (subgenus *Gale*), the Ferrets and Polecats (subgenus *Putorius* proper), the American Ferret (subgenus *Cynomyonax*), and the Minks (subgenus *Lutreola*). This chapter is devoted to the consideration of the species of the first of these sections, after presentation of the characters of the genus at large. The other sections are reserved for succeeding chapters.

The Genus PUTORIUS. (Cuvier.)

< **Mustela**, or **Martes**, of some authors.
× **Viverra** *sp.*, **Lutra** *sp.* of some authors.
= **Putorius**, *Cuvier*, Règne Anim. i. 1817, and of authors generally.
= **Fœtorius**, *Keys. & Blas.* Wirbelth. Eur. 1840.
> **Gymnopus**, *Gray*, Cat. Mamm. Br. Mus. 1842.
> **Lutreola**, "*Wagner*", *Gray*, P. Z. S. 1865, 117. (Type *Mustela lutreola* L.)
> **Gale**, "*Wagner*", *Gray*, P. Z. S. 1865, 118.
> **Neogale**, *Gray*, P. Z. S. 1865, 114. (Type *P. brasiliensis*.)
> **Vison**, *Gray*, P. Z. S. 1865, 115. (Type *P. vison*.)

GENERIC CHARACTERS.—*Dental formula:* I. $\frac{3-3}{3-3}$; C. $\frac{1-1}{1-1}$; Pm. $\frac{3-3}{3-3}$; M. $\frac{1-1}{2-2} = \frac{16}{18} = 34$ (one premolar above and below less than in *Gulo* and *Mustela*). Sectorial tooth of lower jaw (anterior true molar) without an internal cusp. Anteorbital foramen presenting downward-forward (as in *Mustela*; reverse of *Gulo*), a mere orifice, not canal-like, and opening over the last premolar (the opening more anterior in *Gulo* and *Mustela*). Skull as a rule* little contracted at the middle; the rostral portion extremely short, stout, turgid, scarcely tapering, and much more vertically truncated than in *Gulo* or *Mustela*; frontal profile convex, and usually more nearly horizontal than in *Gulo* or *Mustela*. Nasal bones widening forward from an acute base. General outline of skull in profile scarcely arched—sometimes quite straight and horizontal in most of its length. Production of mastoids and auditory bullæ and general prominence of periotic region at a minimum; the bullæ flatter than in *Mustela* or *Gulo*, and scarcely so constricted across as to produce a tubular meatus. Zygomatic arch usually not higher behind than in front, nowhere vertical nor developing a posterior convexity. Depth of emargination of palate little if any greater, or less than, distance thence to the molars. Skull as a whole more massive than in *Mustela*, though smaller.

Size medium and very small (including the smallest species of the whole family). Body cylindrical, slender, often extremely so; legs very short; tail long, terete, uniformly bushy or very slender and close-haired, with a terminal pencil. Ears large, orbicular. Soles commonly furry. Pelage usually close and short, whole-, or oftener, parti-colored; turning white in winter in Northern species. Progression digitigrade. Habits indeterminate—terrestrial, arboreal, or aquatic.

The foregoing characters are drawn up from consideration of the European and North American forms, and may require some qualification, in ultimate details, to cover all the modifications of this extensive genus, containing, as it does, several sections or groups of species, probably of subgeneric value. From *Gulo* or *Mustela* it is at once distinguished by the different dental formula. The skull, as compared with that of its nearest ally, *Mustela*, differs notably in the shortness and bluntness of the muzzle, position and direction of the anteorbital foramen, slight convexity of the upper profile, and other points noted above. There is a decided difference in the character of the auditory bullæ, more readily perceived on comparison than described; the bullæ are usually less inflated—sometimes quite flat, as in *P. vison*; and even when, as in some cases, the inflation of the basal portion is not much less than in *Mustela*, we miss the constriction which in the latter genus produces a well-determined tubular meatus. The skull of *Putorius* is decidedly heavier for its size than that of *Mustela*, in this respect more like that of *Gulo*, though it is comparatively much flattened and otherwise dissimilar from the latter.

The name of the genus is from the Latin *putor*, a stench

* In some species of *Putorius*, however, the constriction is as great as is ever found in *Mustela*.

(*puteo*, to stink), as one of its synonyms, *Fœtorius*, is from *fœtor, fœteo*, of the same signification. The relation of the English *putrid, fetid*, &c., is obvious.

The extensive genus *Putorius* is divisible into several well-marked sections, doubtless of subgeneric value. Three such groups exist in North America. These may be analyzed as follows, in connection with a fourth group, *Putorius* proper, introduced to further elucidate the position and relations of a new subgenus I propose for the reception of the *Putorius nigripes*.

Division of the genus into subgenera.

1. GALE.*—*The Stoats or Ermines, and Weasels.*—Skull smooth, without well-developed sagittal crest. Frontal profile strongly convex and declivous. Pterygoids with small hamular processes, or none. Bullæ auditoriæ nicked at end by orifice of the meatus. Skull moderately abruptly constricted near the middle; postorbital processes slight. Species of small and smallest size, with very slender, cylindrical, "vermiform" body, very long neck, and tail (of variable length) slenderly terete, with terminal pencil, usually black; pelage, including that of the tail, short and close set (the Northern species usually turning white in winter), bicolor, of uniform color above, lighter below. Ears large, high, and orbicular. Palmar pads all separate. Toes scarcely webbed. Habits terrestrial, and somewhat arboreal; not aquatic. Of general distribution in both hemispheres.

2. CYNOMYONAX† (nob., subg. nov.).—*American Ferret.*—Skull developing sagittal crest. Frontal profile scarcely or not convex, strongly declivous. Pterygoids with slight hamular process. Bullæ auditoriæ nicked by orifice of meatus. Sectorial tooth of upper jaw with its outer border nearly straight, developing no decided antero-external process, and the antero-internal process merely a slight spur. Skull abruptly and strongly constricted in advance of the middle, with strongly developed postorbital processes. Last molar of under jaw minute, merely a cylindrical round-topped stump, without trace of cusps or other irregularity of surface. Animal of large size, equalling or exceeding a large Mink, yet retaining the attenuate, elongate and cylindrical body, long neck, large suborbicular ears, slenderly terete black-tipped tail, and close short pelage of *Gale*. Coloration not distinctively bicolor; legs darker than body; peculiar facial marking. Toes not semi-palmate. Palmar pads discrete. Habits terrestrial. No seasonal change of colors. One species known, peculiar to North America.

3. PUTORIUS‡ (proper).—*The Ferrets or Polecats.*—Skull finally developing sagittal crest, and roughened muscular impressions. Frontal profile

---

* *Etym.*—The Greek γαλῆ, a weasel.

† *Etym.*—Greek κυων, dog, μυς, mouse, ὄναξ (or ἄναξ), king.—The genus *Cynomys* (κυων, μυς) is that of the so-called "prairie-dogs", among which the species lives, and upon which it largely subsists.—*Cynomyonax*, "king of the prairie-dogs".

‡ *Etym.*—See above.

convex, strongly declivous. Pterygoids developing large hamular processes. Bullæ auditoriæ nicked by orifice of meatus. Skull scarcely constricted near the middle, where, if anything, it is broader than rostrum; postorbital processes poorly developed. Sectorial tooth of upper jaw as in *Cynomyonax*. Back molar of lower jaw of ordinary size, circular, developing irregularities of the crown. Animals rather large, comparatively stout-bodied, less lengthened, with rather bushy, tapering tail, and low, orbicular ears; pelage long and loose, instead of close-set, variegated above, or there not notably darker than below; do not turn white in winter. Palmar pads separate. Toes not semipalmate. Terrestrial in habits. The species confined to the Old World.

4. LUTREOLA*.—*The Minks.*—Skull of adult developing sagittal crest and muscular impressions. Frontal outline nearly straight and scarcely declivous. Pterygoids with strong hamular process. Bullæ auditoriæ notably less inflated than in the foregoing, prolonged into a somewhat tubular meatus, not nicked at orifice. Constriction of skull and development of postorbital processes intermediate in degree between *Putorius* proper and *Cynomyonax*. Sectorial tooth of upper jaw with its outer border concave, owing to development of a strong antero-exterior spur, which lies out of the axis of dentition, and forms with the antero-interior cusp (present in all *Mustelinæ*) a rather open V, into which the antecedent premolar fits, the antero-internal process developing to a conical cusp. Back lower molar as in *Putorius* proper. Animals of large to largest size in the genus, stout-bodied, rather long and very bushy tail, cylindrico-tapering; pelage moderately loose, but thick, to resist water, very bristly and lustrous, dark-colored, unicolor or only varied with irregular white patches on under parts; no seasonal changes of pelage. Ears very low. Feet semipalmate, natatorial. Palmar pads without hairy intervals. Habits highly aquatic. Species common to both hemispheres.

The first of these subgenera is represented in North America by several species, some of which are not clearly distinguished from their congeners of Europe, while another is specifically identical with an animal which ranges through Central into South America. The second and fourth each contain a single North American species, as far as known, the fourth having a closely allied European congener; while the second, peculiar to America, is the nearest analogue of the third, which has no exact American representative.

The North American species of *Putorius* at large may be determined by the following analysis of subgeneric and specific characters:—

---

* *Etym.*—" Lutreola ", " Little Otter "—diminutive form of the Latin *lutra*, an Otter, which the Mink much resembles. For von Marten's exposition of the word *lutra* in its several forms, and discussion of the philological questions involved, see p. 29.

Analysis of North American species of Putorius.

A. (*Gale.*) Of smallest size (length of head and body under 12 inches), most slender and attenuate body, and longest neck. Ears conspicuous, orbicular. Tail slenderly terete, with the tip usually (rarely in *vulgaris*) black. Toes cleft. Palmar pads separate. Coloration bicolor, in distinct upper and under areas, latter not darker than former, feet not black; or, entirely white, excepting black tip of tail. (Weasels, Stoats or Ermines.)
  a. Head not darker than rest of upper parts, nor variegated with streaks or spots.
    a'. Tail pointed at end, scarcely or not black-tipped, 2 inches or less in length, including hairs; belly white or scarcely tinged with sulphury. . . . . 1. P. VULGARIS.
    b'. Tail with a terminal pencil of black hairs, and over 2 inches long, including hairs.
      a''. Belly pure sulphury-yellow; tail-vertebræ 2-5 inches long, the black tip not confined to the terminal pencil. . . 2. P. ERMINEA.
      b''. Belly tawny, saffron or salmon-yellow; tail 6-7 inches long, the black tip reduced to terminal pencil. . . . 3. P. LONGICAUDA.
  b. Head darker than rest of upper parts, with light stripes or spots; belly as in *b''.* . . . . 4. P. BRASILIENSIS FRENATUS.

B. (*Cynomyonax.*) Much larger; length of head and body over 12 inches; body scarcely stouter, and equally close-haired, and tail very short, slenderly terete, black-tipped. Ears conspicuous, orbicular. Toes cleft. Palmar pads separate. Coloration not bicolor in distinct areas. (American Ferret.)
  Pale brown, nearly uniform, or brownish-white, scarcely darker on the back; feet, end of tail, and broad bar across the face black.
                                  5. P. NIGRIPES.

C. (*Lutreola.*) Size of the last, or rather less; body as stout or stouter. Ears low. Toes semipalmate. Palmar pads fused. Tail uniformly bushy. (Mink.)
  Dark chestnut-brown or blackish, uniform, or only varied by white patches below; tail without differently colored tip.
                                    6. P. VISON.

The Subgenus GALE. (WAGNER.)

This subgenus, which includes the Weasels proper and the Stoats or Ermines, comprises a large majority of the species of *Putorius*, widely distributed over the globe. The leading characters which distinguish it from its nearest allies have already been given (p. 99), together with an analysis of the four species known to inhabit North America. Further details of the skull, teeth, and external form are presented beyond, under head of *G. erminea*, which, as a typical member of the subgenus, may serve as a standard of comparison. We may at once, therefore, proceed to consider the several North American species.

The Weasel.

Putorius (Gale) vulgaris.

PLATE VI. FIGS. 2, 4.

(a. Old World references.)

Mustela, *Variorum* ("*Gesn.* Quad. 1551, 651, f. —.—*Schwenckf.* Theriotroph. 1603, 116.—*Charlet.* Exercit. 1677, 20.—*Rzacz.* Polon. 1721, 235").
Mustela vulgaris, *Aldrov.* Quad. Digit. 1645, 307.—*Sibb.* Scot. Illust. ii. 1684, 11.—*Ray,* Syn. 1693, 195.—*L. S. N.* eds. 2d–5th, 1740–47, 44.—*Klein,* Quad. 1751, 62.—*Jonst.* Theatr. 1755, 152, pl. 64.—*Briss.* Quad. 1756, 241, no. 1.—*Erxl.* Syst. Anim. 1777, 471, no. 12 (synon. much mixed with that of other species).—*Schreb.* Säug. iii. 1778.—*Gm. S. N.* i. 1788, 99.—*Bechst.* Naturg. i. —, 812.—*Turt. S. N.* i. 1806, 61.—*Desm.* Mamm. i. 1820, 179, no. 275; Nouv. Dict. xix. 372; Ency. Méth. pl. 84, f. 1.—*Fr. Cuv.* Dict. Sci. Nat. xxix, 1823, 251, no. 7.—*Is. Geoff.* Dict. Class. x. 213.—*Less.* Man. 1827, 146.—*Fisch.* Syn. 1829, 223.—*Flem.* Br. An. 1828, 13.—*Jen.* Br. Vert. 1835, 12.—*Bell,* Br. Quad. 1837, 141; 2d ed. 1874, 182, f. —.—*Selys-L.* Fn. Belg. 1842, 10.—*Gray,* List Mamm. Br. Mus. 1843, 65.—*Gieb.* Säug. 1855, 782.—*Fitz.* Naturg. Säug. i. 1861, 335, f. 69.—*Gerv.* Cat. Br. Mus. 1862, 93.—*Farwick,* Zool. Gart. xiv. 1873, 17 (albino).
Mustela vulgaris α. **æstiva,** β. **nivalis,** *Gm. S. N.* i. 1788, 99, nos. 11 a, 11 b.
Viverra vulgaris, *Shaw, G. Z.* i. 1800, 420, pl. 98, upper fig.
Mustela (Gale) vulgaris, *Schinz,* Syn. Mamm. i. 1844, 344.—*Gray,* P. Z. S. 1865, 113; Cat. Carn. Br. Mus. 1869, 90.
Putorius vulgaris, *Griff.* An. Kingd. v. 1827, 121, no. 344 (but not same name on p. 120, no. 339).—*Brandt,* Wirb. Eur. N. E. Russl. —, 26.
Fœtorius vulgaris, *Keys. & Blas.* Wirb. Eur. 1840, 69, no. 147.—*Blas.* Wirb. Deutschl. 1857, 231.—*Jäckel,* Zool. Gart. xiv. 1873, 459 (albino).
Mustela nivalis, *Linn.* Fn. Suec. 2d ed. 1761, 7, no. 18; *S. N.* i. 1766, 69, no. 11.—*Müll.* Zool. Prod. 1776, 3, no. 15.—*Erxl.* Syst. An. 1777, 476, no. 14.—*Schreb.* Säug. iii. 1778, pl. 139.—*Hell.* Kon. Vet. Akad. Stockh. vi. 1785, 212, no. 9, pl. 1.—*Less.* Man. 1827, 146.
Mustela gale, *Pall.* Zoog. R.-A. i. 1831, 94, no. 32.
Belette. *Briss.* op. et loc. cit.—*Buff.* Hist. Nat. vii. 225, pl. 29, f. 1.—*Bomare.,* Dict. i. 1768, 262.—*French.*—Marcot, Marcotte, *French.*
Common Weesel, *Penn.* Syn. Quad. 1771, 212, no. 151; Brit. Zoöl. i. —, 95, pl. 7, f. 17.—*Shaw,* op. loc. cit.—Weasel or Weesel, *English.*
Scheenwiesel, *Müll.* Naturs. i. 1776, 276 (= *M. nivalis*).
Wiesel, Kleine Wiesel, *German* (cf. *v. Martens,* Zool. Gart. xi. 1870, p. 276, philological).—Wezel, *Belgic.*—Viesel, Lækatt, *Danish.*—Sneemuus, *Danish* (white).—Snömus, *Swedish* (white).—Ballottula, *Italian.*—Comadreja, *Spanish.*

(b. American references.)

Mustela nivalis, *Forst.* Phil. Tr. lxii. 1772, 373.
Mustela vulgaris, *Harl.* Fn. Amer. i. 1825, 61.—*Maxim.* Reise, ii. 1841, 98.—*Thomps.* N. H. Verm. 1853, 30.—*Hall,* Canad. Nat. and Geol. vi. 1861, 295.
Mustela (Putorius) vulgaris, *Rich.* F. B.-A. i. 1829, 45.
Putorius vulgaris, *Emm.* Rep. Quad. Mass. 1840, 44.—*All.* Pr. Bost. Soc. xiii. 1869, 183; Bull. M. C. Z. i. 1870, 167.
Mustela (Gale) vulgaris *var.* **americana,** *Gray,* P. Z. S. 1865, 113; Cat. Carn. Br. Mus. 1869, 91.
Mustela pusilla, *De K. N. Y.* Zoöl. i. 1842, 34, pl. 14, f. 1.—*Beesley,* Geol. Cape May, 1857, 137.
Putorius pusillus, *Aud. & Bach.* Q. N. A. ii. 1851, 100, pl. 64.—*Bd.* M. N. A. 1857, 159.—*Suckl.* N. H. W. T. 1860, 92.—*Sam.* Rep. Mass. Agric. for 1861 (1862), 154, pl. 1, f. 2, 4.—*Maxim.* Arch. Naturg. 1861, —; Verz. N. Am. Säug. 1862, 49.—*Ross,* Canad. Nat. and Geol. vi. 1861, 441.—*Merriam,* Rep. U. S. Geol. Surv. Terr. 1872, 661 (Idaho).—*Ames,* Bull. Minn. Acad. Nat. Sci. 1874, 69.
Putorius "cicognani", *Rich.* Zoöl. Beechey's Voy. 1839, 10° (err.).
Common Weesel, *Penn.* Hist. Quad. 1781, no. 192; Arct. Zoöl. i. 1784, 75, no. 25.

HAB.—In America, the northern portions of the United States and northward. Europe and Asia, northerly.

SPECIFIC CHARACTERS.—Very small; length of head and body 6 or 8 inches; of tail-vertebræ 2 inches or less; tail-vertebræ about one-fourth or less of the head and body; tail slender, cylindrical, pointed at tip, which is concolor or not obviously black; under parts white, rarely, if ever, tinged with sulphury; coloration otherwise as in *P. erminea*. Caudal vertebræ 15 (*Gerrard*).

General characters and relationships of the species.

To describe the general body-colors of this animal would be to repeat, in substance, most of what is beyond said of *P. erminea*. I find no differences susceptible of intelligible description excepting those given in the foregoing diagnosis, although, as usual in this genus, there is considerable individual variation in the shade of the mahogany-brown upper parts, in the details of the line of demarcation with the white of the under parts, and in the color of the feet, which appear to be indifferently like the back or like the belly. I do not observe, however, in any of the specimens before me, that the under parts are notably tinged with sulphury-yellow, as is frequently or usually the case with *P. erminea*. They are quite purely white.

The points of this animal to which attention should be directed in comparison with its ally, *P. erminea*, are the general dimensions and the color of the tail. This member is both absolutely and relatively shorter than in *P. erminea*; it is cylindrical, very slender, and usually terminates in a point, without the slightest bushy enlargement. In most specimens, as in all the European examples I have seen, there is no black whatever at the end of the tail; on the contrary, the tip is frequently mixed with a few white hairs. In other specimens, however, the end of the tail is dusky, as in No. 6491, from the Yukon (*Kennicott*); while in No. 3316, from Oregon (*Wayne*), the tip is quite blackish. The tail-vertebræ range from rather less than an inch in length to full two inches, if not a trifle more, though the latter dimension seems to be rarely reached; the terminal pencil of hairs from $\frac{1}{4}$ to $\frac{3}{4}$. According to Gerrard, there are fewer (15) caudal vertebræ than in *P. erminea*.

Accounts of authors are surprisingly at variance in assigning dimensions to this animal. De Kay says in one place 12–13 inches (nose to end of tail), but this is probably a slip of the pen, for his detailed measurements amount to 8.80 for head and body and 1.80 for tail-vertebræ; Audubon, 8; Bachman gave 7 inches, the tail-vertebræ 2. Baird gives 6; the tail from 0.83 to 1.60; the head, 1.45; fore foot, 0.58; hind foot, 0.92. The smaller dimensions seem to be nearer the average. The skin from the Yukon, above mentioned, probably well stretched, measures 7.50; tail-vertebræ, 1.25; hind foot, 1.10. The Oregon specimen was apparently about 7 inches; the tail 2. Two skins from British America (4411, Fort Resolution, *Kennicott*, and 4231, Moose Factory, *Drexler*) are notably smaller and shorter-tailed than any others I have seen. They are about 6 inches long, the tail-vertebræ an inch or less, the hind feet about 0.75.* They are also somewhat peculiar in the intensity of a liver-brown shade.

With only such small and dark-colored specimens as these last before us (strictly representing *P. pusillus* of Audubon and Bachman), there might be little difficulty in distinguishing at least an American race; but, as already indicated, such distinctions disappear on examining larger series, and consequently fail to substantiate a geographical race. Whatever minute discrepancies may be noted in comparing certain American with certain other European examples, assuredly these do not hold throughout the series; and, moreover, the differences *inter se* between animals of either continent are as great as any of those which can be detected when the animals of the two continents are compared. Thus, holding in my hands the Yukon specimen and No. 2290, from Leeds, England, I find that I have incontestably the same species. In size and color, these two are much more nearly identical than Nos. 2290 and 2279, the latter being also from Leeds. The Yukon animal has, indeed, a bushy tip to the tail; but, again, the one from Moose Factory has not. A specimen from Scotland (No. 1658) has proved susceptible of overstuffing up to more than 10 inches for length of head and body; but No. 2290 was scarcely 7 inches long. The presence of true *M. vulgaris* on our continent may be considered established. So that the question practically narrows to whether we have not also an additional species. This I cannot admit; for if minute differences of the grade allowed to distinguish a supposed "*pusillus*" be taken into account, we must, to be consistent, also separate from this latter the specimen from Oregon,† with its longer blackish-tipped tail, and so have *three* North

---

* Reliable European writers assign a length of about 8 inches of head and body, the head 1¼, the tail 2. The female is usually an inch, if not more, smaller than the male.

† This furnishes a case parallel with that of *Hesperomys* "*boylii*" and *H.* "*austerus*". There is a strong local influence exerted upon various animals in this region.

American "species" of the *vulgaris* type, namely, *vulgaris* as attested by the Yukon specimen; "*pusillus*", as by the Hudson's Bay example and others; and a nameless Oregon species. We should obviously be reduced to this dilemma in any such attempt to describe *specimens* instead of characterizing species. And in determining our species and races, it is quite sufficient to note the minor variations from a common type without giving the subjects of such variation a name.

Nevertheless, as it is desirable to carry investigations of the characters of animals into minute particulars, the following summary is presented:—

Var. 1. An animal averaging slightly less than *P. vulgaris* of Europe, with the end of the tail blackish. Alaska, &c.

Var. 2. Rather smaller than the last; the tail relatively longer (vertebræ about two inches) and distinctly dusky-tipped. Oregon and Washington Territories.

Var. 3. Very small—about six inches long; tail-vertebræ one inch or less; color darker than in *P. vulgaris*, but tail concolor. Hudson's Bay, &c.

Geographical distribution.

The area over which this species turns white in winter may be approximately deduced from the accounts of various authors. This is nearly coincident with what is now known of the American range of the animal. Mr. J. A. Allen states that it turns in northern New England, but not so far south as Massachusetts, where the change sometimes, but not always, occurs to *P. erminea*. Dr. De Kay denies any change in New York, though I suspect this may not hold for the northern mountainous portions of the State. According to Maximilian, the change takes place in the region he explored, as it doubtless does in all higher latitudes.

The range of the Least Weasel extends entirely across the continent on this hemisphere; but its north and south dispersion are less definite, in the present state of our knowledge. To the northward, Richardson formerly limited its extension to the Saskatchewan; but my specimens, from the Yukon, Fort Resolution, and Hudson's Bay, largely extend the supposed range, and I infer that the animal is generally distributed in British America and Alaska. Audubon's examples were from the Catskills and Long Island; and this author alludes to others from Lake Superior. The Red River and

Upper Missouri regions, Oregon and Washington Territory, are other recorded localities. According to Mr. Allen, it is rather rare in Massachusetts—much more so than *P. erminea*. The total lack of citations of the species from Southern or even Middle districts in the United States is in evidence, though of a negative character, of the geographical distribution at present assigned.

Habits.

Our accounts of the habits of this animal are lamentably meagre; nor can I add to them from personal observation. De Kay says it is by no means a rare animal, but one difficult to capture; that it feeds on mice, insects, young birds, eggs, &c., and possesses all the rapacity characteristic of the tribe. Audubon repeats this, in substance, with the inference that, owing to its small size, it would not be mischievous in the poultry-house, and would scarcely venture to attack a full-grown Norway Rat.

In this dearth of facts respecting the animal in America, we turn to other authors. One of the most particular, and at the same time interesting and apparently reliable accounts, is that given by Thomas Bell (who was evidently familiar with the animal) in the work above cited. Comparing its habits with those of the Stoat, Bell finds them considerably distinct, and believes that the accusations current against the Weasel should mostly be laid rather at the door of the Stoat. He continues:—

"It is not meant to be asserted that the Weasel will not, when driven by hunger, boldly attack the stock of the poultry yard, or occasionally make free with a young rabbit or sleeping partridge; but that its usual prey is of a much more ignoble character is proven by daily observation. Mice of every description, the Field and Water Vole, rats, moles, and small birds, are their ordinary food; and from the report of unprejudiced observers, it would appear that this pretty animal ought rather to be fostered as a destroyer of vermin, than extirpated as a noxious depredator. Above all, it should not be molested in barns, ricks or granaries, in which situations it is of great service in destroying the colonies of mice which infest them. Those only who have witnessed the multitudinous numbers in which these little pests are found, in wheat-ricks especially, and have seen the manner in which the interior is drilled, as it were, in every direction by their runs, can at all appreciate the amount of their depredations; and surely the occasional abduction of a chicken or duckling, supposing it to be even much more frequently chargeable against the Weasel than it really is, would be but a trifling set off against the benefit produced by the destruction of those swarms of little thieves.

"The Weasel climbs trees with great facility, and surprises birds on the nest, sucks the eggs, or carries off the young. It has been asserted that it attacks and destroys snakes; this, however, I believe to be entirely erroneous. I have tried the experiment by placing a Weasel and a common snake together in a large cage, in which the former had the opportunity of retiring into a small box in which it was accustomed to sleep. The mutual fear of the two animals kept them at a respectful distance from each other; the snake, however, exhibiting quite as much disposition to be the assailant, as its more formidable companion. At length the Weasel gave the snake an occasional slight bite on the side or on the nose, without materially injuring it, and evidently without any instinctive desire to feed upon it; and at length, after they had remained two or three hours together, in the latter part of which they appeared almost indifferent to each other's presence, I took the poor snake away and killed it.

"Far different was this Weasel's conduct when a Mouse was introduced into the cage; it instantly issued from its little box, and, in a moment, one single bite on the head pierced the brain and laid the Mouse dead without a struggle or a cry. I have observed that when the Weasel seizes a small animal, at the instant that the fatal bite is inflicted, it throws its long lithe body over its prey, so as to secure it should the first bite fail; an accident, however, which I have never observed when a Mouse has been the victim. The power which the Weasel has of bending the head at right angles with the long and flexible, though powerful neck, gives it great advantage in this mode of seizing and killing its smaller prey. It also frequently assumes this position when raising itself on its hinder legs to look around.

"The disposition which has been attributed to the Weasel of sucking the blood of its prey, has, I believe, been generally much exaggerated. Some persons have positively denied the existence of such a propensity, and my own observation, as far as it goes, would tend to confirm that refutation of the commonly received notion. The first gripe is given on the head, the tooth in ordinary cases piercing the brain, which it is the Weasel's first act of Epicurism to eat clean from the skull. The carcase is then hidden near its haunt, to be resorted to when required, and part of it often remains until it is nearly putrid.

"The Weasel pursues its prey with facility into small holes, and amongst the close and tangled herbage of coppices, thickets and hedge-rows. It follows the Mole and the Field Mouse in their runs; it threads the mazes formed in the wheatrick by the colonies of Mice which infest it, and its long flexible body, its extraordinary length of neck, the closeness of its fur, and its extreme agility and quickness of movement, combine to adapt it to such habits, in which it is also much aided by its power of hunting by scent—a quality which it partakes in equal degree with the Stoat. In pursuing a rat or a mouse, therefore, it not only follows it as long as it remains within sight, but continues the chase after it has disappeared, with the head raised a little above the ground, following the exact track recently taken by its destined prey. Should it lose the scent, it returns to the point where it was lost, and quarters the ground with great diligence till it has recovered it; and thus, by dint of perseverance, will ultimately hunt down a swifter and even a stronger animal than itself. But this is not all. In the pertinacity of its pursuit, it will readily take the water, and swim with great ease after its prey.

"It is, however, sometimes itself the prey of hawks, but the following fact shows that violence and rapine, even when accompanied by superior strength, are not always a match for the ingenuity of an inferior enemy. As a gentleman of the name of Pinder, then residing at Bloxworth in Dorsetshire, was riding over his grounds, he saw, at a short distance from him, a kite pounce on some object on the ground, and rise with it in its talons. In a few minutes, however, the kite began to show signs of great uneasiness, rising rapidly in the air, or as quickly falling, and wheeling irregularly round, whilst it was evidently endeavoring to force some obnoxious thing from it with its feet. After a short but sharp contest, the kite fell suddenly to the earth, not far from where Mr. Pinder was intently watching the manœuvre. He instantly rode up to the spot, when a Weasel ran away from the kite, apparently unhurt, leaving the bird dead, with a hole eaten through the skin under the wing and the large blood-vessels of the part torn through. . . .

"The female Weasel .... brings forth four, or more frequently five young, and is said to have two or three litters in a year. The nest is composed of dry leaves and herbage, and is warm and dry, being usually placed in a hole in a bank, in a dry ditch, or in a hollow tree. She will defend her young with the utmost desperation against any assailant, and sacrifice her own life rather than desert them; and even when the nest is torn up by a dog, rushing out with great fury, and fastening upon his nose or lips."

The signification of the name 'Weasel', or, as it is also sometimes written, 'Weesel', is obscure. Webster states that he does not know the meaning, but observes that the German 'wiese' is a meadow. Von Martens, as quoted on p. 26, discusses the subject in its philological bearings. The name 'Weasel' in strictness should pertain to the present species, as distinguished from its various larger allies, as the Stoats and Ferrets; but it has come to have rather a generic application to the various species of the same immediate group.

The Stoat or Ermine.

Putorius (Gale) erminea.

PLATE VI, FIGS. 1, 3, 5, 6, 7.

(a. *General references.*)

Mus ponticus, quem hodie vocant Hermelam, *Agric.* "De Anim. Subter. 1614, 33".
Mustela candida, *Schwenckf.* "Theriotroph. 1603, 118".
Mustela candida in extrema caudâ nigricans, *Aldrov.* "Quad. Digit. 1645, 310, fig".
Mustela alpina candida, *Wagn.* "Hist. Nat. Helvet. 1680, 180".
Mustela candida s. animal Ermineum recentiorum, *Ray,* "Syn. Quad. 1693, 198".
Mustela alba, *Rzacz.* "Polon. 1721, 235".
Mustela caudæ apice atro, *Linn.* Fn. Suec. 1st ed. 1746, 3, no. 90.
Mustela candida s. ermineum, *Linn.* Syst. Nat. eds. 6th, 7th, 1748, 5, no. 6.
Mustela armellina, *Klein,* "Quad. 1751, 63".
Mustela nivea auribus augustis, caudæ apice nigro, *Hill,* "Hist. Anim. 1752, 543".
Mustela hieme alba, æstate supra rutila infra alba, caudæ apice nigro, *Briss.* Quad. 1756, 242.
Mustela erminea, *Linn.* Mus. Adolph. Frid. 1st ed. 1754, 5; S. N. i. 10th ed. 1758, 46, no. 9; Fn. Suec. 2d ed. 1761, 6, no. 17; Syst. Nat. 12th ed. i. 1766, 68, no. 10.—*Hout.* Natuurl. Hist. iii. 1761, 206, pl. 14, f. 5.—*S. G. Gmel.* Reise, ii. 1770, pl. 23 (*ermineum majus*).—*Müll.* Zool. Dan. Prod. 1776, 3, no. 14.—*Erxl.* Syst. An. 1777, 474, no. 13.—*Schreb.* Säug. iii. 1778, 496, pl. 137, A, B.—*Zimm.* Geogr. Gesch. ii. 1780, 308, no. 205.—*Gm.* S. N. i. 1788, 98, no. 10.—*Herm.* Obs. Zoöl. 45.—*Bechst.* Naturg. i. ——, 797.—*Turt.* S. N. i. 1806, 61.—*Pall.* Zoogr. R.-A. i. 1831, 90, no. 31 (*ermineum*).—*Desm.* Mamm. i. 1820, 180, no. 277; Nouv. Dict. xix, 376; Ency. Méth. pl. 83, f. 2, 3.—*Fr. Cuv.* Dict. Sci. Nat. xxix, 1823, 250.—*Is. Geoff.* Dict. Class. x. 212.—*Less.* Man. 1827, 146.—*Fisch.* Syn. 1829, 222.—*Flem.* Br. An. 1828, 13.—*Jen.* Br. Vert. 1835, 13.—*Bell,* Br. Quad. 1837, 145; 2d ed. 1874, 191, fig.—*Selys-L.* Fn. Belg. 1842, 10.—*Gray,* List Mamm. Br. Mus. 1843, 65.—*Schinz,* Syn. Mamm. i. 1844, 342.—*Schrenck.* Reise Amurl. ——, 40.—*Gieb.* Säug. 1855, 7-1.—*Gerr.* Cat. Bones Br. Mus. 1862, 93.—*Grill,* Zool. Gart. iii. 1862, 225.—*Gray,* P. Z. S. 1865, 111; Cat. Carn. Br. Mus. 1869, 88.

Mustela erminea *a.* æstiva, *b.* hyberna, *Gm.* S. N. i. 1788, 98, nos. 10 *a*, 10 *b*.
Viverra erminea, *Shaw,* Gen. Zoöl. i. 1800, 426, pl. 99.
Putorius erminea, *Griff.* An. Kingd. v. 1827, 122, no. 345.—*Owen,* Br. Foss. Mamm. ——, 116, f. 40, 41, 42 (skull).—*Brandt.* Wirb. Eur. N. E. Russl. ——, 24.
Fœtorius erminea, *Keys. & Blas.* Wirb. Eur. 1840, 69, no. 185.—*Blas.* Wirb. Deutschl. 1857, 228.—*Jäckel,* Zool. Gart. xiv. 1874, 459 (albino).
Armelinus, *Gesn.* Quad. 1551, 852.
Hermelinus, *Scheff.* Lappon. 1673, 343.
Hermellaenus, *Charlet.* Exercit. 1677, 20.—*Jonst.* Theatr. 1755, 153.
Hermine, *Charlev.* Nouv. France, iii. 1744, 134.—*Briss.* Quad. 1756, 243, no. 2.—*Buff.* Hist. Nat. vii. 240; Dict. Anim. ii. 430.—*French.*
Hermelin, *Hall.* Vierf. Thiere, 1757, 455.—*S. G. Gm.* Reise, ii. 192, pl. 23, iii. 370.—*Pall.* Reise, 1771, 129.—*Mart.* Buff. Vierf. iv. 196, pl. 67.—*Müll.* Natursyst. i. 1773, 274, pl. 14, f. 5.—*Stell.* Kamtsch. 1774, 125.—*Martens,* Zool. Gart. xi. 1870, p. 278 (philological).
Hermyn-Wezel, *Houtt.* Nat. Hist. Dieren, ii. 1761, 206, pl. 14, f. 5.—*Belgic.*
Ermellino, *Scatag.* Anim. Quad. ii. pl. 74, fig. from Buffon.—*Italian.*
Stoat, Ermine, *Penn.* Brit. Zoöl. ——, 84.—*English.*
Roselet (summer), *French.*—**Armiño, Armelina,** *Spanish.*—**Armellino,** *Italian.*—**Lekatt,** *Swedish.*—**Gronostay,** *Polish.*—**Gornostai,** *Russian.*

(*b. American references.*)

a. erminea.

Mustela erminea, *Forst.* Phil. Trans. lxii. 1772, 373.—*Harlan,* Fn. Amer. 1825, 62.—*Godman,* Am. N. H. i. 1831, 193.—*Thomps.* N. H. Verm. 1853, 31.—*Hall,* Canad. Nat. and Geol. vi. 1861, 295.
Mustela erminea *var.* **americana,** *Gray,* P. Z. S. 1865, 111; Cat. Carn. Br. Mus. 1869, 89.
Putorius erminea, *Aud. & Bach.* Q. N. A. ii. 1851, 56, pl. 59.—*? Wood.* Sitgr. Rep. 1853, 44 (Indian Territory).—*All.* Bull. M. C. Z. i. 1870, 167 (critical).—*Billings,* Canad. Nat. and Geol. ii. 1857, 455 (biographical).—*Allen,* Pr. Bost. Soc. N. H. xiii. 1869, 183.
Putorius noveboracensis, *De Kay,* Rep. N. Y. Survey, 1840, 18; N. Y. Zoöl. ii. 1842, 36, pl. 12, f. 2 (winter) and pl. 14, f. 2 (summer).—*Emmons,* Rep. Quad. Mass. 1840, 45.—*Bd. M. N. A.* 1857, 166, pl. 36, f. 3 (skull).—*Kenn.* Tr. Ill. State Agric. Soc. 1853–4, 578.—*Ross,* Canad. Nat. and Geol. vi. 1861, 441.—*Maxim.* Arch. f. Naturg. 1861, 220.—*Verz.* N. A. Säug. 1862, 44.—*Gilpin,* Tr. Nov. Scot. Inst. ii. 1870, 15, 59.—*Sam.* Ann. Rep. Mass. Agric. for 1861, 1862, 156, pl. 1, f. 1.—*Ames,* Bull. Minn. Acad. Nat. Sci. 1874, 69.

(*b. cicognani.*)

Mustela (Putorius) erminea, *Rich.* F. B. A. i. 1829, 46.
Mustela erminea, *Thomps.* N. H. Vermont, 1853, 31.
Mustela cicognani, *Bp.* Charlesw. Mag. ii. 1838, 37; Fn. Ital. 1838, sub. *M. boccamela.*—*Wiegm.* Arch. 1839, 423.—*Gray,* Cat. Mamm. Br. Mus. 195.
Putorius cicognani, *Bd.* M. N. A. 1857, 161.—*Suckley,* P. R. R. Rep. xii. pt. ii. 1859, 92.—*Gilpin,* Tr. Nov. Scot. Inst. ii. 1870, 13, 59.—*Sam.* Ann. Rep. Mass. Agric. for 1861, 1862, pl. 1, f. 6.
Mustela richardsoni, *Bp.* Charlesw. Mag. ii. 1838, 38 (based on Richardson).—*Gray,* P. Z. S. 1865, 112; Cat. Carn. Br. Mus. 1869, 90.
Putorius richardsoni, *Rich.* Zoöl. Beechey's Voy. 1839, 10*.—*Bd.* M. N. A. 1857, 164.—*Gray,* Cat. Mamm. Br. Mus. 195.—*Sam.* Rep. Mass. Agric. for 1861, 1862, 155, pl. 1, f. 3, 5, 7.—*Ross,* Canad. Nat. and Geol. vi. 1861, 441.—*Gilpin,* Tr. Nov. Scot. Inst. ii. 1870, 1559.—(?) *Stev.* U. S. Geol. Surv. Terr. 1870, 461 (Wyoming).—*Ames,* Bull. Minn. Acad. Nat. Sci. 1874, 69.
Mustela fusca, *Aud. & Bach.* J. A. N. S. P. viii. pt. ii. 1842, 288.—*De Kay,* N. Y. Z. i. 1842, 35.—*Wagn.* Wiegm. Arch. 1843, Bd. ii. 32.
Mustela (Gale) fusca, *Schinz,* Syn. Mamm. i. 1844, 343.
Putorius fuscus, *Aud. & Bach.* Q. N. A. iii. 1853, 234, pl. 148.
Putorius agilis, *Aud. & Bach.* Q. N. A. iii. 1853, 184, pl. 140.—*Kenn.* Tr. Illinois State Agric. Soc. for 1853–4, 1855, 578 (Illinois).
Putorius kanei, *Bd.* M. N. A. 1857, 172 (Kamtschatka and Siberia).
Putorius erminea *var.* **kanei,** *Gray,* P. Z. S. 1865, 111; Cat. Carn. Br. Mus. 1869, 89.

HAB.—Arctogæa: Europe, Asia, and America, north to the limit of existence of terrestrial Mammals. In America, south to very nearly the southern border of the United States, but no specimens seen from the Gulf States, New Mexico, Arizona, or Southern California. The range meets that of *P. brasiliensis*, which conducts the genus into South America.

SPECIFIC CHARACTERS.—Length of head and body 8-11, oftener 9-10 inches; of tail-vertebræ 2-5 inches, averaging 3½-4, only exceptionably passing the first-named limits. Tail at all seasons brushy, conspicuously black-tipped for ¼-⅗, generally about ⅔, its total length. In summer, dull mahogany-brown above, pale sulphury-yellow below; in winter, in most regions pure white all over except the black end of the tail, tinged in places with sulphury-yellow. Caudal vertebræ 17 or 18 (*Gerrard*).

Discussion of specific characters and relationships.

In entering upon the subject of the Ermines, the following *præmonenda*, which will be attempted to be proven in the course of the article, will assist to an appreciation of the points of the discussion :—

1. The Ermines of Europe, Asia, and America are specifically identical.

2. None of the supposed characters which have been relied upon to separate them have any existence in nature except as peculiarities of individual specimens examined.

3. The American Ermines are of two forms according to size alone, which in the extremes stand widely apart, but which grade insensibly into each other.

4. Within certain limits (to be hereafter defined), length of the whole animal, length of tail, both absolutely and relatively to that of body, and length of the black portion, either absolutely or relatively to that of the tail, are utterly fallacious as a means of specific diagnosis.

5. No question of coloration, of stoutness of body, of shape of ear, of furriness of feet, of character of pelage, and the like, can enter into the question, since such details are proven fortuitous circumstances of sex, age, season, locality, or merely normal individual variability.

I have before me a considerable suite of specimens of the Ermine, taken at various seasons in Great Britain, France, Germany, Sweden, Siberia, and Kamtschatka, together with an immense collection from all portions of North America inhabited by the animal. I may therefore set forth my conclusions without hesitation. The Ermine is the same animal in Europe, Asia, and America. Respecting the various trivial and insignificant distinctions which Gray and others have sought to establish, upon obviously insufficient material and inadequate investigation, it may be stated unequivocally that they fail as bases of specific or even varietal separation. Not that the alleged trifling differences do not exist; I can find them all and others besides: but they occur equally in the specimens from all countries, are not in the least correlated with any supposed geographical limits, and are, in short, an expression merely of the normal individual variability of the animal. As perfect duplicates as I ever examined came from Alaska and Northern Europe respectively: in all those nice points of pelage, shades of color, &c., which the practiced eye recognizes, they are more exactly alike than, for example, several specimens from England and France are among themselves. Every point which has been seized upon to separate an American from the Old World animal is nullified by sufficient series of specimens.

In seeking either resemblances or differences in the nicer minor points, we must not look at the animals as limited by certain continental areas, nor in any way by longitude: experience proves that this would be useless. A creature of thoroughly and conspicuously circumpolar distribution, extending probably as near the pole as any land Mammal, it is modified, when changed at all, by latitude, as expressed in the climate to which it is subjected, state of its food-supply, &c. These points are thoroughly understood in the commercial world by those whose wits are sharpened by their pecuniary interests; and it is surprising that some naturalists have failed to appreciate them.

The existence in North America of the true Ermine being established, there yet remains the question whether there be not also in this country other species of the same type, for we must not hastily assume that, because we have the true Ermine, all our other Stoats must be identical with it.

Throwing out of consideration the quite different *P. longicauda*, three species have of late years been currently recognized. These are the *P. noveboracensis* of De Kay, *P. richardsoni* of Bonaparte (= *agilis* Aud. and Bach.), and *P. cicognani* of Bonaparte (= *fuscus* A. and B.). Of the first-named it may be said, simply, that it is based upon the ordinary United States animal, of dimensions exactly corresponding to an average English specimen, for instance, and not otherwise different. This may be accepted as a convenient standard of comparison for the ordinary United States animal, identical with that from corresponding latitudes in the Old World. The *P. richardsoni* of Bonaparte was originally a mere presumptive attempt to separate the Ermine of America, being based upon *P. erminea* of Richardson, who does not hint at any supposed distinction from the Old World animal, and whose description and measurements indicate identity with ordinary *P. erminea*. Later, the name was adopted by Prof. Baird for specimens from Massachusetts and northward, considerably smaller than the average (8 inches), but with proportions of body and tail much as usual. *P. agilis* of Audubon and Bachman is obviously the same as Baird's *richardsoni*. Specimens from Massachusetts and northward, of about the same size, but shorter tail, were separated by Baird as *P. cicognani*. He compares *richardsoni* with *cicognani* as follows :—" This species is readily distinguished from *P. cicognani* by the longer tail, the vertebræ alone of which are half the length of the body." Measurements of the tail-vertebræ of *P. cicognani* given range from 2.25 to 3.00.

As a matter of fact, I find the tails to present all the several dimensions given by Baird, together with other intermediate dimensions, constituting an unbroken series from the shortest to the longest; and with additional dimensions which connect them as closely with the largest examples of "*noveboracensis*". It will be observed from Baird's tables that the difference among the various examples of "*cicognani*" (2.25 to 3.00 = 0.75) is about the same as that supposed to distinguish *richardsoni* (3.00 as against 4.00). In regard to total size, the same minute gradations are before me, from specimens scarcely 8 inches long to others over 10. The points of relative lengths of the black tip, amount of white on the upper lip, &c., are wholly matters of individual variability, to be thrown out of the discussion. It may be said in brief that the American Ermines are inseparably connected by the most minute intergradations from the smallest and shortest-tailed to the opposite extreme.

This fact ascertained, however, should not blind us to the equally notable fact of the existence of such differences. All the points laid down by Baird are substantiated. There are the larger and smaller Weasels, living side by side, in New York and Massachusetts, for instance—the one scarcely 8 inches long, with the tail-vertebræ under 3, and the other 11 inches, with the tail over 5. And I find the same thing to hold throughout the country to the Arctic Ocean. The *P. kanei* of Baird, a type of which is before me, is merely one of these smaller Ermines from Arctic regions. The author indeed says it is about the size of the *P. cicognani*, " which it otherwise greatly resembles and represents". The point I make is, that it is impossible to draw any dividing line between the extremes. Whatever character, or whatever set of characters, we assume as definitive, is instantly negatived by sufficient material. There *is* no dividing line. The differences may be relegated to the category of individual variability in size; and as such they possess for the zoölogist quite as much interest as if they were "specific characters". To facilitate the recognition and handling of this range of variation, I have above thrown the synonyms in two batches, assorting them as far as practicable; though it must always be remembered that the name refers, in most cases, not to either extreme, but to various intermediates, so that exact location of the names is in the nature of the case impossible. The smaller Stoat may be recognized, by those who desire to give it a name, as *P. erminea* var. *cicognani*.

So far from there being anything remarkable or exceptional in this, it seems that a similar case occurs in Europe. Though I am not cognizant of any species based upon this distinction in size, the specimens before me indicate the same range of variation. Thus, one from France, in winter pelage, and therefore full-grown, is quite as small as typical *cicognani*; for all I can discover, it is as nearly identical with a small Massachusetts skin as if the two had been born in the same litter.*

Those engaged in investigating the points at issue here should not fail to consult, further, Mr. Allen's paper upon the subject above quoted. It will be found an admirable historical summary of the case, an acute analysis of imaginary distinctions, and a logical conclusion. With the exception of the case of *P. longicauda*, which Mr. Allen had not seen, his views are substantially the same as those I have since been led to adopt from my own studies, though I would lay a little more stress upon the actually existing differences than he was inclined to do when arguing solely against that absence of specific distinction in which I wholly agree with him. I wish to here bring out the differences as strongly as he did the resemblances. Since the point at issue is entirely a matter of dimensions, relative and absolute, the following table of measurements is presented without comment as a fair *résumé* of the whole question:—

* "Earum inter Americæ animalia quoque meminit *Charlevoix hist.* de la Nouv. France vol. iii. p. 134. Statura ibi paulo minore sunt. Sic et in Daurie densissimis sylvis occurrent spithamam vix excedentes."—(PALLAS, *Zoog. Rosso-Asiatica* i. 1811, 92.)

MEASUREMENTS OF PUTORIUS ERMINEA.

Measurements of twenty-one alcoholic specimens of PUTORIUS ERMINEA.

Current number.	Original number.	Locality.	Sex.	From tip of nose to— Eye.	Ear.	Occiput.	Tail.	Tail to end of— Vertebrae.	Hairs.	Length of— Fore foot.	Hind foot.	Elbow to end of claw.	Knee to end of claw.	Height of ear.	Nature of specimen.
2313		Illinois	♂ ♀	0.50	1.30	2.00	9.00	4.25	5.25	1.10	1.50	2.00	2.60	0.75	Alc. sum.
2316		Massachusetts					8.50	4.00	5.25						do.
2317		do	♂				7.50	2.25	3.00						do.
6296		Nova Scotia					8.50	4.50	4.50						Alc. win.
6295		do					8.50	4.00	4.50						Alc. sum.
10399		Puget Sound					7.75	3.25	3.50						do.
10704		do	♂ ♀				7.50	3.50	3.50						do.
11233*		Nulato, Alaska					7.75	2.50	3.75						Alc. win.
11236		do					8.00	2.25	4.00						do.
11234		do	♂ ♀ ♂ ♂				8.00	2.60	4.00						do.
11235*		do	♂	0.70	1.50	2.20	8.00	3.25	3.75	1.25	1.75	2.25	3.25	0.80	Alc. sum.
8457		Saint Michaels, Alaska	♂ ♂				8.50	3.00	5.25		1.70				Alc. win.
8457 bis		Fort Churchill	♂				9.00	3.00	5.50						Alc.
		Saint Michaels	♂				9.25	3.00	4.75	1.12					Fresh.
275		Fort Simpson	♂	0.75	1.30	1.95	8.30	3.45	4.50		1.95				do.
361		Great Slave Lake	♂	0.70	1.30	1.40	8.75	4.00	5.70	0.90	1.80				Dry.
564		Fort Resolution	♂	0.78	1.35	1.95	10.00	4.00	5.35	1.40	2.00				Fresh.
1005		Yukon (October)	♂	0.70	1.25	2.03	10.25	3.25	5.30	1.50	2.10				do.
1029		do. (January)	♂	0.90	1.60	2.15	10.25	3.70	5.90	1.40	2.00				do.
1036		do. (December)	♂	0.85	1.30	2.12	10.10	3.50	6.90						do.
1041		do. (February)	♂	0.80	1.40	2.10	9.30	3.50	5.25						do.

* Mammae 6 pairs, abdominal and inguinal.

Note on the skull and teeth.

Skull and teeth.—A description of the cranium and dentition of this typical species will answer well for that of the subgenus *Gale* (see p. 99). The skull, though strong, is smooth in its general superficies, lacking almost entirely the sagittal ridge and roughness of muscular attachment which characterize the crania of the larger forms, like *fœtidus* and *vison* for instance. The forehead is turgid and convex in profile; the muzzle very short, swollen, and nearly vertically truncate. The zygomata are very slender, regularly arched throughout; the anterior root is a thin flaring plate, perforated by a large foramen anteorbitale. The cranium proper is peculiarly cylindrical rather than ovoidal; the postorbital constriction is abrupt, though slight. Supraorbital processes are moderately developed. The palatal emargination is slight; the pterygoids send out a spur to embrace the adjacent foramen, and terminate roundly, without a hamular process, so conspicuous in the larger *Putorii* and in *Mustela*, or with only a slight one. The bullæ auditoriæ are very large, flattish, parallel rather than divergent, and not in the least produced into a tubular meatus; on the contrary, the orifice of the meatus shows from below as an emargination. The glenoid fossæ have so prominent a hinder edge that they seem to present forward rather than downward.

The teeth scarcely furnish occasion for remark, as they present no peculiarities. In a specimen before me, the middle upper premolar of the right side has failed to develop. This is rather a large tooth to thus abort. Among the incisors (much as elsewhere in this subfamily), various irregularities are observable in different specimens, owing to the crowded state of these small teeth. (For cranial and dental peculiarities as compared with *longicauda*, see beyond.)

Description of the external characters.

A general description of this animal, herewith given, necessarily embraces many points shared with its congeners. It may be taken in amplification of the generic characters already given, and serve as a standard of comparison for other species, in the several accounts of which a repetition of non-essential specific characters is by this means avoided.

In general form, the Stoat typifies a group of carnivorous Mammals aptly called 'vermiform', in consideration of the extreme length, tenuity and mobility of the trunk, and shortness of the limbs. This elongation is specially observable in the neck, the head being set exceptionally far in front of the shoulders. The trunk is nearly cylindrical; it scarcely bulges in the region of the abdominal viscera, slopes a little over the haunches, rises slightly about the shoulder muscles, is a little contracted behind these; the neck is but little less in calibre than the chest. The greatest circumference of body is little more than half its length.

The head is shorter than the neck; it is notably depressed, especially flattened on the coronal area and under the throat; it is broad across the ears, whence it tapers with convex lateral outline along the zygomatic region, thence contracting more rapidly to the snout. The bulging of the sides of the head is in great measure due to the bulk of the temporal and masseteric muscles, which form swollen masses meeting on the median vertical line. This also contributes to the flattening of the frontal outline. The width of the head across the ears is about two-thirds its length. The eyes are rather small, situated midway between the nose and ears; they glitter with changing hues, and contribute, with the low forehead and protruding canine teeth, to a peculiarly sinister and ferocious physiognomy. The ample gape of the mouth, thin-lipped, reaches to below the eyes. The nasal pad at the extremity of the muzzle, is entirely and definitely naked; it is obscurely marked with a median furrow. The nostrils are small, circular, with a lateral projection below. The ears are conspicuous, rising high above the short surrounding fur; they are rounded in contour, about as wide across as high above the notch. Most of the auricle is flat and closely furred both sides. The rim completes about three-fifths of the contour. There is a conspicuous lobule reaching half-way up the border behind. The concavity of the vestibule is slight, naked, but hidden by a close-pressed pencil of long, upright hairs from the base of the auricle in front, extending nearly to the top of the ear. The back of the ear is on the occipital cross-line.

The whiskers are few but long, the longest reaching far beyond the head. A few shorter, very slender bristles spring over the eye and on the malar region.

The short forelimb is stout, and not fairly separated from the body much above the elbow. The forearm tapers rapidly to the wrist, causing the feet to appear slender in comparison, though they are really relatively stouter than in many unguiculate

Now, as to the details of coloration, especially the line of demarcation of the two body-colors, we must remember, in the beginning, that we here have an animal which, under ordinary conditions, turns entirely white once every year, and resumes its bicoloration as often; that consequently we must expect to find skins showing every possible step of the transition; and that, moreover, various odd little matters of coloration are certain to appear in different cases. Taken in its perfected summer dress in average latitudes, the animal ordinarily shows a line of demarcation, beginning at the snout, involving the edge of the upper lip, running thence straight along the side of the head and neck to the shoulder; there dipping down the fore edge of the limb to the paw, returning on the opposite border of the limb, running thence nearly straight to the hind leg, dipping down the outside of this also, returning to the perinæum, there meeting its fellow. The tail all around and upper surfaces of the paws are like the back. A slight lowering of this line would leave the end of the muzzle and the whole upper lip dark, as is frequently the case, showing how absurd are any distinctions based on "amount of white on the upper lip". The line also frequently encroaches upon the belly, narrowing the sulphury band. But, as might be anticipated, the chief deviations from this complete summer dress are in the other direction—lessening of dark area. The commonest point here is whiteness of the paws, the dark spurs stopping at the wrist and ankle. Another common state is whiteness of the anal region and under surface of the tail. Frequently light patches reach irregularly up the sides of the head, particularly about the ears. These points may be witnessed in midsummer, and appear to be purely fortuitous—that is, not traces of the regular change.

Coming now to this matter of the change, we find it under several aspects. I am not now speaking of the *mode* of change, but of the appearances presented at different stages. A frequent state of incipient change leaves much of the snout, ears, legs, and tail, sulphury-white, with considerable elevation of the general line of demarcation. This progresses until there may be a narrow median dorsal stripe along the whole length of the animal. In this kind of change, the fur of the dark parts is often found without the slightest admixture of white, the hairs being uniformly as dark as in summer, to the very roots. In other cases, however, with little or no restriction of the general dark area, this insensibly lightens by progressive whitening of the hairs from the roots outward, at first appearing merely paler brown, then white with brown streakiness of uniform character all over. The animal finally becomes pure white except the end of the tail. But this white is generally tinged in places, particularly on the belly and hind quarters, with sulphur-yellow.

Conditions of the change of color.

Much has been said of the mode in which this great change is effected, not only in the case of the Ermine, but of the Arctic Fox, Northern Hare, Hudson's Bay Lemming, and other animals. As I have not personally witnessed the transition, I can only display the evidence afforded in the writings of others. Some contend that the change is rapid and abrupt, resulting in a few hours, simply from lowering of temperature to a certain point. Others argue that the change is gradually accomplished; and of those favoring the latter view, some maintain that the brown coat is shed and a white one grown, while others hold that the extinction of pigment is gradually effected without a renewal of the pelage.

We will first review the evidence adduced by the author of Bell's Quadrupeds (p. 150, *seq.*) :—" The winter change of color which this species so universally assumes in northern climates . . . . is effected, as I believe, not by a loss of the summer coat, and the substitution of a new one for the winter, but by the actual change of color of the existing fur. It is perhaps not easy to offer a satisfactory theory for this phenomenon, but we may perhaps conclude that it arises from a similar cause to that which produces the gray hair of senility in man, and some other animals; of this instances have occurred in which the whole hair has become white in the course of a few hours, from excessive grief, anxiety or fear; and the access of very sudden and severe cold has been known to produce, almost as speedily, the winter change, in animals of those species which are prone to it. The transition from one state of the coat to the other does not take place through any gradation of shade in the general hue, but by patches here and there of the winter colour intermixed with that of summer, giving a pied covering to the animal. . . . . It appears to be established that what ever may be the change which takes place in the structure

of the hair, upon which the alteration of colour immediately depends, the transition from the summer to the winter colours is primarily occasioned by actual change of temperature, and not by the mere advance of the season." The author quotes in support of his views, and as tending to confirm them, the observations of Mr. John Hogg (Loudon's Mag. vol. v.), and details an experiment upon a Lemming which turned white by a few hours' exposure to severe cold.

As a supporter of the view that the change results from renewal of the coat may be cited the eminent naturalist Mr. Blyth, who communicates his conclusions to Mr. Bell in these terms (*op. cit.* 153) :—"Authors are wrong in what they have advanced respecting the mode in which this animal changes its color, at least in autumn; for in a specimen which I lately examined, which was killed during its autumnal change, it was clearly perceivable that the white hairs were all new, not the brown changed in colour."

Once again we have the minute and detailed observations of Audubon and Bachman, made from March 6 to 28, upon an animal they kept in confinement, and which was observed during this period to nearly complete the change from white to the summer colors. These authors agree with Mr. Blyth:—"We have arrived at the conclusion, that the animal sheds its coat twice a year, *i. e.*, at the periods when these semi-annual changes take place. In autumn, the summer hair gradually and almost imperceptibly drops out, and is succeeded by a fresh coat of hair, which in the course of two or three weeks becomes pure white; while in the spring the animal undergoes its change from white to brown in consequence of shedding its winter coat, the new hairs then coming out brown."

This conflicting testimony, which might be largely added to if this were desirable, is perhaps not so difficult to harmonize as it appears at first sight; nor is it in the least required to impugn the credibility of the witnesses of observed facts. I should state in the beginning, however, that it seems to me to be like straining a point to find any analogy between this periodically recurring change in a healthy animal and the tardy senile change coincident with flagging of the vital energies, or with the sudden pathological metamorphosis due to violent mental emotions of a kind to which *feræ naturæ* are not ordinarily exposed. This point aside, I would readily agree with Mr. Bell that subjection to sudden severe cold may materially hasten the change. But it is to be remembered in this connection that the difference in temperature is necessarily coördinated more or less perfectly with the progress of the seasons, so that it becomes in effect merely a varying element in the periodical phenomena. The question practically narrows to this: Is the change coincident with renewal of the coat, or is it independent of this, or may it occur in both ways? Specimens before me *prove* the last statement. Some among them, notably those taken in spring, show the long woolly white coat of winter in most places, and in others present patches—generally a streak along the back—of shorter, coarser, thinner hair, evidently of the new spring coat, wholly dark brown. Other specimens, notably autumnal ones, demonstrate the turning to white of existing hairs, these being white at the roots for a varying distance, and tipped with brown. These are simple facts not open to question. We may safely conclude that if the requisite temperature be experienced at the periods of renewal of the coat, the new hairs will come out of the opposite color; if not, they will appear of the same color, and afterward change; that is, the change may or may not be coincident with shedding. That it ordinarily is not so coincident seems shown by the greater number of specimens in which we observe white hairs brown-tipped. As Mr. Bell contends, temperature is the immediate controlling agent. This is amply proven in the fact that the northern animals always change; that in those from intermediate latitudes the change is incomplete, while those from farther south do not change at all.

The good purpose subserved in the animal's economy—in other words, the design or final cause of this remarkable alteration, is evident in the screening of the creature from observation by assimilation of its color to that of the predominating feature of its surroundings. It is shielded not only from its enemies, but from its prey as well. Another important effect of the whiteness of its coat has been noted. Mr. Bell has clearly stated the case:—"It is too well known to require more than an allusion, that although the darker colours absorb heat to a greater degree than lighter ones, so that dark-coloured clothing is much warmer than light-coloured, when the wearer is exposed to the sun's rays—the radiation of heat is also never greater from dark than from light-coloured surfaces, and consequently the animal heat *from within* is more completely retained by a white than by a dark covering; the temperature therefore of an animal having white fur, would continue more equable than that of one clothed in darker colours, although the latter would enjoy a greater degree of warmth whilst exposed to the sun's influence. Thus the mere presence of a degree of cold, sufficient to prove hurtful if not fatal to the animal, is itself the immediate cause of such a change in its condition as shall at once negative its injurious influence."

The latitudes in which the change occurs in this country include the northern tier of States, and the entire region northward. In this area, the change is regular, complete, and universal. Complete change is also usually effected—but not always—nearly to the southern limits of dispersion in mountainous regions. White winter specimens are the rule in Massachusetts, New York, and Pennsylvania; and I have seen others, pure white, from Illinois, Wyoming Territory, and California (Fort Crook). For the Southern States, from which I have no white examples, I will quote Audubon and Bachman:— "We received specimens from Virginia obtained in January, in which the colours of the back had undergone no change, and remained brown; and from the upper and middle districts of South Carolina, killed at the same period, when no change had taken place; and it was stated that this, the only species of Weasel found there, remained brown through the whole year. . . . Those from the valleys of the Virginia mountains have broad stripes of brown on the back, and specimens from Abbeville and Lexington, South Carolina, have not undergone the slightest change." It may be presumed that in the debatable ground some individuals may change and others not, and that, again, character of successive seasons may make a difference in this respect.

General history and habits of the species.

For the meaning of the name of this animal, we may refer again to Bell:—"The derivation of the word Stoat is very probably, as Skinner has it, from the Belgic 'Stout', bold; and the name is so pronounced in Cambridgeshire and in some other parts of England to the present time. Gwillim, in his 'Display of Heraldrie', gives the following etymology of Ermine:— 'This is a little beast, lesse than a Squirrell, that hath his being in the woods of the land of Armenia, whereof hee taketh his name.'" The latter word is sometimes written in English 'ermin' or 'ermelin'; and the same term occurs in several other languages, as in the French 'hermine', the Italian 'armellino', the Spanish 'armino', Portuguese 'arminho', Dutch 'hermelyn', German, Danish, and Swedish 'hermelin', and Armoric 'erminicq'.* Barbarous nations of the northern portions of the globe would appear to have each their own name for an animal well known to them as an object of the chase and of profit; names of very various signification, according to the different points which attracted their attention. Thus, Pallas enumerates nearly fifty names, most of which have no evident connection with each other. The technical appellation of the animal is derived from *putor*, a bad smell. *Fœtorius*, proposed as a substitute for *Putorius* by Keyserling and Blasius, has the same signification. The name is highly appropriate; for the stench emitted by the animals, of both sexes, is horribly offensive at times, as when under the influence of fear, anger, or the sexual passion; it is only less penetrating and more fugitive than that of the Skunk itself. It may be emitted at pleasure, as in case of the Skunk, and is scarcely perceptible, except at certain periods, when the animal is at rest. The source of the odor is a peculiar fluid contained in special glands situated about the anus, opening upon two conical papillæ, one on each side of the anus, just within the verge of the opening. On slightly everting the anus, these papillæ may be readily perceived; slight pressure will cause them to stand erect, while at the same time the fluid may be caused to spirt several inches in a fine spray, or even trickle in a stream about the parts.†

The female Ermine is provided with these glands, the same as the male. She is much smaller than the opposite sex; but, this and her sexual characteristics aside, she is quite identical. She makes her home in an underground burrow, beneath the stump of a tree, under a pile of rocks, in a decaying log, or the hollow of a tree trunk, and brings forth a large litter. The number at a birth is, however, very variable; four or five may be an average number. They have been found newly born from March to June, according to latitude, but are ordinarily produced in April or May. In northern latitudes, the litter may be born while the female is still in her white pelage, as in the case mentioned by Pallas; he found two young of a white mother, early in May, in the hollow of a tree. The

---

* Compare especially von Martens, *anteà*, p. 22.
† Compare *anteà*, p. 12.

cavity was separated into several compartments, arranged with some care. One of these contained a heap of fresh mice and shrews, another a quantity of the rejected skins, feet, and tails of these animals. The nest was extremely foul. The cry of the young is represented by Pallas to be like that of a newly-born kitten. At the age of ten or twelve days, the little animals were ashy above and white beneath. The mother, courageous in defence of her offspring, could scarcely be driven away, and followed the captor of her brood for a long time. The same author details the methods of capturing Ermine in Siberia—by means of a noose set at the entrance of their burrows, of spring-traps (at least so I understand by *decipulis compressoriis inescatis*), and of a bent stick with slip-knot, set off with a thread crossing their pathway, and placed before a hollow made in the snow where the bait is put. The skins, used for vestments, were sent, he says, chiefly to China, Turkey, and parts of Europe, being little used in Russia, where the tails, the principal ornaments, were reserved by law as the exclusive perquisite of royalty (*privilegio Majestatis reservatæ*). The body was withdrawn from the skin through a single incision across the posteriors; and it is added that not even those tribes "who eat all sorts of nasty things" will consume the flesh, so thoroughly impregnated is it with the fetor. The weight of a male is stated to be from five to eight ounces, more or less; of the female, scarcely four.

Mr. Hogg's observations on the British Stoats, in Loudon's Magazine, v., 718 *et seq.*, as already mentioned, relate chiefly to the changes of pelage as affected by temperature rather than season; but further remarks, bearing upon some of the habits of the animal, will be found interesting:—"Whilst walking along a footpath in a field, one day in the last week of December, 1831, I observed a Stoat, or a Weasel, coming in the same path towards me. I immediately stood still, and, as he approached, I found that he carried his nose in the same relative bearing to the ground, and was in the act of running the scent of some bird, or other small animal, exactly after the manner of a dog 'on scent', and in chase after game. His whole attention being to the ground, with his head down, he did not see me until close to me, when, suddenly catching sight of me, he turned a little aside, stopped short, looked up, and then scampered back along the path, with his tail erected into somewhat of a curve, from the black end of which I was able to distinguish him from a Weasel, and, bounding into a hedge near the path, he there concealed himself; whence he would probably go forth again, when he perceived that all was safe, and would perhaps follow up the scent from which I had disturbed him. I was thus an eyewitness to the fact of a Stoat being able to pursue its prey on scent, and I have little doubt that nature has given the sense of smelling, in a similar degree, to the Weasel and Polecat; which will therefore readily account for their being so destructive to game, and chiefly for their instinct in finding the nests of partridges and pheasants during the breeding season.

" . . . A Stoat does sometimes take to swimming. Walking on a fine evening in the spring, a few years ago, by the banks of the Wear, between Schincliffe Bridge and Old Durham, I noticed an animal swimming in the water; and, making haste to the place, which was just below the same bank whereon I was walking, I saw that it was a Stoat; it then swam gently across the river, which is there both deep and of considerable width, to the opposite bank, where, owing to the thick brushwood, I lost sight of it. In the act of swimming, it lifted its head and neck well out of water, like a dog; and so differed from a water rat, which usually keeps its head close along the surface." That the Stoat readily takes to the water, and swims well, has, however, been long known. Pallas makes this statement: "habitat . . . . necnon circa aquas, in quibus etiam praedam non illibenter quaerit, optime natans", and similar testimony is afforded by the writings of various authors. Audubon, however, says nearly the reverse:—"The Ermine avoids water, and if forcibly thrown into it, swims awkwardly like a cat." But this should be taken with qualification, like the same author's further statement, that the animal "does not, like the Fisher and Pine Marten, pursue its prey on trees, and seems never to ascend them from choice, but from dire necessity, when closely pursued by its implacable enemy, the dog." The Ermine indeed is neither so aquatic as its congener, the Mink, nor so much at home on trees as the Martens; but it has too frequently been observed in such situations to admit the doubt that it both swims and climbs with ease and without reluctance.

The always pleasing pen of Mr. Wm. Macgillivray has furnished us with the following general account of the habits of the Stoat as observed in Great Britain:—"It appears that in

England generally the Ermine is less common than the Weasel; but in Scotland, even to the south of the Frith of Forth, it is certainly of more frequent occurrence than that species; and for one Weasel I have seen at least five or six Ermines. It frequents stony places and thickets, among which it finds a secure retreat, as its agility enables it to outstrip even a dog in a short race, and the slimness of its body allows it to enter a very small aperture. Patches of furze, in particular, afford it perfect security, and it sometimes takes possession of a rabbit's burrow. It preys on game and other birds, from the grouse and ptarmigan downwards, sometimes attacks poultry or sucks their eggs, and is a determined enemy to rats and moles. Young rabbits and hares frequently become victims to its rapacity, and even full-grown individuals are sometimes destroyed by it. Although in general it does not appear to hunt by scent, yet it has been seen to trace its prey like a dog, following its track with certainty. Its motions are elegant, and its appearance extremely animated. It moves by leaping or bounding, and is capable of running with great speed, although it seldom trusts itself beyond the immediate vicinity of cover. Under the excitement of pursuit, however, its courage is surprising, for it will attack, seize by the throat and cling to a grouse, hare or other animal, strong enough to carry it off; and it does not hesitate on occasion to betake itself to the water. Sometimes, when met with in a thicket or stony place, it will stand and gaze upon the intruder, as if conscious of security; and, although its boldness has been exaggerated in the popular stories which have made their way into books of natural history, it cannot be denied that, in proportion to its size, it is at least as courageous as the tiger or the lion."

With a mind preoccupied in contemplation of the exploits of the chase of great *Carnivora*—those grand exhibitions of predatory instincts on the part of some of the strongest beasts, one is apt to overlook, or at least to underestimate, the comparative prowess of some lesser animals. Doubtless, the entomologist would give instances of equal courage and perseverance in pursuit of prey, of vastly greater comparative strength and skill in its capture, and superior destructiveness. Probably the great mass of insect-eating animals—an immense and varied host—are in no whit behind in this respect. And in noting the instincts and predacious habits of the Weasels and Stoats, we observe that, to grant them only equal courage and equal comparative prowess, we must nevertheless accede to them a wider and more searching range of active operations against a greater variety of objects, more persevering and more enduring powers of chase, and a higher grade of pure destructiveness, taking more life than is necessary for immediate wants. The great cats are mainly restricted each to particular sources of food supply, which they secure by particular modes of attack; and, their hunger satisfied, they quietly await another call of nature. Not so, however, with the Weasels. No animal or bird, below a certain maximum of strength, or other means of self-defence, is safe from their ruthless and relentless pursuit. The enemy assails them not only upon the ground, but under it, and on trees, and in the water. Swift and sure-footed, he makes open chase and runs down his prey; keen of scent, he tracks them, and makes the fatal spring upon them unawares; lithe and of extraordinary slenderness of body, he follows the smaller through the intricacies of their hidden abodes, and kills them in their homes. And if he does not kill for the simple love of taking life, in gratification of superlative bloodthirstiness, he at any rate kills instinctively more than he can possibly require for his support. I know not where to find a parallel among the larger *Carnivora*. Yet once more, which one of the larger animals will defend itself or its young at such enormous odds? A glance at the physiognomy of the Weasels would suffice to betray their character. The teeth are almost of the highest known raptorial character; the jaws are worked by enormous masses of muscles covering all the side of the skull. The forehead is low, and the nose is sharp; the eyes are small, penetrating, cunning, and glitter with an angry green light. There is something peculiar, moreover, in the way that this fierce face surmounts a body extraordinarily wiry, lithe, and muscular. It ends a remarkably long and slender neck in such way that it may be held at right angle with the axis of the latter. When the creature is glancing around, with the neck stretched up, and flat triangular head bent forward, swaying from one side to the other, we catch the likeness in a moment—it is the image of a serpent.

In further illustration of the character of the Stoat, I continue with an extract from Audubon, which represents nearly all that has appeared to the point in this country:—

"Graceful in form, rapid in his movements, and of untiring industry, he is withal a brave and fearless little fellow; conscious of security within the windings of his retreat among the logs, or heap of stones, he permits us to approach him within a few feet, then suddenly withdraws his head; we remain still for a moment, and he once more returns to his post of observation, watching curiously our every motion; seeming willing to claim association so long as we abstain from becoming his persecutor.

"Yet with all these external attractions, this little Weasel is fierce and bloodthirsty, possessing an intuitive propensity to destroy every animal and bird within its reach, some of which, such as the American rabbit, the ruffed grouse and domestic fowl, are ten times its own size. It is a notorious and hated depredator of the poultry house, and we have known forty well-grown fowls to have been killed in one night by a single Ermine. Satiated with the blood of probably a single fowl, the rest, like the flock slaughtered by the wolf in the sheepfold, were destroyed in obedience to a law of nature, an instinctive propensity to kill. We have traced the footsteps of this bloodsucking little animal on the snow, pursuing the trail of the American rabbit, and although it could not overtake its prey by superior speed, yet the timid hare soon took refuge in the hollow of a tree, or in a hole dug by the Marmot, or Skunk. Thither it was pursued by the Ermine and destroyed, the skin and other remains at the mouth of the burrow bearing evidence of the fact. We observed an Ermine, after having captured a hare of the above species, first behead it and then drag the body some twenty yards over the fresh fallen snow, beneath which it was concealed, and the snow lightly pressed down over it; the little prowler displaying thereby a habit of which we became aware for the first time on that occasion. To avoid a dog that was in close pursuit, it mounted a tree and laid itself flat on a limb about twenty feet from the ground, from which it was finally shot. We have ascertained by successful experiments, repeated more than a hundred times, that the Ermine can be employed, in the manner of the Ferret of Europe, in driving our American rabbit from the burrow into which it has retreated. In one instance the Ermine employed had been captured only a few days before, and its canine teeth were filed in order to prevent its destroying the rabbit; a cord was placed around its neck to secure its return. It pursued the hare through all the windings of its burrow, and forced it to the mouth, where it could be taken in a net, or by the hand. In winter, after a snow storm, the ruffed grouse has a habit of plunging into the loose snow, where it remains at times for one or two days. In this passive state the Ermine sometimes detects and destroys it.

"Notwithstanding all these mischievous and destructive habits, it is doubtful whether the Ermine is not rather a benefactor than an enemy to the farmer, ridding his granaries and fields of many depredators on the product of his labour, that would devour ten times the value of the poultry and eggs which, at long and uncertain intervals, it occasionally destroys. A mission appears to have been assigned it by Providence to lessen the rapidly multiplying number of mice of various species and the smaller rodentia.

"The White-footed Mouse is destructive to the grains in the wheat fields and in the stacks, as well as the nurseries of fruit-trees. Le Conte's Pine Mouse is injurious to the Irish and sweet potatoe crops, causing more to rot by nibbling holes in them than it consumes, and Wilson's Meadow-mouse lessens our annual product of hay by feeding on the grasses, and by its long and tortuous galleries among their roots.

"Whenever an Ermine has taken up its residence, the mice in its vicinity for half a mile around have been found rapidly to diminish in number. Their active little enemy is able to force its thin vermiform body into the burrows, it follows them to the end of their galleries, and destroys whole families. We have on several occasions, after a light snow, followed the trail of this Weasel through fields and meadows, and witnessed the immense destruction which it occasioned in a single night. It enters every hole under stumps, logs, stone heaps and fences, and evidences of its bloody deeds are seen in the mutilated remains of the mice scattered on the snow. The little Chipping or Ground Squirrel, *Tamias Lysteri* [*sc. striatus*] takes up its residence in the vicinity of the grain fields and is known to carry off in its cheek pouches vast quantities of wheat and buckwheat, to serve as winter stores. The Ermine instinctively discovers these snug retreats, and in the space of a few minutes destroys a whole family of these beautiful little *Tamiæ;* without even resting awhile until it has consumed its now abundant food, its appetite craving for more blood, as if impelled by an irresistible destiny, it proceeds in search of other objects on which it may glut its insatiable vampire-like thirst. The Norway rat and the Common House Mouse take possession of our barns, wheat stacks, and granaries, and destroy vast quantities of grain. In some instances the farmer is reluctantly compelled to pay even more than a tithe in contributions towards the support of these pests. Let however an Ermine find its way into these barns and granaries, and there take up its winter residence, and the havoc which is made among the rats and mice will soon be observable. The Ermine pursues them to their farthest retreats, and in a few weeks the premises are entirely free from their depredations. We once placed a half domesticated Ermine in an outhouse infested with rats, shutting up the holes on the outside to prevent their escape. The little animal soon commenced his work of destruction. The squeaking of the rats was heard throughout the day. In the evening, it came out licking its mouth, and seemed like a hound after a long chase, much fatigued. A board of the floor was raised to enable us to ascertain the result of our experiment, and an immense number of rats were observed, which, although they had been killed in different parts of the building, had been dragged together, forming a compact heap.

"The Ermine is then of immense benefit to the farmer. We are of the opinion that it has been over-hated and too indiscriminately persecuted. If detected in the poultry house, there is some excuse for destroying it, as, like the dog that has once been caught in the sheepfold, it may return to commit further depredations; but when it has taken up its residence under stone heaps and fences, in his fields, or his barn, the farmer would consult his interest by suffering it to remain, as by thus inviting it to a home, it will probably destroy more formidable enemies, relieve him from many petty annoyances, and save him many a bushel of grain."

The same author, alluding to the Weasel's want of shyness, and its ready capture in any kind of trap, continues with a matter that may next interest us—its relative abundance in different localities:—"This species does not appear to be very abundant anywhere. We have seldom found more than two or three on any farm in the Northern or Eastern States. We have ascertained that the immense number of tracks often seen in the snow in particular localities were made by a single animal, as by capturing one, no signs of other individuals were afterwards seen. We have observed it most abundant in stony regions; in Dutchess and Ontario counties in New York, on the hills of Connecticut and Vermont, and at the foot of the Alleghanies in Pennsylvania and Virginia. It is solitary in its habits, as we have seldom seen a pair together except in the rutting season. A family of young, however, are apt to remain in the same locality until autumn. In winter they separate, and we are inclined to think that they do not hunt in couples or in packs like the wolf, but that, like the bat and the mink, each individual pursues its prey without copartnership, and hunts for its own benefit." In Massachusetts, according to Allen, it is comparatively common. I myself saw none in Labrador during my summer visit; but it must be quite abundant, to judge from the number of skins I saw in possession of the natives at various places. According to Richardson, "Ermine-skins formed part of the Canada exports in the time of Charlevoix; but they have so sunk in value, that they are said not to repay the Hudson's Bay Company the expense of collecting them, and very few are brought to the country from that quarter." Nevertheless, it would appear that the Ermine is much more abundant in British America generally than it is in the United States. Over three-fourths of the large miscellaneous collection of skins we have examined in the preparation of this article came from this country and from Alaska. The writer last mentioned speaks of it as "common", and adds that it often domesticates itself in the houses of the fur traders, where it may be heard the live-long night pursuing the white-footed mouse. Up to a certain limit of latitude it would appear to increase in numbers to the northward. The abundance of an Ermine, either the present or succeeding species, on the Missouri is attested by the regalia of ceremony of some of the Indian tribes—picturesque costumes decorated with the tails, in rude imitation of royal fashion.

Like a majority of thoroughly predacious animals, the Ermine is somewhat nocturnal; that is to say, it is active and successful in the dark. Nevertheless, it is too often abroad in the daytime, either in sport or on the chase, to warrant our reckoning it among the truly nocturnal Carnivores. In the choice and construction of its retreats we see little evidence of burrowing instincts, or, indeed, of any considerable fossorial capacity. It retreats beneath stone heaps, under logs and stumps, in hollows of trees, and also in true underground burrows, though these, it should be observed, are usually those made by Rodents or other burrowers whom it has driven off or destroyed. Nevertheless, there is evidence that the animal sometimes digs. Thus Captain Lyon, as rendered by Richardson, states, that he observed a curious kind of burrow made by Ermines in the snow, "which was pushed up in the same manner as the tracks of moles through the earth in England. These passages run in a serpentine direction, and near the hole or dwelling place the circles are multiplied, as if to render the approach more intricate." Audubon has a passage of similar effect:—"We have frequently observed where it had made long galleries in the deep snow for twenty or thirty yards, and thus in going from one burrow to another, instead of travelling over the surface, it had constructed for itself a kind of tunnel beneath."

Accounts of different writers indicate a great variation in the number of young produced at a birth—from two to twelve. We may safely assume that these are unusual extremes, the average litter being five or six. As in case of the Miuk, the rutting season is early; in the United States, during a part of February and March. Young have been noted, toward the southern extreme of the range of the species, before the end of March; but most are produced in May or late in April. Without definite information respecting the period of gestation, we may surmise this to be about six or seven weeks. Information is also wanting of the length of time that the young nurse or require to have food brought them by the parents.

*On the distribution of the Ermine in the Old World.**

Georgi (*loc. cit.* [*i. e.*, Geogr. Phys. Besch. iii. 6], p. 1539) indicates, with regard to the distribution of the Ermine in Russia, the southern temperate, and the cold regions almost to the Arctic Ocean. He mentions, as special localities, the Polish-Russian and Dnieper governments, Curland, Livonia, and Ingermannland, also Finland, the governments on the Volga and its tributaries, and also the governments of Archangel, Wiburg, Wologda, Perm, those of the southeast to Bucharia; Siberia from the Ural to the Jenisei, Dauria, the Lena River, Kamtschatka, and finally the Kurile and Aleutian Islands; and calls attention to its abundance in Siberia. Pallas (Zoogr. R.-A. i. p. 92) gives the Ermine as inhabiting not only the whole of Europe, and Asia to India, but also asserts its extension into America, remarking, however, its absence from the Kurile and Aleutian Islands. According to Wosnessenski's observations, communicated to me

---

* Translated from Brandt's article already quoted with reference to *Gulo luscus.*

personally, and contrary to the opinion of Pallas, Ermines are met with on the Aleutians and Behring Islands, where they are hunting the Mice and Shrew-mice marching after the food-provisions of man. The same author also speaks of their frequent occurrence in Kamtschatka and on the coasts of the sea of Ochotsk. Von Saritschew (Reise, i. p. 92) observed Ermines on the middle course of the Indigirka, and von Wrangel (Reise, ii. p. 238) near Werchojansk, in latitude 67°, longitude 33°; Gebler (Katun. Gebirge, p. 85) mentions their existence in West Siberia; Eversmann in the governments of Kasan and Orenburg; Lehmann (Reise, Zoöl. App. by Brandt, p. 302) names, besides Orenburg, the country of the Bashkirs and Fort Spask; Hohenacker (Bull. Nat. Hist. Moskou, 1837, 2, p. 137) enumerates them among the Mammals of the countries of the Caucasus; Nordmann mentions their appearance in Bessarabia, Ekaterinoslaw, and Asia Minor (Demidoff, Voy. iii. p. 17); and Czernay (Bull. Nat. H. Moskou, 1851, p. 274) in the governments of Charkow and Ekaterinoslaw. Kessler calls them frequent inhabitants of all the four governments of the District of Kiew. Briucken (Mém. p. 47) and Eichwald (Skizze, p. 237) number them among the animals of Lithuania. Their appearance in Curland is mentioned in the Description of the Province of Curland by v. Derschau and v. Keyserling (p. 130), and also by Lichtenstein (Bull. Nat. H. Mosc. 1829, p. 289). According to a communication from Fischer, Ermines are met with in Livonia only in certain localities and a very few places (Naturg. v. Livland, p. 144). Their frequent appearance near St. Petersburg I am able to attest by many years' experience. In Finland, they are mentioned by Sadelin (Fauna Fenn. p. 10, and the Förteckning öfver Sällskapets p. Fauna Fenn. Samlingar, p. 7). Oseretskowski indicates them also on the coast of Lapland. Schrenk (Reise, i. pp. 66, 97) reports them on the Pinega River and in the District of Mesen. From the latter region, a specimen was received by the Academical Museum through the kindness of Mr. Bystrow, inspector of schools (see my report in the Scientific Bull. of the Acad. of Sciences of St. Petersburg, v. x). In the government of Wologda there are said to be collected annually from 5,000 to 10,000 skins (v. Baer Beitr. vii. p. 251). Sujew (Pallas, Trav. iii. p. 87) numbered them among the inhabitants of the lower Obi, and this was lately confirmed by Ermann (Reise, i. p. 562). The Ural expedition brought back with them a male specimen killed on June

7, 1847, on the Wischera, in latitude 62°. The summer coat of this individual agreed substantially with that of other Ermines killed at the same season in other regions. The tail of the above mentioned animal shows a white ring before its black end, very likely only an individual peculiarity. The balls of the feet and the joints of the toes were distinctly visible. Von Hoffmann informed me that the Ermines follow the Lemmings to the Arctic Sea.

The many Ermines killed near St. Petersburg are always brown in summer but white in winter, which, Pallas says, is also the case with those living near the Caspian Sea (Zoögr. i. p. 93). In my memoir on the periodical change of pelage of animals of the Weasel kind (Bull. Sc. Cl. Phys. Mat. v. ix. n. 12, Mélanges Biolog. i. p. 185), I mentioned the capture of an Ermine in the brown or summer pelage, in the month of November, on the island of Oesel, and doubted then the likelihood of such an occurrence. The following communication, however, from Dr. Moritz, of Tiflis, respecting *Mustela vulgaris*, permits me to believe that many individuals do not change when the winter is mild. The northern limits of the animal's distribution in the Russian possessions are the shores of the Arctic Ocean; the southern limits include the whole region south of the Caucasus; the western, Poland; and the eastern, Kamtschatka and East Siberia. Concerning the value of the fur, and the yearly proceeds, see v. Baer, Beitr. viii. p. 183.

The Long-tailed Weasel.

Putorius longicauda.

? Mustela longicauda, *Bp.* Charlesw. Mag. N. H. 1838, 38 (based on long-tailed variety of *P. erminea* from Carlton House, *Rich.* F. B. A. i. 1829, 47, in text).—*Gray*, List Mamm. Br. Mus. 195.
? Putorius longicauda, *Rich.* Zoöl. Beechey's Voy. 1839, 10*, (in text; same as foregoing).
Putorius longicauda, *Baird*, M. N. A. 1857, 169 (Nebraska, Montana).—*Suckley*, P. R. R. Rep. xii. pt. ii. 1859, pp. 93, 114 (Milk River).—*Hayd.* Trans. Am. Philos. Soc. xii. 1862, 142.—(?) *Ross*, Canad. Nat. and Geol. vi. 1861, 441.—*Coues & Yarrow*, Zoöl. Expl. W. 100 Merid. 1875, 591 (Colorado and New Mexico).
Putorius culbertsoni, *Bd. MSS.* Mus. Smiths. (labels of nos. 4320, 4325).
Hermelin des oberen Missouri, *Maxim.* Verz. N.-A. Säug. 1862, 46, pl. "8", f. 8 (penis-bone).

HAB.—Region of the Upper Missouri and its tributaries; Minnesota, Dakota, Montana, Nebraska, Wyoming; also Colorado, Utah, New Mexico, Arizona. North apparently to Carlton House; west probably to the Pacific.

SPECIFIC CHARACTERS.—Size of *P. erminea*; tail absolutely and relatively longer, with hairs ¼ to ⅜ the head and body. Below tawny or buffy, with a salmon (not sulphury) tinge abruptly defined against white of cheeks and chin. Black on end of tail almost reduced to the terminal pencil alone, scarcely ¼ the whole length. Color entirely white in winter. Male, total length of head and body, 10.50; tail-vertebræ 6.75; tail with hairs 8.50. Female, total length of head and body, 8.50; tail vertebræ, 5.75; tail with hairs, 6.75.

Description. *

The size is entirely within the range of that of *P. erminea*, and there is little to note in this respect, excepting the greater length of the tail; the general build, however, appears stouter than is usual in *P. erminea*, the muzzle blunter. The tail is remarkably long—not that it is entirely beyond the maximum of that of *erminea*, but that when shortest it is about at such maximum, and that its normal average is beyond the average of that of *P. erminea*. The two animals being of substantially the same size of body, the tail is relatively longer—including the hairs it is three-fourths to four-fifths the length of the body and head. The black on the tail is normally restricted to the minimum—or even beyond this—of ordinary *erminea;* it occupies scarcely anything more than the terminal pencil alone, extending less than half an inch on the vertebræ. The upper parts are much as in *erminea*, but there is a peculiar olivaceous cast, owing to admixture of some green in the brown—not that any green shows as such, but it gives a particular tone to the parts. Below, and on both sets of paws, the color is a rich and beautiful buffy-yellow mixed with salmon color, quite different from the clear sulphury of *P. erminea*. This color is abruptly displayed against the pure white of the chin and cheeks. The female is considerably smaller than the male, as usual in this genus, but is not otherwise different. (This particular specimen is much lighter than her mate, but such distinction will not hold.) The following measurements were carefully taken in the flesh :—

---

* From a pair in the Coues collection, killed in August, 1874, in Northwestern Montana.

Measurements of two specimens, ♂ and ♀, of PUTORIUS LONGICAUDA.

Original number.	Locality.	Date.	Sex.	From tip of nose to—				Tail to end of—		Length of—		Elbow to end of claws.	Knee to end of claws.	Height of ear.	Width of ear.	Girth of chest.	Nature of specimen.
				Eye.	Ear.	Occiput.	Tail.	Vertebræ.	Hairs.	Fore foot.	Hind foot.						
4505*	Chief Mountain Lake, Montana, 49° N...	Aug. 17, 1874	♀	0.75	1.65	2.00	8.50	5.75	6.75	1.20	1.75	2.10	3.00	0.80	0.80	3.75	Fresh.
4586	Chief Mountain Lake, Montana, 49° N...	Aug. 23, 1874	♂	0.80	1.90	2.30	10.50	6.75	8.45	1.30	2.00	2.50	3.10	0.95	1.10	4.50	...do.

* Killed up a tree with a stick. Pupil oblique, elliptical. Pores of anal glands opening on papillæ, one on each side of edge of anus. Odor strongly fœtid.

General account of the species.

The subject of the present article differs notably in the above particulars from the common type of Ermine. It is probably the same as the *longicauda* of Bonaparte, though it must be observed that we have no assurance of this. It is the *longicauda* of Baird. In case Bonaparte's animal should prove not the same, the present must be called P. *culbertsoni* Baird, MSS.

After dwelling at the length I have upon the variability in the length of tail of *P. erminea*, and on the extent of the black pencil in that species, it may seem inconsistent to introduce such features in a specific diagnosis. But it will be observed that the character of the member is something over and above that shown by *P. erminea* in any of its interminable variety, and that I use it in combination with another peculiarity, the color of the under parts. Taken together, these seem perfectly distinctive; at any rate, I find the same features preserved throughout a considerable series of specimens, without the slightest intergradation with *P. erminea*. The specimens are distinguishable at a glance. While I make no doubt that this animal is an offshoot from *P. erminea*, yet the differentiation is complete, and no intermediate specimens are known; while, for that matter, it is doubtless true that all of the species have come from an original stock. This particular offshoot is a step toward those members of the genus which extend into tropical America. This is evident in the coloration of the belly, very little increased intensity of which would assimilate it to the rusty and orange-brown shades prevailing further south.

Besides the types of my description, I have examined a dozen or more additional specimens—those recorded by Baird in his work, and others since received at the Smithsonian. None show any gradation with *P. erminea*. In No. 4325, from old Fort Union, "Nebraska" (now Montana), the tail-vertebræ (the tail has not been skinned) measure 6.50 inches, with the hairs about 8.00. No. 4320, from Fort Laramie, Wyoming, taken in December, 1859, is pure white; the black tip under 1.50 long; the vertebræ (unskinned) are about 6.00; the specimen in its winter dress is readily recognized by these features as pertaining to *P. longicauda*. Another specimen is in winter dress from Fort Clarke. There are several from Utah. A skin, too defective for satisfactory identification, but probably belonging here, is in the collection from Puget's Sound—a locality which, if substantiated, would considerably extend the known range of the species. It is the individual which formed the basis of *P. "richardsoni"* in Dr. Suckley's report above cited.

Since writing the foregoing, I have examined the skull of this species in comparison with that of *P. erminea*, and I find remarkable distinctions. Coördinately with the shorter and broader head, the skull shows differences of shape as well marked as those subsisting between *M. martes* and *foina*. An example of *P. noveboracensis* measures 1.90 by 0.95; a female specimen of *longicauda* 1.80 by 1.03, the resulting difference in contour being obvious. The cranial portion proper of *P. longicauda* is much more expanded and ovate, the width there (0.90) being half the total length, while the same measurement of *P. noveboracensis* (0.80) is much less—only about half the length of the skull exclusive of the rostral portion. The skull of *P. longicauda* is notably more constricted behind the orbits. The zygomata are much more obliquely offset from the skull. The anteorbital foramina are narrowly oval and very oblique. There is a remarkable inward obliquity of the last upper molar, different from anything I have seen in *P. ermineus*. I have not seen a male skull; it will be found larger by about a fourth of an inch in length. We may tabulate the cranial characters of the two species as follows:—

P. erminea.—Zygomatic width of skull one-half its length. Cranial width much less than half the total length. Width of skull at point of greatest constriction half the zygomatic width. Anteorbital foramina large, subcircular. Set of back upper premolar nearly vertical.

P. longicauda.—Zygomatic width of skull about three-fifths its length. Cranial width half the total length. Width of skull at point of greatest constriction about two-fifths the zygomatic width. Anteorbital foramina small, very obliquely oval. Set of back upper premolar obliquely inward.

I think that after all the relationships of this species are closest with *P. frenatus*, notwithstanding the absence of the facial markings peculiar to the latter. It shares with *P. frenatus* the rusty-reddish or salmon-colored under parts, well contrasting with the clear sulphury-yellow of *P. erminea*. Moreover, southern examples, such as those from New Mexico, show a decided approach to *P. frenatus* in darkening of the color of the head. This is sometimes so decided, that were white spots present in these cases, the specimens would unhesitatingly be referred to *P. frenatus*; and we know that in Central American and Mexican skins the facial markings of *P. frenatus* are not seldom extinguished.

The Long-tailed Stoat is the characteristic form of the genus throughout the region of the Missouri and its tributaries. While I am not assured that it inhabits this country to the exclusion of *P. erminea*, I may state that I never met with the latter in any of my travels, and that I have not seen specimens from fairly within this region, though some from its confines are before me. It is the Weasel of the Rocky Mountains too, for a corresponding extent, and, as above indicated, very probably reaches to the Pacific. Mr. Ridgway informs me that he found a specimen, which he satisfactorily identifies from memory of its creamy-yellow under parts, in the Wahsatch Mountains, near Salt Lake City, Utah. I have also seen the species in the mountains of Colorado.

I found the animal to be quite numerously represented in Northern Montana, on the boundless prairies of the Upper Missouri and Milk River, living in burrows underground along with the Gophers (*Spermophilus richardsoni*), Badgers, and Kit Foxes. In these treeless domains, it occupies as its home the deserted burrows of the Gophers. I once surprised a family of five or six in such a retreat; I could hear them spitting angrily below, but did not succeed in my endeavor to dislodge them. This was late in July; the young were well grown at this period. Later in the season, at Chief Mountain Lake, one of the headwaters of the Saskatchewan, on the eastern base of the Rocky Mountains, latitude 49°, several specimens were secured. Here the species was living on wooded ground; indeed, one of my specimens was caught up a small tree, and killed with a stick. It climbed and leaped among the branches with ease and agility, much like a Squirrel. Skins were in demand by the men of our party for the manufacture of tobacco-pouches; they made very pretty ones, and many were killed for this purpose.

The specimen mentioned above from the Wahsatch Mountains was found dead by Mr. Ridgway in the nest of a *Buteo swainsoni*. This shows that the animal, despite its ferocity and activity, may fall a victim to the rapacity of the larger Hawks. The individual had its neck torn, and was already partly eaten by the two strong and voracious young Buzzards which occupied the nest. The nest contained also the remains of a Chipmunk, and of a Black-headed Grosbeak (*Goniaphea melanocephala*).

When irritated, this species diffuses a fœtid odor quite as strong as that of the Mink.

The Bridled Weasel.

Putorius (Gale) brasiliensis frenatus.

a. brasiliensis (Sewast.).

Mustela brasiliensis, *Sewast.* Mem. Acad. St. Petersb. iv. 1813, 356, pl. 4.—*Fisch.* Syn. 1829, 222.—*Burm.* Abh. Nat. Ges. Halle, ii. 1854, 46.—*Gerr.* Cat. Bones Br. Mus. 1862, 94.—*Gray,* Ann. Mag. N. H. xiv. 1874, 374.
Mustela (Putorius) brasiliensis, *D'Orbig.* Voy. Amér. Mérid. ——, pl. 13, f. 3 (skull).
Mustela (Gale) brasiliensis, *Schinz,* Syn. Mamm. i. 1844, 346.
Mustela (Neogale) brasiliensis *var.* brasiliana, *Gray,* P. Z. S. 1865, 115 (type of *Neogale*); Cat. Carn. Br. Mus. 1869, 92.

b. æquatorialis Coues.

Mustela aureoventris, *Gray,* P. Z. S. 1864, 55, pl. 8 (Quito; very young) (not *M. auriventer* Hodgs.).—*Gray,* P. Z. S. 1865, 115 (Ecuador and New Grenada; adult); Cat. Carn. Br. Mus. 1869, 92.
Putorius brasiliensis *var.* æquatorialis, *Coues* (merely as a substitute for Gray's preoccupied name).
? Mustela macrura, *Tacz.* P. Z. S. 1874, 311, pl. 48 (Central Peru).
? Mustela affinis, *Gray,* Ann. Mag. N. H. xiv. 1874, 375 (New Granada).

c. frenatus (Licht.).

Mustela frenata, *Licht.* Darstellung . . . Säug. 1827-34, pl. 53 (Mexico).—*Aud. & Bach.* J. A. N. S. P. viii. pt. ii. 1842, 291.—*Gray,* Zoöl. Voy. Sulphur, 1844, 31, pl. 9 (head).—*Tomes,* P. Z. S. 1861, 287 (Guatemala).—*Gray,* List M. Br. Mus. 1843, 65.—*Gerrard,* Cat. Bones Br. Mus. 1862, 94.
Mustela (Gale) frenata, *Wagn.* Suppl. Schreb. Säug. ii. 1841, 234.
Putorius frenatus, *Bach.* J. A. N. S. P. viii. 288.—*Aud. & Bach.* Q. N. A. ii. 1851, 71, pl. 60 (Texas to Monterey and southward).—*Bd.* M. N. A. 1857, 173, pl. 19, fig. 5 *a* ; Mex. B. Surv. ii. pt. ii. 1859; Mammals, 19, pl. 17. figs. 1 and 2, *a–e*.—*Coues,* Am. Nat. i. 1867, 352.
Mustela xanthogenys, *Gray,* Ann. Mag. N. H. xi. 1843, 118; Zoöl. Voy. Sulph. 1844, 31, pl. 9; (ref. to *Pall.* Zoog. 92); List M. Br. Mus. 1843, 66; P. Z. S. 1865, 115; Cat. Carn. Br. Mus. 1869, 93; Ann. Mag. N. H xiv. 1874, 375.—*Gerr.* Cat. Bones Br. Mus. 1862, 94.
Putorius xanthogenys, *Bd.* B. N. A. 1857, 176, pl. 19, f. 3 *a* (California).—*Newb.* P. R. R. Rep. vi. 1857, 42 (San Francisco).
Putorius mexicanus, *Berlandier.* MSS. ic. incd. 4 (Tamaulipas and Matamoras).
'Comadreja' of the *Mexicans.*
(Compare *Mustela javanica* Seba, Thes. i. pl. 48, fig. 4 = *M. leucogenis* Schinz, Syn. i. 344; not Japanese.—Cf. Pall. Zoog. R.-A. i. 1811, 92, footnote.)

HAB.—Southern Texas to California. Up the Pacific side to San Francisco, Fort Crook, and probably Astoria, Oregon. South to Guatemala. Var. *æquatorialis,* thence to Ecuador. Var. *brasiliensis,* Brazil.

SPECIFIC CHARACTERS.—Size and proportions of *P. erminea ;* top of head notably different in color (darker) from the back, and blotched with white; chin white; other under parts more or less strongly tinged with tawny-yellow or orange-brown; tail tipped with black. No seasonal change of pelage.

General account of the species.

In respects of size and form, this species scarcely differs from *P. erminea*. The pelage appears to be coarser, thinner, and more glossy than it usually is in the *P. erminea*, evidently as a consequence of the more southern habitat of the animal. The palms, soles, and ears are rather more scantily haired. There are no indications that the animal turns white in winter.

The pattern of coloration is as usual in this genus, with the addition of the peculiar head-markings, to be presently described. The upper parts are of a mahogany-brown, as in the summer coat of other Stoats, but differ in the shade much as polished mahogany differs from that wood in the rough, being darker and richer in tint. This color deepens insensibly into blackish on the head. The darkest examples before us, from Guatemala, are almost chocolate-brown, and quite black on the head. This intensity of coloration is quite coincident with decrease of latitude; and the northernmost examples, from California, are much paler—of a lighter and more yellowish-brown than the average of *P. erminea*. There is a similar parallelism in the color of the under parts. Aside from the chin and throat, which usually remain quite purely white, the under parts range from a tawny-yellow to a rich orange-brown. In running through the series from California to Guatemala, I have seen nothing quite like this in any of the northern Stoats, in which any yellowish of the under parts which may exist is sulphury; the only approach to it being a salmon shade on *P. longicauda*. In the orange-bellied Guatemalan skins, moreover, the line of demarcation from the white of the throat is quite abrupt; in others, the transition is by insensible degrees. The light color of the under parts runs down both fore and hind limbs to the feet; but the tops of the feet are indifferently colored like the belly or like the back; at any rate, I find specimens varying in this respect, without finding any clue to a rule which might determine this condition. The tail is like the back all around; it blackens insensibly at the tip, for a shorter distance than is usual in northern Stoats; the defined black only occupying, on an average, about an inch of the end in addition to the terminal pencil of hairs, which is about another inch longer.

Measurements of two specimens of Putorius brasiliensis frenatus.

Current number	Locality	Sex	From tip of nose to—				Tail to end of—		Length of—		Nature of spec.
			Eye.	Ear.	Occiput.	Tail.	Vertebræ.	Hairs.	Fore foot.	Hind foot.	
2329	Matamoras, Mex	♂	1.00	1.75	2.25	11.00	6.75	2.00	1.20	1.80	Alc.
2323	....do...........	♀	0.60	1.50	2.20	8.25	4.75	5.30	1.10	1.66	Alc.

The facial white markings of this species deserve special consideration. Upon the most cursory examination, one may satisfy himself of their irregular, indeterminate character, and would expect to find them, as they really are, variable to the last degree. They are similar in this respect to the white on the chin and about the lips and along the belly of the Mink, or on the chest of the Marten, and of a part with the variation above mentioned in the color of the paws of the present species. They appear to be, in fact, simply an exaggeration and permanent retention of certain white markings that occur in *P. erminea* (unassociated with beginnings or remains of the winter dress). In several European examples of *P. erminea*, I find a little white coronal or two white supraciliary or auricular spots, and a wholly variable extent of white upon the cheeks. The usual pattern in *P. frenatus* is this: a triangular or quadrate white frontal spot just between the eyes, and a broad oblique white stripe on each side of the head. In addition, there may or may not be an occipital white spot between the ears. The frontal spot is usually isolated from the white stripes, but may fuse with them, completing the "bridle". It is sometimes reduced to a mere nasal stripe, with corresponding reduction of the pre-auricular markings. In a specimen from Fort Crook, California, which I refer to this species, there are only a few white hairs on the muzzle, and a slight patch at the base of the ear. But the malar stripe, on the variations of which *P. xanthogenys* chiefly rests, is still more unstable in character; for its width and the outline it forms with the black of the cheeks are wholly indeterminate.

I am inclined to think that this animal has become fairly differentiated from an original stock which comprised *P. erminea*, although traces of a former connection may still subsist on the confines of its present habitat. The Fort Crook specimen above mentioned imperfectly represents the species. It is, morover, associated in that locality with an animal which turns perfectly white in winter, and is in other respects inseparable from northern Ermines. Thus, No. 2839, from Fort Crook (*Feilner*), is pure white underneath, has the head like the back in color, and both of the usual undressed mahogany-brown; yet it shows the white frontal spot and has a decided trace of the malar stripe. It is accompanied by No. 3872, pure white all over. Still, the white markings of No. 2839 may be remains of a seasonal change, or merely like the similar appearances that some specimens of the European *P. erminea* present. I refer these two specimens to *P. erminea*, but include the other from Fort Crook, No. 3830, in the beginning of the *brasiliensis* series. I refer to this species, with some hesitation, a very young Stoat from Astoria, Oregon (No. 3520, June 19, 1858, *J. Wayne*). Although the head is not darker than the back, and no head-stripes are apparent, the belly shows strongly the characteristic fawn color of *frenatus*.

There remains the discussion of the relationships of the South American forms. Although I have not specimens from Brazil or Ecuador, the sufficient descriptions of authors enable me to speak with confidence respecting them. There is evidently but one series of linked forms. We have already seen that *frenata* begins in Upper California, as *xanthogenys*, which is merely the northernmost palest form, between which and true *frenata* (City of Mexico, &c.) there is no difference requiring recognition by name. In Guatemala, *frenata* already assumes the rich coloration that culminates further south in *brasiliensis*. Gray, indeed, who usually subdivides altogether too much, does not attempt to separate *frenata* and *brasiliensis* except varietally. *P. æquatorialis* (as I call what Gray named *M. aureoventris*, a term preoccupied to all intents and purposes by *auriventer* of Hodgson) was originally described from a very young animal ("length 6 inches, tail 4½"—the adult is afterward described as length 12, tail 8), without facial markings, but the adult has the auricular blotch, though the frontal spot appears to be extinguished. It is described as very darkly and richly colored, the under parts and ear-spot "golden-yellow"; the coloration of the plate is almost precisely that of Guatemalan specimens before me. But that the facial markings may be completely extinguished, as a matter of individual variation, is shown by a specimen before me from Costa

Rica. It is very dark and richly colored, with the merest trace of white markings behind one eye—not both.

I find nothing in the ascribed characters of *Mustela macrura* Taczanowski forbidding its reference to the Middle American series; nor is there anything in Dr. Gray's brief and unsatisfactory account of *M. affinis* incompatible with the characters of the present species in their now ascertained range of variation.

There is nothing peculiar in the relationships of the various forms as here advanced. It is paralleled in the cases of other Mammals and many Birds, and, in fact, might have been predicated.

We are in possession of no special information upon the habits of the Bridled Weasel, which, however, may be presumed to differ little, if at all, from those of its allies. Dr. Newberry represents it as abundant about San Francisco.

CHAPTER V.

MUSTELINÆ—Continued: THE AMERICAN FERRET.

The subgenus *Cynomyonax*—Subgeneric characters—*Putorius* (*Cynomyonax*) *nigripes*, the American or Black-footed Ferret—Synonymy—Specific characters—Habitat—General account of the species—ADDENDUM: On the species of the subgenus *Putorius*—*P. fœtidus*, the Polecat or Fitch—Synonymy—Description—*P. fœtidus* var. *furo*, the Ferret—Synonymy—Remarks—Ferret breeding and handling—*P. fœtidus* var. *eversmanni*, the Siberian Polecat—Synonymy—Remarks—*P. sarmaticus*, the Spotted Polecat—Synonymy and remarks.

I HAVE been obliged to establish a new subgenus for the reception of the singular *Putorius nigripes* of Audubon and Bachman, which curiously combines some features of both *Gale* and *Putorius* proper, with others peculiar to itself. As indicated by the name of American or Black-footed "Ferret", it is the strict analogue in this country of the European Ferret, or Polecat, with which it agrees so closely in some respects that I was at first inclined to refer it to the subgenus *Putorius* itself. But further examination has satisfied me that the sum of its peculiarities ranks as high, at least, as that characterizing other admitted subgenera of *Putorius*. I have already concisely contrasted its characters with those of other sections of the genus (p. 99), and shall devote this chapter to further consideration of the remarkable animal.

The Subgenus CYNOMYONAX. (COUES, 1877.)

The dental formula of this subgenus is the same as that of the genus *Putorius* at large (pm. $\frac{3-3}{3-3}$).

The details of the dentition agree most closely with those of subgenus *Putorius*, though peculiar in one respect. The back lower molar is a mere cylindrical stump, with hemispherical crown, too small and weak to develop the little cusps seen plainly in *P. fœtidus* and *P. vison*. The inferior incisors, in the specimen examined, are so crowded that the middle one on each side sets entirely back of the line of the rest, exactly as in a specimen of *P. fœtidus* before me. The dentelure of the upper jaw might be described in terms identical with those applicable to *P. fœtidus*, though the back molar seems to be rather weak. *P. fœtidus* and *C. nigripes* both differ from *L. vison* in the character of the upper sectorial tooth, which in *vison* develops, as elsewhere described, an antero-exterior process, wholly wanting in the other subgenera.

The skull of *Cynomyonax* differs notably from that of *Gale*, and agrees with those of *Putorius* proper and *Lutreola* in its size, relative massiveness, and development of ridges and depressions. It is nevertheless at once distinguished by the extreme degree of constriction behind the orbits, where the width of the cranium is *much* less than that of the rostrum. (In *L. vison*, the constriction is moderate; in *P. fœtidus*, there is scarcely any.) Coincidently with this narrowing of the skull near the middle, the postorbital processes are better developed than they are in either of the two genera last named, and the postmolar production of the palate is extremely narrow. The interpterygoid emargination is comparatively shallow as well as narrow, not nearly reaching half-way to the molars; the palate ends (in the specimen examined—it may not in others) transversely instead of with strongly concave or even acute emargination. The pterygoids, as in *Gale*, do not develop decided hamular processes (conspicuous in *P. fœtidus* and *L. vison*). The bullæ auditoriæ, as in both *Gale* and *P. fœtidus* (they are notably flatter in *L. vison*), have considerable inflation, with scarcely a tubular prolongation and nick at end. In brief, the skull combines the size, massiveness, and roughness of *Putorius* proper and *Lutreola*, with other characters rather of *Gale*, and some peculiarities of its own.

In external details, *Cynomyonax* is similarly interrelated to *Gale* and *Putorius* proper, though nearer the former (*Lutreola* being more specialized in adaptation to aquatic habits than either of the other subgenera). Though of such large size, *Cynomyonax* retains the attenuate body, long neck, very short legs, slim tail, large orbicular ears, and close-set pelage of a true Stoat. On the other hand, the pattern of coloration, excepting the black-tipped tail, is different, and more like that of the Ferrets in some respects, while it is entirely peculiar in others.

It is interesting to observe that this single American analogue of a special Old World group occurs in the western portion of the country, furnishing another among many instances of the closer relationships of the Western than of the Eastern fauna with that of the other hemisphere.

American or Black-footed Ferret.

Putorius (Cynomyonax) nigripes.

PLATE VII.

Putorius nigripes, *Aud. & Bach.** Q. N. A. ii, 1851, 297, pl. 93 (Lower Platte River).—*Bd. M. N. A.* 1857, 180 (from the foregoing).—*Gray*, P. Z. S. 1865, 110; Cat. Carn. Br. Mus. 1869, 88 (its validity queried).—*Coues*, Am. Sportsman Nov. 20, 1874 (call for specimens).—*Ames*, Bull. Minn. Acad. 1874, 69 (presumptively attributed to Minnesota).

HAB.—Region of the Platte River, and other portions of the central plateau. Has been found in Kansas, Nebraska, Wyoming, Montana, and Colorado; north to Milk River, Montana.

SPECIFIC CHARACTERS.—Above pale brownish, mixed with a few blackish-tipped hairs, especially on the lower back; below nearly white; hairs everywhere white at the roots; general color-aspect brownish-white; a broad stripe across forehead, the feet and the end of the tail, black. Length 19 inches; tail-vertebræ 4, with hairs 5¼; fore leg 4. Skull 2.60 long; rather under 0.50 broad at point of greatest constriction (zygomatic width unknown).

General account of the species.

Until very recently, nothing was known of this remarkable animal beyond what was given by Audubon. The original of his figure, if ever preserved, does not appear to have been examined by other naturalists. Doubt has been cast upon the existence of such an animal,† and the describer has even been

---

* *Digest of the original description.*—Dentition strictly as in *Putorius* (teeth 34). Form elongate; forehead arched and broad; muzzle short; ears short, broad at base, triangular, closely furry both sides; feet covered with hair on both surfaces. Tail narrowly cylindrical. Pelage finer than that of the Mink or Pine Marten, and even shorter (relatively) than that of the Ermine; the outer hairs few, short, and coarse. All the pelage white at the roots; the bases of the longer hairs with a yellowish tinge, their ends broadly reddish-brown; soft under fur white, with a yellowish tinge, giving the animal on the back a yellowish-brown appearance, in some parts approaching to rufous; on the sides and rump, the color is a little lighter, gradually fading into yellowish-white. Nose, ears, sides of head, throat, under surface of neck, belly, and under surface of tail white; a shade of brownish on the chest between the fore legs. A broad black patch on the forehead, enclosing the eyes and reaching near the tip of the nose; legs to near the shoulders and hips brownish-black; end of tail black for about two inches.

Type procured by Mr. Alexander Culbertson on the lower waters of the Platte. Stated to inhabit the wooded parts of the country to the Rocky Mountains, and to be perhaps found beyond that range, though not observed by any travellers from Lewis and Clarke to the present day. Habits said to resemble those of the Ferret of Europe. "It feeds on birds, small reptiles and [other] animals, eggs and various insects, and is a bold and cunning foe to the rabbits, hares, grouse and other game of our western regions.'

† Dr. J. E. Gray, for instance, with characteristic sagacity, queried it amidst a number of purely nominal species he admitted without question.

suspected of inventing it to embellish his work. I have, therefore, the greater pleasure in being able to present a full account of the species.

The first specimen known after the type was a fragment of a skin which for some years lay unrecognized in the National Museum at Washington. According to my present recollection, the object being not at hand, it consists of a squarish piece of the skin of the lower back, with the tail attached.

A second specimen (No. 11932) lately reached the same museum, but unfortunately in very defective state of preservation. It was procured from some point on the Platte River, and presented by Mr. J. W. Munyon, or Munyou. The skull was smashed to pieces. I was only enabled to determine that the animal had thirty-four teeth, and was therefore *Putorius*, not *Mustela*, and that its relationships appeared to be with the European *P. fœtidus* group.*

Being so short of the necessary material when I began to study this group, I caused an advertisement of my wants, with a description of the species, to be inserted in several of the sporting newspapers, and extensively copied by papers of the region inhabited by the species. This had the gratifying result that in a short time the required specimens were received at the Smithsonian Institution; and my thanks are due to several gentlemen who kindly interested themselves in the matter.

The third specimen (counting the above-mentioned scrap of skin as one) was brought to Washington by Dr. F. V. Hayden, Director of the United States Geological Survey. This one was taken by Dr. Law, in the valley of the Cache La Poudre River, near the northern border of Colorado. It was in better condition than Mr. Munyon's, but still defective, having lost part of the tail and most of the head, which had been shattered by a rifle-ball. The length of this individual was about eighteen inches to the root of the tail. Dr. Hayden informed me that it had been shot at the mouth of a prairie-dog hole, of which it had taken possession, and that its stomach contained remains of one of these quadrupeds. He also spoke to me of another individual, kept for some time in confinement at Greeley, Colo-

---

* The whitishness about the mouth and ears of *P. fœtidus*, contrasting with dark parts, gives somewhat the appearance of a stripe across the face, which is perfected in *C. nigripes*, in which the face-markings recall those of the Grison, *Galictis vittata*.

rado, which had also been secured in a prairie-dog town; and represented the species as being not at all rare, though very difficult to obtain, owing to the facilities for its retreat into the safe recesses of the burrows of the Marmots.

Shortly afterward a fourth specimen came to hand from Fort Wallace, Kansas, where the animal is said to be called the "prairie-dog hunter", from the habits indicated in the preceding paragraph. This individual, sent by Mr. L. H. Kerrick, fairly well mounted, was the first I had seen with the head and tail complete and in good preservation.

Another specimen, from Wyoming or the contiguous portion of Colorado, was sent to Prof. Baird by my friend, Capt. James Gilliss, of the Army, then stationed at Cheyenne Depot, Wyoming. This one I think I have not seen.

Still another specimen, important as extending the known geographical distribution of the species, was very recently received at the Smithsonian Institution, from Mr. C. Cavileer, of Pembina, Dakota. This was procured on Milk River, Montana.

I am informed by Prof. Baird that two living specimens were sent from some part of the West to New York, one of which died *en route*, and was probably thrown away; of the ultimate disposition of the latter I do not know.

Mrs. M. A. Maxwell, a well-known naturalist and taxidermist, of Boulder, Colorado, who made a remarkably fine Centennial exhibit of the animals of Colorado at the late International Exposition, at Philadelphia, procured two or three specimens in the vicinity of Denver. They were taken on the prairie land in dog-towns. These specimens, very nicely prepared, I had the pleasure of inspecting when Mrs. Maxwell's collection was on exhibition in Washington, during the winter of 1876–77. One of them had been "drowned out" of a prairie-dog hole, and kept for some time in captivity. It became, I am informed, quite tame, though it was furious when first captured.

The skull from which the foregoing cranial and dental characters were drawn up was sent from Nebraska by the late Mr. W. F. Parker, formerly editor of the "American Sportsman", in which one of my advertisements was inserted. I do not know whether or not it was accompanied by a skin. It is No. 14530 of the National Museum.

No. 11932 shows the characteristic black facial stripe, black feet, and black end of the tail. The general light brownish-white color and character of the pelage are peculiar. The coat is very short and close, the individual hairs appearing scarcely longer than those of a Stoat; there is nothing of the length of pelage of either Mink, Polecat, or Marten. The fur is everywhere, even on the darkest part of the back, white at the roots, and on the under parts it is entirely white, excepting a faint brownish discoloration. There is a stronger tinge of pale brown on the back, and a certain dorsal area shows blackish-brown tips of the hairs, not strongly pronounced enough, however, to materially alter the general cast of the parts. The tail has nothing of the bushy character seen in a Mink or Polecat, being cylindrical, close-haired, scarcely enlarged at the terminal brush, and relatively as slender as that of a Stoat. As far as can be judged, this specimen agrees closely with the dimensions assigned by Audubon.

Dr. Hayden's specimen, in better order, corresponds closely in coloration with that just described: it is dingy whitish all over, with a slight brownish cast on the upper parts, and a dorsal area of sparse dark brown streakiness. All four paws are quite black; on the fore legs, these black stockings run up to the shoulder all around the limb, except on the outer surface, where a pale line extends down from the body. On the hind limbs, the black is more restricted, soon fading into smoky-brown below the knee. A line along the soles is whitish. There is a curious blackish stripe through the umbilicus. The feet are remarkable for the great length of the numerous bristles on the toes, projecting far beyond and almost hiding the claws; the palms and soles are densely furry. The specimen equals a very large Mink in size.

The Kansas specimen affords some additional characters, especially relating to the general shape. The body seems proportionally as slender and the neck as long as in an Ermine. The tail-vertebræ are only about five inches long, decidedly less than one-third of the length of the head and body, which is apparently some eighteen or nineteen inches, but is perhaps stretched. The circumference of the body is about seven inches. The slender tail has no enlarged terminal brush. The physiognomy and general aspect is rather that of an overgrown Weasel than of a Mink or Ferret. The ears are very prominent, perhaps even more so than those of a Stoat, and are not perfectly orbicular, having an obtuse point at the highest part of the border; they measure, in their present state, 1.10 above notch, 0.70 above head. The longest whiskers (black) reach to the back of the ear; others grow on the chin, the cheeks back of the angle of the mouth, and on the forehead. The brownish-black mask is well contrasted with nearly white surroundings, except on the forehead, where the dingy brownish of the upper parts extends to it. The ears are mostly white, with a dark touch at the lower front border. The dingy brownish of the upper parts is a little stronger than in either of the two other specimens here described. The blackish tip of the tail is about one and a half inches long.

I made no written memoranda of my examination of Mrs. Maxwell's specimens, but remember that they presented nothing requiring special comment, being fairly illustrative of the characters here detailed.

Audubon's figure is unmistakable, and gives a very fair idea of this interesting animal. As remarked by Prof. Baird, it is singular that so conspicuous a species should have so long eluded the observation of the many explorers who have traversed the region it inhabits, and where, apparently, it is by no means rare. Its retiring habits, and the nature of its resorts, doubtless tend to screen it. In the summer of 1876, I conducted a natural history party through the region supposed to be its centre of abundance, where Dr. Hayden's and Captain Gilliss's specimens were secured; but I failed to obtain a sight of it, though I was in the midst of prairie-dog towns, and continually on the watch for this particular animal. The geographical distribution above assigned will probably require to be considerably enlarged.

ADDENDUM TO CHAPTER V.

On the species of the extralimital subgenus PUTORIUS.

In further illustration of the genus *Putorius*, I wish to introduce a notice of the extralimital species of the subgenus *Putorius*, which, as already said, includes the Fitches, Ferrets or Polecats.*

No representatives of this particular group are indigenous to America, but the Ferret is extensively bred, in confinement or semi-domestication, for the purpose of hunting rats, rabbits, &c.

---

* The untechnical reader must not confound the proper use of the term "polecat" for the Ferret group, with its frequent application in this country to the Skunks (*Mephitinæ*).

The Subgenus PUTORIUS.

For the characters of this group see *anteà*, p. 99. The nonpareil note on p. 74 is equally applicable to the synonymy of the following extralimital species of the subgenus.

1. **Putorius fœtidus.**—*Polecat or Fitch.*

PLATE VIII.

Mustela putorius, *L. S. N.* i. 10th ed. 1758, 46, no. 5 ; S. N. ii. 1766, 67, no. 7.—*Schreb.* Säug. iii. 1778, 485, pl. 131.—*Gm.* S. N. i. 1788, 97.—*Bechst.* Naturg. i. 479.—*Pall.* Zoog. R.-A. i. 1811, 87.—*Desm.* Mamm. i. 1820, 177, no. 271.—*Fr. Cuv.* Mamm. ii. 34.—*Flem.* Br. An. 1828, 14.—*Jen.* Br. Vert. 1835, 11.—*Bell*, Br. Quad. 1837, 156, fig.; 2d ed. 1874, 203.—*Selys-L.* Fn. Belg. 1842, 9.—*Blainv.* Compt. Rend. xiv. 1842, 210 seq. pls. —.—*Schinz*, Syn. 1844, 339.—*Gieb.* Odont. 33 ; Säug. 1855, 779.
Viverra putorius, *Shaw*, Gen. Zoöl. i. 1800, 415.
Fœtorius putorius, *Keys. & Blas.* Werb. Eur. i. 1840, 68, no. 143.—*Blas.* Wirb. Deutschl. 1857, 222, f. 125.—*Chatin*, Ann. Sci. Nat. 5th ser. xix. 1874, 98 (anatomical).
Mustela fœtida, "*Klein.*"
Putorius fœtidus, *Gray*, List Mamm. Br. Mus. 1843, 64 ; P. Z. S. 1865, 109 ; Cat. Carn. Br. Mus. 1869, 87.—*Gerr.* Cat. Bones Br. Mus. 1862, 92.
Putorius verus, *Br.*—*Brandt*, Bem. Wirb. Nord. Eur. Russl. 26.
Putorius communis, "*Cuv.* R. A."—(*Gray.*)
Putorius typus, "*F. Cuv.*"—(*Gray.*)
Putorius vulgaris, *Griff.* Cuv. R. A. v. 1827, 120, no. 339 (no. p. 121, no. 344).—*Fitz.* Naturg. Säug. i. 1861, 328, f. 68.
Putois, *Buff.* Hist. Nat. vii. —, 199, pl. 123.
Polecat, Fitch, Fitchet, Fitchew, Foumart, Fulmart or Fullmart, *English.*—*Penn.* Brit. Zoöl. i. 89, pl. 6.
Iltis, *German*, cf. *v. Martens*, Zoöl. Gart. xi. 1870, 275 (philological).
Wicha, Madrui, *Selys-L. l. c.*

Form stout ; ears short and rounded ; tail rather bushy, cylindric-tapering, about one-third the head and body ; fur very long and loose on most parts of the body (the well-known "fitch" of commerce), yellowish-brown, overlaid with glossy blackish-brown, the tail, legs, and chest mostly blackish ; head dark, the ears, a space in front of them, lips and chin, usually white. Varies interminably in proportion of the yellowish and blackish. Length about 16 inches ; tail 5½ ; head 2¼ ; ears ½. The name Polecat is probably a contraction of Polish cat. Foumart, &c., are merely Foul Mart, in distinction from the *Mustelas*, or "Sweet" Marts, the odor in this species being much more disgusting. The animal inhabits Europe.

1a. Var. **furo.**—*The Ferret.*

Mustela furo, *L. S. N.* i. 1766, 68, no. 8.—*Schreb.* Säug. iii. 1778. —, pl. 133.—*Gm.* S. N. i. 1788, 97.—*Desm.* Mamm. i. 1820, 178, no. 273.—*Jen.* Br. Vert. 1835, 12.—*Fisch.* Syn. 1829, 219.—*Bell*, Br. Quad. 1837, 161, fig.—*Schinz*, Syn. 1844, 340.—*Fr. Cuv.* Mamm. ii. 22.—*Gieb.* Odont. 33, pl. 19, f. 5 ; Säug. 1855, 780.
Viverra furo, *Shaw*, Gen. Zoöl. i. 1800, 418.
Fœtorius furo, *Chatin*, Ann. Sci. Nat. 5th ser. xix. 1874, 98 (anatomical).
Mustela putorius var., *Flem.* Br. An. 1828, 14.
Putorius vulgaris var. furo, *Griff.* Cuv. R. A. v. 1827, 120, no. 339 a.
Putorius fœtidus var. furo, subfuro, *Gray*, P. Z. S. 1865, 110 ; Cat. Carn. Br. Mus. 1869, 87.
Ferret, *English.*—*Penn.* Brit. Zoöl. i. —, 91.
Furet, Furet Putoire, *French.*—*Buff.* Hist. Nat. vii. —, 209, pl. 26.
Frett, Frettel, Frettchen, *German.*—*Blas.* Wirb. Deutschl. 1857, 225.

This is the well-known tame Ferret, now only recognized in a state of domestication. It is smaller and slenderer than the Fitch, yellowish-white or white, with pink eyes. This is an excellent example of a "variety," properly so called, in distinction from a geographical race. The root of all the various vernacular names (there are many others than those above given) seems to be the Latin *fur*, a thief. There may also, as has been suggested, be a relationship with the Latin *riverra*, by which name the present, among other species, seems to have been known to the ancients.

The rearing of Ferrets seems to be a growing industry in this country, though still not practiced to the extent it is in Europe. The following article, entitled "Ferret Breeding and Handling", by Mr. F. Mather, appeared in the "American Sportsman" (newspaper) of November 28, 1874:—

"I have had several inquiries of late from readers of the SPORTSMAN concerning the breeding, management and hunting of ferrets, together with invitations to write it up. It appears somewhat singular that no one has done this before, at least I do not remember to have seen anything on the subject in any American paper, and this fact causes me to comply with the request more readily than I should have done had others with more experience volunteered to publish it.

"Practical details having been asked for, we will consider them as they are in our day, and not stop to trace their origin nor where first used. We have two varieties, the brown or 'fitch-ferret,' and the white one. The latter is probably an albino, as its eyes are pink; but it breeds true to color every time, possibly a 'sport,' as the florists say, that has been perpetuated. The white ones seem to be in most favor for some unknown reason, judging from the inquiries that I have received. I keep both kinds, and have them mixed, and don't see any difference in hunting qualities, and can only account for the preference on the ground that the white ones are thought to be the prettiest pets. Having no strong local attachments, they require to be constantly confined, although instances have been known where they were at liberty and did not go away; still, as they are just as good for chickens as for rats and rabbits, it is best not to trust them too far. Two or three animals may be kept in a common shoe-box with slats or wire-cloth fronts, a box for a nest in one corner, and a drawer containing coal-ashes or earth in another. This should be emptied often and renewed; they will make all their muss here and will then keep clean and healthy. A cellar is not a good place for them—too damp and cold; a yard or wood-shed is better. I have a ferret-yard made for the purpose, built of hemlock boards; it is sixteen feet long by six wide; the sides are four feet high, the boards running up and down to prevent climbing; it is also floored to prevent digging. I have in this at present eighteen ferrets, but could accommodate fifty, as they only foul one corner. A tin spout conveys milk into the feeding-pan, and meat is thrown over. Their nest is a box with a cover; it is full of straw, and a hole in one side is the door. One-half will be covered this winter to keep the snow off.

"They will shiver in the summer, and it is not good to keep them in too warm a place if they are expected to hunt in the snow; but a small box of straw where they can huddle up together and so keep warm is sufficient. I saw three ferrets last summer in a small box that was sheet-ironed inside (the owner thought that they could gnaw like rats), where the poor things had lived for a month in their filth. It was horrible enough to breed a pestilence; in fact it did breed one for the ferrets. I told the owner so, but he thought not. Mr. Bergh should have seen that!

"They will keep very clean if they have a chance, but will drag food into their nests and store it if they have too much at any time. This can be partially guarded against, in the case of bones and large pieces, by making the entrance as small as possible. We feed skimmed milk, beef-heads, and other meats; salt meat is said to produce scurvy. Milk fattens a ferret very rapidly, and they are apt to get too fat on this diet.

"My menagerie is run upon economical principles, and it is a hard specimen of either animal or vegetable that does not find a consumer. A kingfisher hovering over the ponds is apt to tumble in; he is then skinned, and if the fish are not hungry it goes to the ferrets. A fat woodchuck goes in—there's provision for several days. A hen-hawk "towering in his pride of place" over my young fowl, often finds a lasting repose in the ferret-yard; while the refuse of fresh fish is also eagerly devoured. Chicken heads often afford an occasional variety. I have a pair of mink who will eat as much in one day as two ferrets would in three; they will devour the entire carcass of a muskrat in twenty-four hours.

"In handling a ferret take it with the hand around the ribs, and if it struggles let its fore legs go through the fingers; they do not like to be held below the ribs. Do not handle young ferrets until nearly grown; do not handle a female about to have a litter. Their period of gestation is about forty-five days I think,—can't speak positively on this point,—it may be a few days more or less; they usually have a litter of from five to ten in May or June. I have heard of their having two litters in one season, but it has never occurred with me. When about to have young, put each female in a box by herself, and don't let the young run with the old male until they can use their teeth to defend themselves. I prefer to say male and female, though some call them dog-ferret and bitch; and last winter a man in Buffalo said, 'In the hold country we calls 'em a 'ob and a gill.'

"In handling wild ferrets put on a pair of leather gloves and pick them up, rub their heads and pat them, and in a few days you can take your bare hand. Now a word about 'trained' ferrets. That is all humbug. A ferret that is tame and well handled will go into a hole and go to the bottom of it, and come out if it finds nothing; hunting is their nature, and a little fasting stimulates them wonderfully. Those kept for rats are generally worked with dogs; and although I have often run the rats out of my barn with them, it is especially for rabbit hunting that I keep them. They are of no use for the large white rabbit, but I used to find that the little gray fellow that abounds in the vicinity of Honeoye Falls had a very unsocial way of sitting in his hole under ground and declining to come out and have fun, but since I have used ferrets he has changed his habits. With the sneaking method of netting the rabbit at the mouth of the hole when driven out with a ferret I have no kind of sympathy, but as 'Molly Cotton' clears the hole with a ten-foot bound after passing a ferret, and keeps going faster if possible, often into a thicket, it *is* sport to stop her. Some prefer a very small ferret, as they use them without a muzzle and they cannot hold a rabbit as a large one does; but I prefer a good stout fellow, and if he is disposed to kill a rabbit (their dispositions vary) I muzzle him. A muzzle is made of a small piece of leather shaped like a letter T, a little wider at the bottom however; a string is put in each end of the top and one in each lower corner, the leather is put under his chin and the top piece tied around his nose; the other two strings are tied behind his ears. Some have the lips pierced, and after healing they are tied shut. I have never tried this, nor breaking the teeth, which latter practice is brutal. The ferret can be carried in a bag, with drawn string, strung over the shoulder, or in a tightly-buttoned coat pocket. As the animal enjoys its short liberty when hunting and, however tame it may be, does not want to go into the bag again while you move on; it is always best to let it get ten feet or more from the hole before you attempt to pick it up or it may dodge back and refuse to come out. In this case tie a rabbit on a stick and put it down and the ferret will follow out.

"If the leather muzzle don't work, or gets lost, you can improvise one with a string by making a loop that will not get larger or smaller, and put it over his nose and then tie behind his ears, taking care to have the knot under his throat and the last tie on top of his head.

"In England I believe they use small bells on their rat-ferrets to tell their whereabouts. In conclusion I would say, if you use ferrets for rats don't trust a strange dog with them, and if for rabbits don't stand in front of the hole."

1b. Var. eversmanni.—*Siberian Polecat.*

Mustela putorius?, *Licht.* Eversm. Reise, 23; ref. to *Pallas*, i. Zoog. 89, *note*, but not *Mustela sibirica* Pall. *ibid*. p. 90.
Mustela putorius *var.* eversmanni, *Fischer*, Syn. 1829, 219.
Mustela eversmanni, *Less.* Man. 1827, 144, no. 379.—*Schinz*, Syn. 1844, 339.
Putorius eversmanni, *Gray*, P. Z. S. 1865, 109; Cat. Carn. Br. Mus. 1869, 87.
Mustela putorius, *Blyth*, "J. A. S. B. xi. 281."—(*Gray.*)
Mustela putorius thibetanus, *Hodgs.* "J. A. S. B. xxiii, 1849, 446."—(*Gray.*)

This is the Asiatic Polecat, which appears to have been first noted by Pallas, in text of p. 89 of the Zoographia, from Siberia. This is to be carefully distinguished from the *Mustela sibirica* of Pallas, p. 90, a very different animal, elsewhere noticed in the present work. It is apparently but little if any different from *P. fœtidus*, to which Blasius assigns it without query. Certain cranial differences adduced by Gray may require confirmation. I have seen no specimens of the supposed species.

2. Putorius sarmaticus.—*Spotted Polecat.*

Mustela sarmatica, *Pall.* Itin. i. 1771, 453; Spic. Zool. 1780, xiv. 79, pl. 4, f. 1; Zoog. R.-A. i. 1831, 89.—*Erxl.* Syst. Anim. 1777, 460, no. 6.—*Schreb.* Säug. lii. 1778, 490, pl. 132 (from Güldenstädt).—*Zimm.* Geogr. Gesell. ii. 1780, 305, no. 201.—*Gm.* S. N. i. 1788, 97, no. 15.—*Turt.* S. N. i. 1806, 60.—*Desm.* Mamm. i. 1820, 178, no. 204; Nouv. Dict. xix. 371; Ency. Méth. pl. 82, f. 4.—*Fr. Cuv.* Dict. Sci. Nat. xxix. 1823, 252, no. 9.—*Is. Geoff.* Dict. Class. x. 212.—*Fisch.* Syn. 1829, 220.—*Less.* Man. 1827, 145.—*Schinz*, Syn. 1844, 340.—*Gieb.* Säug. 1855, 780.
Viverra sarmatica, *Shaw*, Gen. Zoöl. i. 1800, 430 ("Sarmatia Weesel").
Fœtorius sarmaticus, *Keys. & Blas.* Wirb. Eur. 1840, 68, no. 142.—*Blas.* Wirb. Deuts. 1857, 226.
Putorius sarmaticus, *Griff.* Cuv. R. A. v. 1827, 121, no. 343.—*Gray*, List Mamm. Br. Mus. 1843, 64; P. Z. S. 1865, 110; Cat. Carn. Br. Mus. 1869, 88.
Mustela peregusna, *Güld.* N. Comm. Petrop. xiv. 1769, 441, pl. 10 (*peregusina* is also found).
Mustela præcincta, *Rzaczynski*, Hist. Nat. Pol. 1736, 328.
Vormela, *Gesn.* Quad. 1551, 768.
Pereguzna, Perewlaska, Przewlaska, Parælasia, *Pall.* Itin. *l. c.*
Tigeriltis, Gefleckte Iltis, *German.*
Sarmatier, *Müll.* Natura. Suppl. 1776, 33.
Peronasca, *Buff.* Hist. Nat. xv.
Putois de Pologne, *Cuv.* R. A. i. 148.
Marte à ceinture, *Less. l. c.*

This remarkably distinct species is black, on the upper parts brown spotted with yellow, the ears and a frontal band white. It inhabits Eastern Europe, Poland, and Russia.

CHAPTER VI.

MUSTELINÆ—Continued: The MINK.

The subgenus *Lutreola*—Subgeneric characters and remarks—*Putorius vison*, the American Mink—Synonymy—Habitat—Specific characters—Description of external characters—Measurements—Variation in external characters—Variation in the skull—Comparison with the European Mink—Notice of allied Old World species, *P. lutreola* and *P. sibiricus*—General history and habits of the Mink—"Minkeries".

WE come now to consider a particular modification of the genus *Putorius*, in adaptation to an aquatic mode of life. Both the foregoing subdivisions of the genus comprehend terrestrial and more or less arboreal species; the present one, *Lutreola*, consists of species which are scarcely less aquatic than the Otters themselves; and the consequent modifications, both in cranical and external characters, are decided.

The Subgenus LUTREOLA. (WAGNER.)

The leading peculiarities of this section have been already pointed out (p. 100), and contrasted with those of *Gale, Cynomyonax,* and *Putorius* proper.

The skull of the Mink bears out the general points of "build" which distinguish *Putorius* at large from *Mustela*—such as the short, turgid, truncate rostrum, comparatively shallow interpterygoid emargination, position of anteorbital foramen, &c. As might be expected from consideration of the habits of the animal, a resemblance to the cranium of an Otter is better marked in this than in other sections of the genus, the bullæ auditoriæ, in particular, being notably flattened, and the whole upper outline of the skull being straightened. In its own genus, the resemblances of the skull are with that of *Putorius* proper and of *Cynomyonax*, rather than with that of *Gale*. In addition to the absolutely much greater size in *Lutreola*, the massiveness of the skull, with the strong flaring sagittal and lambdoidal crests defining temporal fossæ, contrasts strongly with the smooth condition of the parts in *Gale*. In *L. vison*, there is, in addition to the comparative flatness of the auditory bullæ, some constriction and outward prolongation of the meatus, which is not seen in *Gale* or *Cynomyonax*, and scarcely indicated in *Putorius* proper. The frontal outline is nearly straight, and but little sloping (much as in *Lutra*). The pterygoids develop strong hamular processes, also seen in *Putorius* proper, but which are weak or wanting in *Gale*. There is much constriction of the skull near the middle, and the postorbital processes are well developed.

The dentelure of *Lutreola* is probably the strongest to be found in the genus *Putorius* at large, and there is reason to suppose that it reaches a maximum in the large North American species of this subgenus. In *L. vison*,* the teeth, aside from the lesser number of premolars, are singularly like those of *Mustela martes*, as a matter of superficial resemblance; and the superiority in size and strength over those of *Putorius* proper, or of *Cynomyonax* (not to mention *Gale*), is very evident on comparison. In the American, if not in other species of *Lutreola*, the following points may be specially noted:—

The back upper molar is of relatively large size, conspicuously exceeding that of *Putorius fœtidus* or *Cynomyonax nigripes* in relative as well as absolute bulk. The inner moiety is much larger than the outer;† its free border is nearly circular; it is divided from the outer by a strong constriction; the outer is somewhat trefoil-shaped. The inner moiety presents a raised rim and a central tubercle; the outer has a corresponding tubercle, but the border is divided into two prominences, making three in all on this half of the tooth. The posterior upper premolar (sectorial tooth) shows certain characters not shared by any American species of the genus at large. There is developed, at the antero-*external* corner of the tooth, a decided process or spur, only less in size than the ordinary antero-*internal* one; and the projection of this gives to the outer border of the tooth a decidedly concave outline.‡ This process, together with the internal one, gives the fore end of the tooth a V-like re-

---

*I have not been able to examine the teeth of any Old World species of this subgenus.

†This is not the case either with *P. fœtidus* or *C. nigripes*, but is scarcely a subgeneric character, for it is said not to occur in the European species of *Lutreola*.

‡There is a trace of this process in *P. fœtidus* and *C. nigripes*, but it is not sufficiently developed to render the outer border of the tooth concave, nor to make a V-reëntrance at the fore end.

entrance, into which the antecedent premolar is set. Moreover, the antero-internal process, instead of being a mere heel or spur standing off from the tooth, as in the other subgenera here compared, develops into a strong, conical, acute cusp, sometimes with two points. The back lower molar, contrary to what might have been expected, is absolutely not larger than that of *P. fœtidus*,* and therefore smaller relatively to the general development of the teeth. The anterior lower molar (sectorial tooth) develops on the inner side a slight but unmistakable supplementary tubercle, like that so evident in *Mustela*, but smaller; the other species of *Putorius* which I have examined have no trace of this lobe, or a mere rudiment. And, in general, it may be said of the molars and premolars of *Lutreola*, that their various cusps are better developed than in most, if not all, other sections of *Putorius*.

The details of external form of *Lutreola* are so fully given beyond in the description of *L. vison* that they may be here omitted. There is but one species known to inhabit North America, very closely related to the Mink of Europe. The alleged differences between the two are presented further on, in concluding a discussion of their affinities.

The American Mink.

Putorius (Lutreola) vison.

PLATE IX.

Mustela vison, *Briss.* Quad. 1756, 246, no. 6 (from Canadian specimen, same as described by Buffon and Pennant).—*Schreb.* Säug. iii. 1778, 463, pl. 127 *b*.—*Gm.* S. N. i. 1788, 94.—*Turt.* S. N. i. 1806, 58.—*Cuv.* R. A. i. 1817, 150.—*Harl.* Fn. Amer. 1825, 63.—*Less.* Man. 1827, 149.—*Maxim.* Reise, i. 1839, 213.—*Blainv.* Ostéogr. Mustela, pl. 13 (teeth).—*Thomps.* N. H. Verm. 1853, 31.
Mustela (Martes) vison, *Desm.* Mamm. i. 1820, 183.—*Griff.* Cuv. R. A. v. 1827, 124.
Mustela (Putorius) vison, *Rich.* F. B.-A. i. 1829, 48, no. 16.
Mustela (Lutreola) vison, *Wagn.* Suppl. Schreb. ii. 1841, 241.
Lutra vison, *Shaw*, Gen. Zoöl. i. 1800, 448 (based on the *Vison* of Buffon).
Putorius vison, *Gapp.* Zoöl. Journ. v. 1830, 202.—*Emmons*, Rep. Quad. Mass. 1840, 43.—*De K.* N. Y. Z. i. 1842, 37, pl. 11, f. 1 (animal), pl. 8, f. 3, A, B (skull).—*Aud. & Bach.* Q. N. A. i. 1849, 250, pl. 33.—*Kenn.* Tr. Ill. State Agric. Soc. for 1853-4, 1855, 578.—*Beesley*, Geol. Cape May, 1857, 137.—*Baird.* M. N. A. 1857, 177, pl. 37, f. 2, 3 (skulls).—*Newb.* P. R. R. Rep. vi. 1857, 42.—*Coop. & Suckl.* N. H. W. T. 1860, 93, 115.—*Billings*, Canad. Nat. and Geol. ii. 1857, 448.—*Ross, op. cit.* vi. 1861, 29.—*Maxim.* Verz. Am. Säug. 1862, 52.—*Sam.* Am. Rep. Mass. Agric. for 1861, 1862, 157, pl. 1, f. 8.—*Gilpin*, Tr. N. Scotia Inst. ii. 1870, 12, 59.—*Ames*, Bull. Minn. Acad. Nat. Sci. 1874, 69.—*Coues & Yarrow*, Zoöl. Expl. W. 100 Merid. v. 1875, 60.—*Allen*, Bull. U. S. Geol. Sur. vol. ii. no. 4, 1876, 326 (skull).
Mustela lutreola, *Forst.* Phil. Trans. lxii. 1772, 371.—*Sab.* Frank. Journ. 1823, 652.—*Fisch.* Syn. 1829, 221 (partly).—*Godm.* Am. Nat. Hist. i. 1831, 205.—*Hall*, Canad. Nat. & Geol. vi. 1861, 295.

---

* As elsewhere stated, in *Cynomyonax* this tooth is singularly minute.

CHARACTERS OF PUTORIUS VISON. 161

Putorius lutreolus, ["*Cur.*"] *Allen,* Bull. M. C. Z. i. 1869, 175 (critical); ii. 1870, 169 (Florida).—*Allen,* Pr. Bost. Soc. N. H. xiii. 1869, 183.
Putorius lutreolus *var.* **vison,** *Allen,* Bull. Ess. Inst. vi. 1874, 54, 59, 62.
Mustela (Lutreola) lutreola *var.* **americana,** *Schinz,* Syn. Mamm. i. 1844, 347.
Vison lutreola, *Gray,* List Mamm. Br. Mus. 1843, 64 (partly).—*Gerr.* Cat. Bones Br. Mus. 1862, 92 (partly).
Mustela canadensis, *Erxl.* Syst. An. i. 1777, 455 (mixed with synonymy of another species, but clearly referable here from the description, which can only apply to the Mink. See *Bd.* M. N. A. text on p. 151).
Mustela canadensis *var.* **vison,** *Bodd.* Elench. An. i. 1784, 86 (after Buffon).
Mustela winingus, *Barton,* Am. Phil. Tr. vi. 1800, 70 (no descr. ; St. Louis, Mo.).
Mustela minx, *Turt.* S. N. i. 1806, 58.—*Ord,* Guthr. Geog. 2d Am. ed. ii. 1815, 291, 298.
Mustela lutreocephala, *Harl.* Fn. Amer. 1825, 63.
Vison lutreocephala, *Gray,* P. Z. S. 1865, 116 ; Cat. Carn. Br. Mus. 1869, 94.
? Mustela rufa, *H. Smith,* Jard. Nat. Lib. xiii. 1842, 189.
Putorius nigrescens, *Aud. & Bach.* Q. N. A. iii. 1853, 104, pl. 124 (not in orig. ed.).—*Baird,* M. N. A. 1857, 180.—*Gilpin.* Tr. N. Scotia Inst. ii. 1870, 12, 60.
Mink, *Smith's* Virginia, 1624.—*Kalm,* Itin. iii. 22.
Mink, Common Mink, American Mink, *Authors* and others.
Minx, *Lawson,* Carol. 1709, 121.—*Brickell,* Nat. Hist. North Car. 1737, 118.—*Penn.* Arct. Zoöl. 1784, 87, no. 35.
Otay. *Sagard-Théodat,* Hist. Canad. 1636, 748 (ed. of 1866, iii. 680).
Foutereau, *La Hontan,* Voy. i. 1703, 81. Also of *French Canadians.*
Vison, *Buff.* Hist. An. xiii. 1765, 304, pl. 43 (based on specimen in Mus. Aubry, as were the descrs. of Briss. & Penn.).—*Bomare,* Dict. iv. 1768, 615.—*Penn.* Hist. Quad. 1781, no. 203 ; Arct. Zoöl. i. 1794, 78, no. 29.
Visone, *Scatoglia,* An. Quad. iv. 1775, pl. 155, f. 2 (from Buffon).
American vison, *Gray,* P. Z. S. 1865, 116.
Lesser Otter. *Penn.* Hist. Quad. 1781, 228.—*Forst.* Phil. Trans. lxii. 1772, 371.
Jackash, *Hearne,* Journ. ——, 376.
Shakweshew or Atjackashew, *Cree Indians* (= "*Jackash*").
Mountain-brook Mink, *Aud. & Bach. l. c.*
Little Black Mink, *Bd. l. c.*
Mountain Mink of *Hunters.*

HABITAT.—North America, at large. North to the Arctic coast, but not abundant north of Fort Resolution.

SPECIFIC CHARACTERS.—Larger and stouter than the Stoats; ears shorter: tail uniformly bushy, nearly as in *Mustela;* feet semipalmate; color dark chestnut-brown; tail, and usually a dorsal area, blackish; chin white, the edges of the upper lip rarely also white, the throat, breast, and belly often with irregular white patches. Length 15-18 inches; tail-vertebræ 6-8.

*Description of external characters.**

This animal, with the essential characters of dentition, &c., of *Putorius,* differs notably from the typical Stoats and Weasels (*Gale*) in its larger size and much stouter form, in which respects it approaches the true Martens. It shares with these the uniformly enlarged, bushy, and somewhat tapering tail, instead of a slenderly terete tail with enlarged bushy tip, as in the Stoats. The tail-vertebræ are one-half (more or less) as

---

* From numerous specimens in the Smithsonian Institution from all parts of North America.

11 M

long as the head and body; the terminal pencil is only as long as the hairs of the tail in general. Unlike the Martens, the Mink has small low ears, smaller than those of the Weasels. The ears are scarcely longer than the adjacent fur, though they overtop it a little, as the fur lies flat; they are rounded, and well furred both sides. The general shape of the head—long, low, flat, subtriangular—is as in other *Putorii*. The small eye centres over the angle of the mouth, half-way between the nose and ear. The whiskers are in four or five series, the longest reaching opposite the occiput; they are stiff and strong; other bristles grow over and behind the eyes, on the cheeks, and on the middle of the chin; similar bristles are usually seen upon the wrists and ankles. The extremity of the snout is protuberant and definitely naked. The feet are broad; the hinder have a slightly oblique set; the fore have ten balls, the hind nine, as in other *Putorii* (five digital pads at the ends of the digits, five palmar, and four plantar). The palmar and plantar pads are not separated by hairy spaces (except the hindmost outer palmar one), there being only a crease between them. Ordinarily, the pads are conspicuously naked, but in northern and some winter skins they must be searched for amidst the overgrowing hair. This is a purely fortuitous circumstance. The palms and soles are always furry around the pads. On the top of the feet, the hairs reach to or rather beyond the ends of the nails. The digits are all webbed at bases for a considerable distance, especially the middle ones. The third and fourth fingers are subequal and longest; the second and fifth not so nearly equal, and both much shorter; the first is quite short. The toes of the hind feet have almost the same relative proportions. The pelage consists of a dense, soft, matted under fur, mixed with long, stiff, lustrous hairs, on all parts of the body and tail. The gloss is greatest on the upper parts; on the tail the bristly hairs predominate. Northern specimens have the finest and most glistening pelage, though the long hairs are the stoutest; in southern specimens there is less difference between the under and over fur, and the whole pelage is coarser and harsher.

In color, the Mink ranges from a light dull yellowish-brown, not very different from that of a Marten, or of some styles of the European *P. fœtidus*, to a rich blackish chocolate-brown. These extremes (which will be presently considered) aside, the animal is ordinarily of a rich dark brown, scarcely or not paler below than on the general upper parts; but a dorsal area is usually the darkest, and the tail is quite blackish. A strong mark of the species is the white chin; this is rarely absent, but still its indeterminate character is shown in the fact that its extent and posterior contour are wholly irregular. As generally found, it occupies the whole under jaw about as far back as the angle of the mouth. It is sometimes prolonged as an irregular streak down the throat; sometimes it is indicated only by a few specks, or it may be altogether absent. This white seldom invades the upper lip; that it sometimes, however, does so is attested by the specimens before me, one of the differences claimed from the European *P. lutreola* being thus obviously negatived. Besides the white on the chin, there are often, perhaps usually, other white patches on the under parts, particularly on the chest, between the fore legs, and on the lower belly between the hind legs. These markings are wholly indeterminate in extent and contour. To recount their vagaries would be futile. In very rare instances, the tail is tipped with white.

Measurements of seven specimens of PUTORIUS VISON.

Current number	Original number	Locality	Sex	From tip of nose to— Eye	Ear	Occiput	Tail	Tail to end of— Vertebræ	Hairs	Length of— Fore foot	Hind foot	Elbow to end of claws	Knee to end of claws	Height of ear over notch	Width of ear	Nature of specimen
2402	....	Steilacoom, W. T.	..	1.00	....	....	15.00	7.00	8.25	1.75	2.50	3.50	4.75	0.80	0.80	Alcoholic.
1154	....	Wisconsin	..	0.90	2.10	2.75	16.50	7.50	8.75	1.80	2.50	....	....	1.00	0.90	do.
....	263	Yukon (Lockhart)	♂	1.10	....	3.50	19.50	6.50	7.50	....	2.60	....	....	....	....	do.
....	264	....	♂	1.15	2.00	3.10	15.10	7.20	10.40	....	2.70	....	....	....	....	do.
....	1595	....	♀	1.20	2.20	3.20	16.00	7.70	8.20	1.95	2.70	....	....	....	....	do.
....	1641	....	♂	1.20	2.15	3.10	18.00	8.00	9.00	....	2.80	....	....	....	....	do.
....	....	....	♂	....	....	3.30	17.40	8.50	9.50	....	....	....	....	....	....	do.

Variation in external characters.

In the extensive series of Minks before me, two extremes of size and color are apparent. One of these, represented by a few skins from Washington Territory and the Upper Missouri, is rather larger than any others I have seen—some 18 or 20 inches long, exclusive of the tail. (But the ordinary dark Mink has been found over 20 inches in length.) They are remarkably light-colored, pale dull yellowish-brown all over, the tail but little darker, with the usual white marks on the chin and elsewhere underneath. Such specimens are noted by Prof. Baird, p. 179, in text. Although by no means to be overlooked in any formal account of the species, the fact that this style shades insensibly into the ordinary state shows that it is merely one phase of individual variation, which need not be recognized by name. The other extreme has been described and figured as *Putorius nigrescens* by Audubon and Bachman,* as above.

---

* In order to set forth fully the characters claimed for this supposed species, the following digest of the original description is given:—

Smaller than *P. vison;* teeth in the under jaw larger than the corresponding teeth in the upper jaw; feet less deeply palmated than in *P. vison;* ears broader and longer; fur softer and more glossy. Color dark brownish-black.

In form, in dentition, and in the shape of the feet, this species bears a strong resemblance to a stout Weasel; the head is broad and depressed, and shorter and blunter than that of *P. vison*. Ears large, oval, and slightly acute, covered on both surfaces with fur; legs rather short and stout; feet small and less webbed than in *P. vison*. The callosities under the toes are more prominent than in that species, and the palms scarcely half as long. Toes covered with short hairs almost concealing the nails, and the hairs between the toes leaving only the tubercles visible. Fur blackish-brown from the roots to the tips; whiskers and ears blackish-brown; a white chin-patch (not shown in the figures); under surface of body a shade lighter and redder than the back; tail blackish-brown, blackening on the end. Length of head and body 11 inches; tail-vertebræ 6, with hairs 7; soles $2\frac{1}{6}$; ear $\frac{1}{2}$.

Mountain Mink of hunters. From Pennsylvania, New York, New England, and Canada, and supposed to be more northerly than *P. vison*.

"We have had abundant opportunities of comparing many specimens [with *P. vison*]. We have seen some with their teeth much worn, and females which from the appearance of the teats had evidently suckled their young. They were all of the size and colour of the specimen above described, and we can no longer doubt that the latter is a distinct species from *P. vison*. The comparison in fact is not required to be made between these species, but between the present species and *P. lutreola* of Europe. . . . . We had no opportunity of placing this little species by the side of the European. We are inclined to believe, however, that distinctive marks will be found

It consists in the combination of small size and dark colors. The specimens representing it are a foot or little more in length, and of a rich blackish chocolate-brown; the white on the chin and elsewhere is found as usual. It has been claimed that this cannot be merely a young "Mink", on the ground that it has been found breeding. Hunters and trappers practically recognize as distinct a "Mountain Mink" of this character, the differences which result in the enhanced value of the pelt appealing to them strongly. But, in any event, the specimens before me establish one fact, namely, that it is impossible to draw any dividing line between "*P. nigrescens*" and the common Mink. They melt into each other insensibly. The question is narrowed to whether the supposed species is a reasonably marked variety, or whether it is merely a fortuitous state under which the Mink may anywhere present itself. The latter is my present view. It is certain that young Minks are darker than the old ones, and that the animal increases in stature for some time after it is "mature", *i. e.*, in possession of reproductive powers. The fact that the small blackish individuals are found breeding is therefore by no means conclusive. Nor is the supposed "*nigrescens*" characteristic of any particular faunal area.

In this connection, the remarks of Mr. B. R. Ross in the paper above cited have much practical pertinence, and his opinion, based upon long experience, is entitled to weight. Speaking of the ordinary Mink, he remarks:—"The color of its pelt varies greatly. In winter its shades range from a dark chestnut to a rich brownish black. The tint of all the body is uniform, except that the belly is sensibly lighter, and that there is a series of white blotches, running with greater or smaller breaks from the end of the chin to some distance below the forelegs, and again continued with more regularity from the middle of the belly to the anus. In some skins these markings are of small extent, but I have never seen them entirely wanting. There are commonly spots under either one or both of the forelegs, but not invariably. I have remarked that the coloration of this animal, as well as of the Otter and Beaver, grows

---

in the small rounded feet and short tarsus of our present species, in its longer and rather more pointed ears, its shorter head and longer lower incisors, together with a more general resemblance to our common weasel (*P. erminea*) in summer dress."

lighter as it advances in years, and that the white blotches or spots are of greater size and distinctness in the old than in the young. The fur of a young Mink (under three years) when killed in season is very handsome; its color is often an almost pure black. The skin is thin and pliable, approaching nearly to the papery consistency of that of the Martin. When aged, the hide is thick and the color more rusty. The summer pelage is short, but tolerably close, and is of a reddish brown color, and the tail, though still possessing black hairs, shews distinctly the under-fur of a decidedly rusty hue. Its feet are rather pointed and not large. Its legs are short but muscular, and its track in the snow is easily distinguished from that of the Martin, whose longer and well-covered paws do not sink so deeply. Indeed, when the snow is at all deep and soft, the Mink makes a regular furrow, similar to that made by an Otter under like circumstances, though of course smaller. . . . .
I am strongly inclined to the opinion that there is only one species of Mink on this continent, and consider it highly probable that the *P. Nigrescentes* of Aud. & Bach. are merely common Minks under three years of age. I have seen numbers of skins here of exactly the same color, size, and furring as those described under that head in Prof. Baird's work on North American Mammals, which were simply young *P. visones*. This gentleman also states that the American species of Mink never has the edge of the upper lip white. I have never seen *the whole* of that part so colored, but in one specimen now on my table there is a white spot beneath the nostrils."

To the above account of the variations in pelage must be added another source of change in specimens, namely, the fading by long exposure to the light. Some mounted individuals which have been in the Smithsonian museum for about twenty years are now bleached to a dingy white nearly all over.

The time that the Mink requires to attain full stature is seen from the foregoing. As usual in this genus, the female averages considerably smaller than the male.

Variation in the skull of the Mink.

Having already given the principal characters of the skull in treating of the subgenus *Lutreola*, it only remains to note the variation presented by the present species.

Skulls of *L. vison* ordinarily range from 2.35 × 1.35 to 2.75 × 1.65,* but the extreme limits of variation are considerably further apart than these. Mr. J. A. Allen † has tabulated and discussed the variations according to geographical distribution. I present his article in full :—

"Eighteen skulls from the northern parts of the continent, mainly from Alaska, average 2.66 in length and 1.58 in width, the extremes being, length, 3.02 and 2.30; width, 1.90 and 1.40. Thirteen skulls from the highlands of Northeastern New York average 2.40 in length and 1.34 in width, the extremes being, length, 2.60 and 2.17. Three skulls from Pennsylvania (undoubtedly males) average 2.49 in length and 1.48 in width. In the northern series, the sex of the skull is given by the collector, whence it appears that the twelve males have an average length of 2.81, and the six females an average length of 2.48, showing a considerable sexual variation in size. Yet the smallest males (2.64 and 2.63) fall below the largest female (2.68), if the skulls are all correctly marked. None of the other females, however, exceed 2.55, and only three of the males fall below 2.70. In the New York series, the sex is not indicated; but, judging from the proportion of the small to the large skulls, the sexes are about equally represented in the two series, but in the New York series there is a very gradual decline from the largest to the smallest. The northern series of eighteen is selected from a series of twenty-three; the New York series of thirteen from a series of thirty. In each case only very old skulls were chosen, the immature specimens in each case being thrown out in order to have a fair basis for comparison. The immature and middle-aged specimens greatly predominate in the New York series, owing, doubtless, to the species being more closely hunted there than in the more unsettled districts of the far north.

"Taking these two series as a basis for a general comparison, there is indicated a considerable decrease in size from the north southward, amounting to 0.26 in length and 0.24 in width, or about one-tenth of the average size of the New York

---

* A skull of the common Ferret, *P. fœtidus* var. *furo*, before me, is almost exactly of the former dimensions. Tame Ferrets' skulls I have examined show a curious depression of the cranial portion—even a concavity of the upper profile, which I have not observed in *P. fœtidus*. A skull of the latter measures 2.60 × 1.55.

† Bull. U. S. Geol. and Geogr. Surv. Terr. vol. ii. no. 4, 1876, pp. 327, 328.

series. A single specimen, marked 'Brookhaven, Miss.', and another marked 'Tuscaloosa, Ala.', however, have a length respectively of 2.60 and 2.80, the former equaling the largest New York specimens, and the latter nearly equaling the average size of the males of the northern series, while a single male skull from Fort Randall, D. T., 2.90 in length, is the second in size of the whole series; one Fort Yukon specimen only being larger! Other specimens from the Upper Missouri region, however, are much smaller, as are other specimens from Prairie Mer Rouge, La., indicating that the specimens above mentioned are much above the average for their respective localities.

"*Measurements of thirty-seven skulls of* PUTORIUS VISON."

Catalogue-number.	Locality.	Sex.	Length.	Width.	Remarks.
6530	Fort Yukon, Alaska	♂	3.02	1.90	Very old.
8709	Alaska	♂	2.82	1.64	....do.
8797	......do	♂	2.83	1.62	...do.
8796	......do	♂	2.75	1.61	....do.
8707	......do	♂	2.73	1.62	....do.
8703	......do	♂	2.75	1.57	....do.
8702	......do	♂	2.68	1.62	....do.
8798	......do	♂	2.64	1.55	... do.
8648	Alaska (Kadiak)	♂	2.63	1.52	....do.
8708	Alaska	♀	2.68	1.58	...do.
6531	......do	♀	2.55	1.50	...do.
8704	......do	♀	2.45	1.45	...do.
8706	......do	♀	2.32	1.40	...do.
8705	......do	♀	2.30	1.40	....do.
3284	Nelson River	♂	2.86	1.62	....do.
4309	Fort Simpson	♂	2.70	1.51	....do.
8132	......do	♂	2.90	1.79	....do.
4305	......do	♀	2.55	1.46	....do.
12915	Fort Randall		2.90	1.61	....do.
3730	Essex County, New York		2.60	1.48	Old.
3824	......do		2.60	1.38	...do.
1169	......do		2.40	1.32	Old. *P. 'nigrescens'* A. & B.
3085	......do		2.40	1.38	....do.
3084	......do		2.40	1.31	....do.
3823	......do		2.32	1.32	Old.
3822	......do		2.30	1.23	....do.
2242	Saranac Lake, New York		2.47	1.37	....do.
2243	......do		2.40	1.30	....do.
2241	......do		2.35	1.31	....do.
2244	......do		2.20	1.18	....do.
2250	......do		2.40	1.48	....do.
2267	......do		2.17	1.20	....do.
1847	Pennsylvania		2.50	1.48	....do.
4834	......do		2.50	1.48	....do.
4835	......do		2.47	1.48	....do.
1894	Tuscaloosa, Ala		2.80	1.61	do.
11315	Brookhaven, Miss		2.60	1.50	....do."

Comparison with the European Mink.

I have only been able to compare my series of American Minks with one European specimen, which, being mounted, does not permit examination of the teeth. But as we have seen that the curious difference in the character of the molar and last premolar of *Mustela martes* and *M. americana* holds good, there is reason to presume that the same difference may constantly obtain in the Minks, as held by Dr. Gray. In this case, very curiously, it is the American animal which has the larger molar, not the European. This could never have been predicated by analogy; it illustrates the constantly recurring lesson of the danger of this mode of reasoning in zoölogy, and the necessity of appeal to observed facts in every case. All the many skulls of American Minks examined (about forty) show the massive last molar with an inner moiety very much larger than the outer, as against the opposite which is alleged of the Old World species. A discrepancy in average size of the American and European Minks is obvious; but the difference is within the range of variation of the former. The white upper lip, the rule in the European species, is the rare exception in the American. As far as external differences go, it would be impossible to separate the two forms specifically; we could only predicate a geographical race upon the *average* superior stature and *generally* dark upper lip of the American form. Attending only to these superficial details, Mr. Allen * came to the justifiable conclusion of the specific identity of the two animals; but had his able and pertinent discussion embraced consideration of the dental peculiarities, his views would doubtless have been materially modified. I am unable to endorse his general statement (*loc. cit.*) respecting the *Lutreola* group, that "we have here again but one circumpolar and widely dispersed species, with possibly two continental or geographical races which may be more or less easily recognized". For aside from the question of *P. vison*, the *P. sibiricus* (see foot-note), which Mr. Allen would bring into the same connection, is an entirely different species, to judge from the single excellent specimen before me. In justice to this writer, however, I should not omit to add that since his examination of the skulls he has presented *P. vison* as a distinct species.

The comparative diagnosis of *P. lutreola* and *P. vison* would be as follows:—

P. lutreola.†—Back upper molar small, quadrate, transverse, the inner moiety scarcely larger than the outer [*fide* Gray]. Averaging smaller; upper lip normally white.

---

* Bull. Mus. Comp. Zoöl. i. 1869, pp. 175-177—an article important as a contribution to the present discussion, and as satisfactorily showing that the external characters supposed to distinguish two species do not hold.

† I introduce short notices of the two Old World species allied to *P. vison*, as further contributions to the history of the group.

P. vison.—Back upper molar large, with great constriction across the middle, making an hourglass-shape, the inner moiety of which is nearly twice as large as the outer [40 specimens seen]. Averaging larger; upper lip normally dark.

Putorius (Lutreola) lutreola.—*European Mink.*

Viverra lutreola, *L.* Fn. Suec. 2d ed. 1761, 5, no. 13.—*Pall.* Spic. Zool. xiv. 1780, 46, pl. 21, f. a.; Zoog. R.-A. i. 1831, 80, no. 23.
Mustela lutreola, *L.* S. N. i. 1766, 66, no. 3 (Finland).—*Schreber,* Säug. iii. 1778, 462, pl. 125.— *Lepech.* Itin. i. ——, 176, pl. 12.—*Gm.* S. N. i. 1788, 94, n. 3.—*Turt.* S. N. i. 1806, 58.—*Nilss.* Skand. Fn. 11, 152.—*Less.* Man. 1827, 147.—*Falck,* Act. Soc. Sc. Fenn. ii. 1847, 523.—*Gieb.* Säug. 1855, 484.
Mustela (Lutreola) lutreola, *Schinz,* Syn. Mamm. i. 1844, 346.
Lutra lutreola, *Shaw,* G. Z. i. 1800, 443.—*Glog.* N. Act. Acad. Nat. Curios. xiii. 501.
Putorius lutreola, *Griff.* Cuv. R. A. v. 1827, 122, no. 347.—*Brandt,* Bem. Wirb. Nord. Eur. Russl. 1856, 27.—*Anjubault,* Bull. Soc. Agric. Sarthe, xiii; Rev. Mag. Zool. 1863, 77 (see *Brehmer,* Arch. Vereins Mecklenb. 1863, 291; *Taragon,* Rev. Zool. xv. 357; *Heinzel,* Verb. Ntrf. Vereins Brünn. i, 1862, 18).
Fœtorius lutreola, *K. & B.* Wirb. Eur. 1840, 69, no. 148.—*Blas.* Wirb. Deutschl. 1857, 234, no. 5 (der Nörz).—*Struck.* Arch. Nat. Mecklenb. xiii. 1859, 139.—*Krause,* Peterm. Geog. Mitth. 1866, 425.
Vison lutreola, *Gray,* List Mamm. Br. Mus. 1843, 64 (includes both species); P. Z. S. 1865, 117 Cat. Carn. Br. Mus. 1869, 94.—*Gerr.* Cat. Bones Br. Mus. 1862, 92 (includes both species).—*M. Schmidt,* Zool. Gart. 1865, 168, fig.
Lutra minor, *Erxl.* Syst. An. i. 1777, 451, no. 3 (mixed with *P. vison*).
Mank, Nurek, Tuhcurl, Nœrza, Norz, Nörz, Nærz, Nurtz, *Authors.*
Kleine Fischotter, Sumpfotter, *Germ.*
Nörz, *Martens,* Zool. Gart. xi. 1870, 278 (philological).

The characters of this species are sufficiently indicated in the text above. Gerrard gives the caudal vertebræ as 17.

Putorius (Lutreola?) sibiricus.—*Siberian Mink.*

Mustela sibirica, *Pall.* Itin. ii. app. 701; Spic. Zool. 1780, xiv. 89, pl. 4, fig. 2; Zoog. R.-A. i. 1831, 90, pl. 7.—*Erxl.* Syst. 1777, 471, no. 11.—*Schreb.* Säug. iii. 1778, 495, pl. 133 B.— *Zimm.* Geogr. Gesch. ii. 1780, 306, no. 202.—*Gm.* S. N. i. 1788, 98, no. 16.—*Turt.* S. N. i. 1806, 61.—*Desm.* Mamm. i. 1820, 177, no. 272; Nouv. Dict. xix. 369.—*Fr. Cuv.* Dict. Sci. Nat. xxix. 249.—*Is. Geoff.* Dict. Class. x. 212.—*Gray,* List Mamm. Br. Mus. 1843, 66.— *Gerr.* Cat. Bones Br. Mus. 1862, 94.—*Gieb.* Säug. 1855, 781.
Viverra sibirica, *Shaw,* Gen. Zoöl. i. 1800, 431.
Putorius sibirica, *Griff.* Cuv. R. A. v. 1827, 122, no. 346.
Vison sibirica, *Gray,* P. Z. S. 1865, 117; Cat. Carn. Br. Mus. 1869, 94.
(?) "Mustela italsi, *Temm.* Fn. Jap. 34, pl. 7, f. 2 (*natsi* by misprint on plate)."—(*Gray.*)
Putois de Sibérie, *Cuv.* R. A. i. 148.
Chorock, "*Sonnini's* Buffon, xxxv. 19."
Kulon, *Tartars.*
Kulonnok, Chorok, *Russian.*

(No. 1451, Mus. Smiths., from the Bremen Museum.) This animal is a *Putorius* (teeth 34), and may come near the Minks, as the toes appear to be extensively semi-palmate and the ears are very short. The general aspect, however, is that of a Ferret or Polecat, *P. fœtidus,* like which species, and like *P. nigripes,* it has dark facial markings contrasted with white surroundings. The tail is long and bushy, about as it would be in a Marten (*Mustela*) of the same size. The color is peculiar—a uniform, clear, rich, fulvous or tawny brown ("buff" or "fawn" color), scarcely paler below, the tail

General history and habits of the Mink.

The history of the American Mink, to which we will confine our attention, begins at an early date, long before Linnæus conferred precision upon zoölogical writing by establishing the binomial nomenclature. Says Sagard-Théodat, in 1636, referring to the Hurons:—" Ils ont vers les Neutres une autre espece d'animaux nommez Otay, ressemblant à un escurieux grand comme un petit lapin, d'un poil tres-noir, & si doux, poly & beau qu'il semble de la panne. Ils font grands cas de ces peaux desquelles ils font des robes & couuertures, où il y en entre bien une soixantaine qu'ils embellissent part tout à l'entour, des testes, & des queuës de ces animaux qui leur donnent bonne grace, & rendent riches en leur estime." Early in the seventeenth century we find the animal unmistakably indicated under the name of *Mink* or *Minx*.* The derivation of these words—or rather of this term, for the two are obviously the same—is from the original Swedish *maenk*, applied to the *P. lutreola* of Europe. The term *otay* had long been in use at that time, and *foutereau* was an Early French designation, used, for instance, by La Hontan (1703) for " a sort of small amphibious weasels". Of the meaning of the term *vison*, generally adopted since Buffon as the specific designation, I have only to remark, on the authority of von Martens, its apparent relation with *weasel*, through *veso*. The word *jackash*, sometimes found, is obviously a rendering by an English tongue of the Cree name, which is given by Richardson as *Shakweeshew* or *Atjackashew*.

"The Minx", says Lawson, about the beginning of the last century, "is an animal much like the English Fillemart or

---

throughout the same. Throat and soles of feet whitish. Forehead, cheeks, region around eyes, and naked nasal pad blackish-brown; end of snout all around (isolating the dark nose-pad), edge of upper lip, and chin white. Length apparently about 15 inches; tail-vertebræ 6 or 7; hairs at the end full 3 inches longer.

Pallas says his animal is peculiar to Farther Siberia, from the Yenisei River eastward to the sea, to the 60° parallel, but is not found in Kamtschatka nor in the Tschuctschi region. Gray attributes it to the Himalayas, China, Japan, and Formosa, quoting Temminck, as above.

* The identity in form with the English *minx* may possibly be more than fortuitous. *Minx* was a name of a female puppy, and subsequently signified a pert, wanton girl, doubtless through the same association of ideas that caused the vulgar name of a she-dog to become a shameful term of reproach for a lewd woman. There is something in the forward, prying, and spiteful nature of the animal to render *minx* applicable.

Polcat. He is long, slender, and every way shaped like him. His Haunts are chiefly in the Marshes, by the Seaside and Salt-Waters, where he lives on Fish, Fowl, Mice and Insects. . . . These are likewise found high up in the Rivers, in whose sides they live; which is known by the abundants of Fresh-Water Muscles Shells (such as you have in England) that lie at the Mouth of their Holes. This is an Enemy to the Tortoise, whose Holes, in the Sand, where they hide their Eggs, the Minx finds out, and scratches up and eats,"—with more in the same quaint style.

Buffon described "Le vison" in 1765 from a Canadian specimen in M. Aubry's museum, the same apparently that served as the basis of Brisson's earlier and Pennant's subsequent account. Pennant indeed has also his Minx or Lesser Otter, but this is simply because he did not recognize that this was the same as his *vison*.

Since these earlier authors, the Mink, a very common animal of this country, has been frequently mentioned by writers, and taken its place in all the systematic works. It has served as the basis of several nominal species, but these have occasioned little if any confusion, the zoölogical characters of the animal being well marked. The only question, indeed, is as to its relationships with the European *P. lutreola*. For many years a specific distinctness was seldom doubted, but of late the opinion has tended the other way. The Mink has been placed alternately in the genera *Mustela* and *Putorius*, partly owing to a varying acceptation of these names by authors, partly to a misconception of its dental characters. It is a true "Weasel", with 34 teeth, not a Marten, which has 38. It is of larger size, stouter form, and bushier tail than an average species of *Putorius*, approaching in these respects to the Martens, *Mustela*. In those points in which it is modified for its eminently aquatic mode of life, namely, the half-webbing of the toes, short ears, and the close-set, bristly, glistening pelage, it makes an approach toward the Otters. In fact, the specific term *lutreola*, "little otter", applied to the European form by Linnæus, is highly appropriate. The non-essential modifications which the animal presents have been unnecessarily made by Dr. Gray the basis of a subgenus *Vison*.

The peculiar odor which the animals of this genus have in common attains in this large and vigorous species a surpassing degree of fetor, though of the same quality. No animal of this country, except the Skunk, possesses so powerful, penetrating, and lasting an effluvium. Its strength is fully perceived in taking the animal from a trap, or when the Mink is otherwise irritated. Ordinarily the scent is not emitted to any noticeable degree; it is under voluntary control, and the fact that the Mink spends most of its time in the water is another reason why its proximity, even in numbers, is not commonly perceived by smell. Both sexes possess the scent-bags; they lie in the perinæum, one on each side of the rectum, and open upon a papilla on either side of the anus, just within the edge of the external orifice. As usual, the apparatus pertains primarily to the sexual relations, and, in fact, can have no other office of consequence, since the effluvium is not powerful enough to deter pursuit on the part of a determined enemy, as is the case with the intolerable emanations of the Skunk. Its service seems to be that of attracting the sexes. It is used with advantage, like the castoreum of Beavers, by trappers, to increase the efficacy of their bait. It belongs to the class of musky odors, which, in minute quantities, are not disagreeable to most persons; and, indeed, a moderate amount of mink scent is to me less undesirable than the ineffably rank odor of a he-wolf for instance. The former is special and peculiar; the latter seems to convey all that is obscene in the nature of the animal.

The distribution of the Mink in this country is scarcely limited. In a word, it is found in suitable places throughout North America. Sir John Richardson found it on Mackenzie's River as far north as 66°; "and there is every reason to believe that it ranges to the mouth of that river, in latitude 69°". Audubon says that he has seen it "in every State in the Union", and remarks its abundance in the salt marshes of the Southern States. Although he could at that time only speak at second hand of its occurrence in regions west of the Rocky Mountains, I have sufficient evidence in the way of specimens that it is there equally well represented. Its essentially aquatic nature leads it to seek, in general, well-watered sections, and it will never be found far away from water, except it be caught during the journeys it makes from one stream or pool to another. Nevertheless, I have found it in great plenty along the watercourses of some of the driest portions of the interior of the continent, as in Dakota and Montana.

The very scarcity of water in such regions is one cause of the apparent abundance of certain aquatic animals in spots, as around the pools and along the few streams; they become aggregated in a few places rather than generally dispersed over the country, so that their numbers appear greater than they really are. In the region last mentioned, there was scarcely any water, running or stagnant, even if enduring for only a part of the year, the muddy banks of which were not dotted with numberless tracks of Mink, Muskrats, and Meadow Mice. All around the permanent pools, the entrances to the burrows of the first named were to be found. The holes were noticed more or less nearly at water-level, according to the state of evaporation of the water; they were generally dug in a rather steep part of the bank, and from the entrance of the burrow a "way" led far out into the pool.

Whilst encamped for a month or more in the autumn of 1873, on Mouse River, in Northern Dakota, a friend with me procured a large number of Minks without difficulty. In addition to our steel traps, we built numerous deadfalls, and were equally successful with both means. The Minks were not at all wary about the traps. Any contrivance by which a small log could be made to fall against another on touching a trigger, the bait being covered so that the animal could only reach it from the desired position, sufficed perfectly well. Such a trap may be built, where there is wood, with a hatchet and pocket-knife in a few minutes. We set them at intervals for several miles along the stream, wherever, judging from the number of tracks, we were most likely to be successful. They were placed as near as convenient to the water's edge, baited with a duck's head or breast, and scented with the Mink's odor. In setting the steel traps, we placed them in the "ways" leading into the burrows, and in very shallow parts of the stream, where a little water rippled over pebbly shingle. It was found best, on the whole, not to bait the trap itself, but to build a little box of flat stones, with a narrow entrance, at which the trap was set, the bait being placed further in. The Mink of this region seemed to me rather smaller and darker than average, and they rarely showed white along the chest or belly.

The tenacity of life of the Mink is something remarkable. It lives for many hours—in cases I have known for more than a day and night—under the pressure of a heavy log, sufficient to hold it like a vice, and when the middle of the body was pressed perfectly flat. Nay, under one such circumstance which I recall, the animal showed good fight on approach. When caught by a leg in a steel trap, the Mink usually gnaws and tears the captive member, sometimes lacerating it in a manner painful to witness; but, singular to say, it bites the part beyond the jaws of the trap. This does not appear to be any intelligent attempt to free itself, but rather an act of the blind fury excited by consciousness of capture. Some have averred that it is an instinctive means of lessening pain, by permitting a flow of blood from the portion of the limb beyond the point of seizure; but this seems to me very problematical. The violence and persistence of the poor tortured animal's endeavors to escape are witnessed in the frequent breaking of its teeth against the iron—this is the rule rather than the exception. One who has not taken a Mink in a steel trap can scarcely form an idea of the terrible expression the animal's face assumes as the captor approaches. It has always struck me as the most nearly diabolical of anything in animal physiognomy. A sullen stare from the crouched, motionless form gives way to a new look of surprise and fear, accompanied with the most violent contortions of the body, with renewed champing of the iron, till breathless, with heaving flanks, and open mouth dribbling saliva, the animal settles again, and watches with a look of concentrated hatred, mingled with impotent rage and frightful despair. The countenance of the Mink, its broad, low head, short ears, small eyes, piggish snout, and formidable teeth, is always expressive of the lower and more brutal passions, all of which are intensified at such times. As may well be supposed, the creature must not be incautiously dealt with when in such a frame of mind.

The gun is not often used to procure Mink, not only because of the injury to the pelt which would ensue, but because its use is difficult and unsatisfactory. I have never secured one in this way, though I have more than once fired at them swimming in the water. If on the lookout, as they usually are, they may dive at the flash, and evade the shot. They immediately disappear likewise if only wounded; and even if killed outright, which is not often the case, they sink, and are not likely to be recovered. Shots at a Mink on land but rarely offer; I do not remember to have had but a single one, and then the animal escaped me.

From what has gone before, the prime characteristic of the Mink in comparison with its congeners may be inferred: I mean its amphibious mode of life. It is to the water what the other Weasels are to the land or the Martens to the trees. It is as essentially aquatic in its habits as the Otter, Beaver, or Muskrat, and spends perhaps more of its time in the water than it does on land. In adaptation to this mode of life, the pelage has that peculiar glossiness of the longer bristly hairs and felting of the close under fur which best resists the water, much as in the cases of the other animals just mentioned. Were not fashion so notoriously capricious, Mink pelts would maintain a conspicuous place in the fur marts of the world; certainly few surpass them in richness of color, gloss, and fineness. Yet they have been found under some circumstances not to repay cost of transportation, although it should be added, at times the price they fetch shows them to be better appreciated. The darkest colored samples are regarded as the most valuable—such as those coming from the so-called *Putorius nigrescens*. As in other cases, the quality of the fur depends largely upon season, and other varying circumstances. Nova Scotian pelts have been regarded with particular favor. On this subject, the following extract from Dr. Gilpin's article above quoted is given :—" This fur once valueless has steadily increased in price, till last winter [1865] not seldom five dollars was paid for a single skin. Our Indians trap but very little now. The idle boys about the villages take many. The farmer, indignant at his slaughtered fowl yard, adds a few more. In every land and every village, there is a social gipsey who loves sport and hates work ; who fishes, and fowls, and traps, eats his own trout or poached salmon or moose meat, taken out of season, and exchanges his little pile of fur for tea and tobacco at the country store. Many come from this source. Thus a gathering pile collects and dangles at the country store. The owner packs and sends them to the Halifax market, where of late years it has become the habit for the fur dealers to tender in writing for them. About six thousand are annually exported from Nova Scotia proper."

Coincidentally with the aquatic habitat, the food of the Mink is somewhat modified, in comparison with that of the land species of the genus. It is probably our only species which feeds habitually upon reptiles, fish, molluscs, and crustaceans—more particularly upon frogs, fresh-water bivalves, crawfish, and the like. Nevertheless, it is not confined to such diet, but shows its relationships with the terrestrial Weasels in a wide range of the same articles of diet as the latter secure. It is said to prey upon Muskrats—a statement I have no hesitation in believing, though I cannot personally attest it. A recent writer,* in an article which I would quote were it written in a style suited to the present connection, narrates an incident which may be here briefly related, as showing that the Mink is a formidable enemy of the Muskrat, though yielding to the latter in weight. Whilst snipe-hunting on a marshy island below the Kickapoo Rapids of the Illinois River, the writer noticed an object, which appeared like a ball some six or eight inches in diameter, rolling toward the water; and soon ascertained that it was a Mink and a Muskrat clinched together, and so completely covered with mud as not to have been at first recognized. At his approach, the Mink released its hold and made its escape; but the Muskrat was already dying of severe wounds in the head and neck, from which the blood was flowing profusely. The Muskrat had evidently been captured and overcome in fair fight by broad daylight, and the Mink would have devoured its victim had not the hunter interfered. It is also destructive to our native rats and mice—the *Arvicolas, Hesperomys, Sigmodon,* and *Neotoma;* it is known to capture Rabbits, especially the *Lepus palustris*, its associate in many marshy or swampy tracts; while its not infrequent visits to the poultry-yard have gained for it the hearty ill-will of the farmer. Various marsh-inhabiting birds are enumerated in the list of its prey, among them the rails and several smaller species; and we may presume that it does not spare their eggs. But most birds are removed from its attack; for the Mink is not a climber, at least to any extent. In respect to poultry, its destructiveness seems to result rather from the regularly repeated visits of an animal that has located in the vicinity than the wholesale slaughtering sometimes accomplished by the Ermine. According to those who have excellent opportunity of judging, the Mink does not as a rule kill more than it eats. Still, the opposite case has been recorded. Its modes of hunting offer nothing peculiar. Like the Weasel and Stoat, it has been known to pursue its prey by scent.

The Mink often annoys hunters by stealing the game they have shot before they have an opportunity of bagging it. An incident related by a recent anonymous writer in "Forest and Stream" is in point, and furthermore illustrates the wonderful energy and perseverance sometimes displayed by the Mink in

---

* M. A. Howell, jr. "The trapper not the only enemy of the Muskrat." <Forest and Stream of Dec. 21, 1876.

securing its food. Speaking of a duck-shooting excursion, during which some of the birds that had been killed were not recovered till next day, the writer goes on to say:—"The first spot which claimed attention, was where our 'hen mallard' had 'struck hard pan.' Here was a sight! feathers and blood marked the scene of a terrific struggle for what remained of a duck's life. Here, for at least ten feet in circuit, the snow, grass and twigs, were whipped into a confused mass, here and there besprinkled with blood, and quite as often decorated with feathers; then there was a trail, leading directly to the river bank, and out upon the ice; the trail thence proceeded up the bank of the river on the ice for about half a mile, when it disappeared directly in line of a hole in the bank, where we discovered the bird half buried, head foremost, into a hole about one-half the size of the body, frozen stiff. When discovered we worked, not without difficulty, at the extrication of the bird. It required all our force to draw it out, when, as it broke from its fastenings, two large minks suddenly appeared, and darted back into their retreat, the last we saw of the varmints after a half hour of close watching. The ground along the shore was rough, covered with heavy grass, brush, drift wood, and many willows. Here the natural obstacles precluded the possibility of such a trip by land, and the little piece of engineering practiced by this one mink, in capturing and conveying home its prize was truly marvellous. That there was but one mink, the trail bore direct evidence throughout its entire length from the scene of the struggle. As we followed the line, we could easily trace the wide trail of the mallard, as it was dragged bodily along over the fresh snow, and the deep penetration of its claws into the new ice, spoke volumes of the force exerted by that small animal in the completion of so severe an undertaking, and the excessive amount of *mink power* expended in the completion of a successful foraging expedition. Here and there throughout the line of trail were frequent halting places, where our mink had stopped for *a rest*. Every time there appeared numerous tracks around the body of its victim, as though pleased to inspect its trophy before the next heat, and then as the distance shortened, the strokes of its tail at regular intervals of march, marked upon the snow upon either side of the trail the determined intention of the animal to go through with its *meat* before it was too cold to squeeze into a small space, where the sharp frost would soon fix it permanently. When drawn out, we found that a couple of 'square meals' had been made from the head, neck and breast, and enough left for several days to come."

This account of the Mink's theft called forth shortly afterward in the same paper the following instance of its stealing fish; the editor, Mr. Charles Hallock, remarking that he had known Minks to carry off fish weighing no less than twelve pounds:—
"We were spending our vacation in the woods of Maine, fishing, and traveling about for a good time in general. One day we came across an old dam made to flood a piece of lowland. As this looked like a good place to fish we stopped, seated ourselves upon the edge of the dam, and cast in our line. The fish were quite plenty, and as fast as we caught one we threw it behind us upon the scaffolding. After a dozen or so had been caught, I thought I would light my pipe, pick up the fish and put them in the shade, and I started to do so. I accomplished the first object, but upon looking for the fish I could not find a single one. I thought that my chum must have removed them, and was playing a joke upon me, but on mentioning it to him he was as much surprised as I was. They could not have fallen through the cracks, nor leaped over the side without our knowing it. Where were they? That was the question. He returned to fish, and I seated myself upon the bank to digest the subject. Presently he caught another fish and threw it upon the boards. Immediately I saw a Mink run out from a hole near by, snatch the fish and carry it off. This explained the mysterious disappearance of the others."

The movements of the Mink on land, though sufficiently active, lack something of the extraordinary agility displayed by the more lithe and slender-bodied Weasels, as a consequence of the build of its body; while, for the same reason, it does not pursue the smaller animals into their extensive underground retreats, nor so habitually prowl about stone heaps and similar recesses. It is altogether a more openly aggressive marauder, though not less persistent and courageous in its attacks. It appears to be more perfectly at home in the water, where it swims with exactly the motions of an Otter, and in fact appears like a small specimen of that kind. It swims with most of the body submerged—perhaps only the end of the nose exposed—and progresses under water with perfect ease, remaining long without coming to the surface to breathe. This may be partly the reason of its long survival under the pressure of a deadfall.

The Mink is not properly a migratory animal. In most sections it remains permanently where it takes up its abode. In others, however, it may be forced to remove at times, owing to scarcity or failure of its food-supply, such as may ensue from the freezing of the waters in northern parts. Under such circumstances, it may perform extensive journeys overland. Trappers have indeed spoken to me of a "running" time with the Minks, but I cannot satisfy myself that reference is here had to anything more than periods of sexual activity, when the animals are hunting mates. I do not think that whatever "migration" may take place is anything more than casual.

The rutting season begins early—generally in February—and April is for the most part the month of reproduction. Five or six young are ordinarily produced at a birth. Litters have been found in the hollow of a log, as well as in the customary burrows.

The Mink has been frequently tamed, and is said to become, with due care, perfectly gentle and tractable, though liable to sudden fits of anger, when no one is safe from its teeth. Without showing special affection, it seems fond of being caressed, and may ordinarily be handled with perfect impunity. The following account of the semi-domestication of Minks on an extensive scale will be read with interest, not alone for its novelty, but also because it gives some precise information respecting the reproduction of the species.

"Minkeries."

The Mink appears to be the only species of its genus which has been systematically reared and trained for ratting in this country as the Ferret is in Europe. The relationship of the two animals at once suggests the feasibility of an experiment, which has been tried with complete success, as we learn from an interesting article lately published in "Forest and Stream" (October 22, 1874—apparently taken from "Fancier's Journal and Poultry Exchange" of October 15, 1874). I reproduce the passage in substance.

Mr. H. Resseque, of Verona, Oneida County, N. Y., has frequently exhibited at fairs two tame female Minks, which he hands to the by-standers to be caressed and passed from one to another. The animals were perfectly gentle, submitting to be handled, but it was noticed that they kept their eyes on their keeper, to whom they would frequently extend their paws like a child wishing to be taken to its parent. Seven years ago, Mr. Resseque came in possession of a live wild Mink, and through her progeny his stock has on some occasions amounted to ninety individuals, besides the numerous specimens disposed of. At the late Albany County fair, his "minkery" was one of the novel features.

Mr. Resseque's minkery consists of twelve stalls, each twelve feet square, of stale soil, and surrounded with a fence and some special precautions to prevent the escape of the animals. In each stall is placed a dry-goods' box for the home of the female; it has two openings for ingress and egress, opposite each other, besides a door on top to allow of inspection and cleaning. The animals are fed on sound, fresh meat, as they do not relish tainted flesh. In summer it is given to them daily, but in cold weather a large quantity is thrown in at once and allowed to freeze, the Minks helping themselves at pleasure. In February, their allowance is shortened, to get them into condition for breeding. Mr. Resseque claims that this slight degree of fasting makes them more lively and playful, and it is á part of his plan to imitate nature as closely as possible—their supply of food, in the wild state, being restricted at this season.

In the minkery, the sexes are not allowed to run together except during the month of March, which is considered the running season in a state of nature. If allowed together for a longer period, the male teases and annoys the female. At this time, the males fight desperately, and if not soon separated one always gets the mastery. The females come in heat with great regularity, all being ready for the male within ten days; and the period of excitement lasts about four days. One male serves six females. The females reproduce when one year old. The duration of gestation scarcely varies twelve hours from six weeks. There is but one litter annually. The litters run from three to ten in number; the young are born blind, and remain so for five weeks. When newly born, they are light-colored, hairless, and about the size and shape of a little finger. By the time the eyes are open, they are covered with a beautiful coat of glossy hair. The young females develop sooner than the males, attaining their stature in ten months, while the males are not full-grown until they are a year and a half old. It is noted that in every litter one or the other sex predominates in numbers, there being rarely half of them males and the other half females. If taken in hand when their eyes are first open, they are readily tamed; they should not subsequently be allowed to remain with the mother or in each others' society. By continual petting and handling, they become like domestic ratters, and have all the playfulness of the young of the feline tribe. They may be handled, without fear of their sharp teeth, but they prove extremely mischievous, their scent leading them to food not intended for them. Their fondness for bathing will prompt them to enter a tea-kettle or any open vessel; and when wetted they will roll and dry themselves in a basket of clothes fresh from the laundry, or even upon a lady's dress, occasioning much inconvenience.

Minks are not burrowing animals in a state of nature, but freely avail themselves of the holes of Muskrats and other vermin. They cannot climb a smooth surface, but ascend readily where there is roughness enough for a nail-hold. The grown male will weigh about two pounds; the female is heavier than she looks, averaging between one and a half and one and three-fourths pounds. These tame Minks make excellent ratters, hunt vigorously, and soon exterminate the troublesome pests. Rats will make off on scenting them; they are so bewildered in flight that they give no battle, but yield at once; and the Mink severs the main vessels of the neck so quickly and skilfully that an observer would scarcely imagine the deed had been done.

When wild Minks are confined with the tame ones, the latter always prove stronger than the former, and come off victorious in the contests that ensue. They have been observed to beat off a cat that imprudently invaded the minkery in quest of food. So completely domesticated are the animals that a person may enter the inclosure with impunity, and observe the animals playing about him like kittens.

Mr. Resseque states that he finds ready sale for his Minks— in fact, that he cannot supply the demand. His prices are $30 per pair—$20 for a female, $10 for a male, and $25 for an impregnated female. It is to be hoped that this novel branch of industry will be perpetuated and extended. There are plenty of Minks in this country, the services of which are available without difficulty for the purpose of destroying vermin, and in the aggregate their good services would have a very decidedly appreciable result. They have a great advantage over terrier dogs in being able to enter any ordinary rat-hole and drive their prey from its hidden resorts.

From the "Forest and Stream" of July 2, 1874, the following article is extracted in further illustration of this branch of industry:—

"Messrs. Phillips & Woodcock, of Caneadea, New York, commenced two years ago the business of breeding mink for their fur. A correspondent of the Buffalo Express describes the 'Minkery' in the following terms:—

"'The "Minkery," designed to accommodate one hundred minks for breeding, consists first of an enclosure about forty feet square, made by digging a trench one foot deep, laying a plank at the bottom, and from the outer edge starting the wall, which consists of boards four feet high, with a board to cap the top, projecting upward eight or ten inches to prevent their climbing over. Within this enclosure is a building 14 by 24, supplied by running water, from which the mink catch living fish, that are often furnished, with the greatest delight.

"'The building is constructed by an alley three feet wide around its circumference. Within are two rows of cells four feet deep and two and a half wide, each having a door ventilated at the top and bottom with wire screens, as is also the front entrance, what the proprietors call the anteroom, four by four feet, which must be fastened within every time the building is entered, to prevent the escape of the imprisoned animals. On entering the main hall, which the minks have access to (when not rearing their young), they present a very playful group.

"'The person feeding them is often mounted, for their food and their tenacity of hold is so strong that they may be drawn about or lifted without releasing their hold upon the food. The nest of the female is very peculiarly constructed with grass, leaves, or straw, with a lining of her own fur so firmly compacted together as to be with difficulty torn in pieces. The aperture leading to the nest is a round opening, just sufficient to admit the dam, and is provided with a deflected curtain, which covers the entrance and effectually secures her against all invasion when she is within. About the middle of March the females are separated from the males until the young are reared. The necessity for this arises from the fact that the males seem inclined to brood the young almost as much as the dam, when both are permitted to remain together.

"'The expense of feeding these animals is almost nominal, being supplied pretty much entirely from the usual offal of a farm yard, with occasional woodchucks and game in general.

They eat this food with equal avidity after decomposition has taken place, devouring every particle of flesh, cartilage, and the bones. The flesh and bones entire of the woodchuck are consumed often at a single meal. While the expense of keeping is thus trivial, the profitable yield of the animal is comparatively immense, it being considered a moderate estimate or claim that the mink with her increase will equal the avails of a cow.'"

We find in Audubon and Bachman several paragraphs upon the same subject, which will be transcribed:—" The Mink, when taken young, becomes very gentle, and forms a strong attachment (?) to those who fondle it in a state of domestication. Richardson saw one in the possession of a Canadian woman, that passed the day in her pocket, looking out occasionally when its attention was roused by any unusual noise. We had in our possession a pet of this kind for eighteen months; it regularly made a visit to an adjoining fish-pond both morning and evening, and returned to the house of its own accord, where it continued during the remainder of the day. It waged war against the Norway rats which had their domicile in the dam that formed the fishpond, and it caught the frogs which had taken possession of its banks. We did not perceive that it captured many fish, and it never attacked the poultry. It was on good terms with the dogs and cats, and molested no one unless its tail or foot was accidentally trod upon, when it invariably revenged itself by snapping at the foot of the offender. It was rather dull at midday, but very active and playful in the morning and evening and at night. It never emitted its disagreeable odour except when it had received a sudden and severe hurt. It was fond of squatting in the chimney corner, and formed a particular attachment to an armchair in our study.

"The latter end of February or the beginning of March, in the latitude of Albany, N. Y., is the rutting season of the Mink. At this period the ground is usually still covered with snow, but the male is notwithstanding very restless, and his tracks may everywhere be traced, along ponds, among the slabs around sawmills, and along nearly every stream of water. He seems to keep on foot all day as well as through the whole night. Having for several days in succession observed a number of Minks on the ice hurrying up and down a millpond, where we had not observed any during the whole winter, we took a position near a place which we had seen them pass, in order to procure some of them. We shot six in the course of the morning, and ascertained that they were all large and old males. As we did not find a single female in a week, whilst we obtained a great number of males, we came to the conclusion that the females, during this period, remain in their burrows. About the latter end of April the young are produced. We saw six young dug from a hole in the bank of a Carolina rice-field; on another occasion we found five enclosed in a large nest situated on a small island in the marshes of Ashley river. In the State of New York, we saw five taken from a hollow log, and we are inclined to set down that as the average number of young the species brings forth at a time."

CHAPTER VII.

Subfamily MEPHITINÆ: The Skunks.

General considerations—Cranial and dental characters—The anal armature—Division of the subfamily into genera—Note on fossil North American species—The genus *Mephitis*—*Mephitis mephitica*, the Common Skunk—Synonymy—Habitat—Specific characters—Description of external characters—Description of the skull and teeth—Variation in the skull with special reference to geographical distribution—Anatomy and physiology of the anal glands and properties of the secretion—Geographical distribution and habits of the Skunk—History of the species—ADDENDUM: on hydrophobia from Skunk-bite, the so-called "rabies mephitica".

General considerations.

A CONCISE diagnosis of this subfamily will be found on p. 10, where the characters of the group are contrasted with those of the other North American subfamilies.

The subfamily is confined to America, its nearest Old World representatives being the African *Zorillinæ*. It is a small group, of only two or three genera and perhaps not more than four or five really good species, among the great number of nominal ones indicated by authors. More precise knowledge than we now possess will be required to fix the number of species, especially in the genus *Conepatus*. No more than three species are known to inhabit North America north of Mexico, each one typical of a different genus or subgenus. There is a Mexican species of *Mephitis* proper, apparently perfectly distinct from *M. mephitica*. One North American and Mexican species of a second allied subgenus, *Spilogale*, and one or several North, Central, and South American species of the very different genus, *Conepatus*, complete the list as far as known.

In entering upon the *Mephitinæ*, we pass to a group quite

different from the *Mustelinæ* in general external appearance as well as in structural characters. The closest relationships of the Skunks are with the Badgers (subfamily *Melinæ*); the affinities of these two being so well marked that some authors have combined them in the same subfamily. The Skunks and Badgers agree in many points of external conformation; in fact, *Conepatus mapurito*, one of the Skunks, is almost as much of a Badger, to all outward appearance. They are terrestrial animals, of more or less perfected fossorial habits; the walk is plantigrade; the fore claws are enlarged, straightened, and well fitted for digging. The general form is very stout; the legs are short, and the body consequently low; the tail is more or less bushy, and the whole pelage is loose. The physiognomy is somewhat hog-like, especially in the Badgers and in *Conepatus*, owing to the production and enlargement of the snout. These animals neither climb trees nor swim in the water; their gait is comparatively slow and lumbering; their retreats are burrows in the ground, dens in rocks or logs, or sometimes the shelter afforded by out-of-the-way nooks in human habitations. Some of the species hibernate.

Cranial and dental characters.

There is also a singular cranial character by which the Skunks and Badgers may be collectively distinguished from any other North American *Mustelidæ*. The conduit of the posterior nares is completely separated into right and left passages by a vertical bony septum, which extends to the hind end of the palate. In all the other *Mustelidæ* treated in this work, the posterior nares are thrown into one channel by total lack, posteriorly, of any such partition.

Nevertheless, the structural characters of most weight in classification are abundantly sufficient to mark off *Mephitinæ* and *Melinæ* as groups differing from each other as much as most other subfamilies of the *Mustelidæ* do. Reference to the tables of characters already given (pp. 7, 8) will show this. Here I may recall some of the leading peculiarities of the *Mephitinæ*.

The skull of any Skunk may be known at a glance, on comparison with that of any other Musteline animal, by the depth of the emargination between the pterygoids, which is always much greater than the distance from the end of this emargination to the molars. The post-molar portion of the bony palate in *Mephites* and *Spilogale* is *nil*, or almost so; that is, the palate ends nearly or exactly opposite the posterior border of the last molar. In *Conepatus*, the palate reaches a little farther back, but still not nearly half-way to the ends of the pterygoids. In other North American *Mustelidæ*, the palate usually extends half-way or more to the extremities of the pterygoids. The cranium of the *Mephitinæ* is further peculiar in the periotic region. The auditory bullæ themselves are small, and but moderately inflated at the base, with well-marked constriction of a tubular meatus;* while the parts lying behind the bullæ are unusually expanded, presenting a flattish and more or less horizontal large surface, which widely separates the paroccipital processes from the bullæ.† In *Lutrinæ* and *Enhydrinæ*, the paroccipitals are remote from the bullæ, but there is no such inflation of the mastoid region as is witnessed in some of the *Mephitinæ*, as in *Spilogale*, where the swelling of the mastoid cells results in a convexity of the parts only less than that of the bullæ themselves. The anteorbital foramen is remarkably small, circular, canal-like, and occasionally divided into several smaller openings. The postorbital processes are small or obsolete; the postorbital constriction of the skull is comparatively slight. The glenoid fossa is shallow, presenting much forward as well as downward, and never locks the condyle of the jaw, as so often happens in *Melinæ*. The coronoid process of the mandible is variable in *Mephitinæ*, for while in *Mephitis* and *Spilogale* it is erect and conical, as usual in *Mustelidæ*, in *Conepatus* it takes a backward slope, and is obtusely falcate, as in *Enhydrinæ*.

The teeth of *Mephitinæ* are also diagnostic in the combination of a large quadrate back upper molar with pm. $\frac{3-3}{3-3}$ or $\frac{2-2}{3-3}$ (the latter formula peculiar to *Conepatus*, but not always obtaining, even in that genus).‡

The detailed descriptions of the skull and teeth given beyond under heads of the several genera of *Mephitinæ* render further account unnecessary here. I would, however, advert to the extraordinarily high rate of variability inherent in the crania of these animals. In other groups, *genera* might very well be

---

* In *Melinæ*, the inflation of the bullæ is at a maximum for the family.

† In *Melinæ*, and also in *Mustelinæ*, the paroccipitals are close to, or in contact with, the bullæ.

‡ *Melinæ*, with pm. $\frac{3-3}{3-3}$, have a perfectly triangular back upper molar; *Lutrinæ*, with quadrate back upper molar, have pm. $\frac{4-4}{3-3}$.

established upon differences which are here nothing but fortuitous individual variations, or even the progressive changes with age during the life of the same individual. A Skunk's skull is as variable in shape as its pelage is in color. (Compare Plate X with XI, or Plate XIII with XIV, and see what extraordinary differences skulls of the same species may show.)

The general pattern of coloration, and the colors themselves, are likewise diagnostic of this subfamily, as all the species are black and white.

The anal armature.

No general sketch, however cursory, of leading features of this subfamily should fail to note the point which renders the Skunks infamous, makes their very name an opprobrious epithet, and almost forbids its use in the ordinary conversation of the polite. The matter is so notorious that comment may be confined to the zoölogical aspects of the case, including a refutation of various absurd notions still current among the vulgar. Special interest attaches to the subject, since it seems probable that there is some occult connection between failure of the supply of the fluid and a state of the system in which the saliva of the animals is capable of inoculating a disease similar to hydrophobia.

It was supposed for many years that the intolerably offensive fluid was the animal's urine, voided by an ordinary act of micturition, but with malice prepense. Its wide diffusion was sometimes fancied to be secured by means of the bushy tail, which, charged with the liquid, served as a mop to flirt it around. The obvious difficulties in the way of anatomical investigation long kept the facts in the case concealed.

The fluid is the secretion of certain glands situated in the perinæum, on each side of the rectum. So far from being peculiar to Skunks, similar glands exist throughout the *Mustelidæ*, and are, in fact, among the characteristic structures of the family. In the *Mephitinæ*, however, they reach the maximum of development, and their secretion acquires qualities which make it the most penetrating, diffusible, and intolerable of animal effluvia. The anatomical structure is fully described beyond; here I need only advert to some leading features.

Each gland is a secretory sac enveloped with a muscular tunic, and furnished with a duct to convey the secretion; the orifice of this duct is upon a papilla, which is situated on the side or the anus, just within the verge. Contraction of the muscular investment compresses the sac, and causes the fluid to spirt from the anal pore; the action is precisely that of a syringe with compressible bulb. The Skunk is as cleanly as any other animal, and the peculiar action observed at the moment of the discharge prevents the wetting of the fur. Forcible erection of the tail is accompanied by a tension of the perinæum, and an eversion of the anus, most favorable to forcible, unimpeded, and direct evacuation of the contents of the sac. The operation is wholly under the voluntary control of the animal, and seems to be chiefly resorted to in self-defence, although there is reason to suppose that the evacuation must recur at intervals simply to avoid over-distension of a continually secreting organ with its own products. Ordinarily, however, the Skunk is not more odorous than many other animals; it may even be captured, under some circumstances, without provoking an emission; nor do the horrible possibilities of the stench always render the flesh of the animal uneatable. In contemplating this singular provision of nature for the protection of an otherwise inoffensive and almost defenceless creature, we cannot but admire the simplicity of the means employed. Some little further development of glands common to the *Mustelidæ*, and some inscrutable modification of the operations of the secretory follicles, which gives a peculiar character to the fluid elaborated, result in means of self-preservation as singular as it is efficacious, habitual reliance upon which changes the economy of the animal and impresses its whole nature.

Division of the subfamily into genera.

There are two strongly marked generic types of the *Mephitinæ*, one of them susceptible of subdivision into two subgenera. In a former paper,* in which the skulls and teeth of the *Mephitinæ* were described, I allowed three full genera, following Dr. Gill;† but I am now rather inclined to consider *Spilogale* as only a subgenus. It certainly differs much less from *Mephitis* proper than *Conepatus* does, and the degree of differentiation seems to me to accord closely with that subsisting, for example, among the subdivisions of the genus *Putorius*.

---

* Bull. U. S. Geol. & Geogr. Surv. Terr. 2d ser. no. 1, 1875, p. 12.
† Arrang. Fam. Mamm. 1872, 66.

The divisions of *Mephitinæ* are expressed in the following diagnoses:—

A. Teeth 34; pm. $\frac{3-3}{3-3}$. Dorsal outline of skull not in one continuous curve. End of muzzle truncate vertically, or with little obliquity. Palate ending opposite last molar (more or less exactly). (Periotic region varying with the subgenera.) Coronoid process of jaw conical, erect, its fore and hind borders converging to a vertical apex in advance of condyle. Angle of mandible not exflected. Snout not notably produced nor depressed. Nostrils lateral. Tail very long and very bushy. Soles comparatively narrow, hairy at least in part. North, Middle, but probably not South American. *Genus* MEPHITIS.*

  a. Skull not depressed, the dorsal outline irregularly convex, highest over the orbits. Zygomata moderately arched upward, highest behind. Postorbital processes usually obsolete. Mastoid processes flaring strongly outward, much beyond orifice of meatus. Periotic region not particularly inflated. Size large. Colors massed in large areas . . . . *Subg. Mephitis.*

  b. Skull depressed, the dorsal outline approaching straightness, particularly over the orbits. Zygomata strongly arched upward, highest in the middle. Postorbital processes well developed. Mastoid processes slight, scarcely produced beyond orifice of meatus. Periotic region peculiarly inflated by development of mastoid sinuses, the under surface swollen, and giving a *quasi* appearance of a second bulla auditoria behind the real one . . . . *Subg. Spilogale.*†

B. Teeth normally 32; pm. $\frac{2-2}{3-3}$, sometimes, however, $\frac{3-3}{3-3}$, from presence of an additional minute premolar,‡ corresponding to the anterior one of *Mephitis*. Dorsal outline of skull one continuous curve, more or less regular, from occipital protuberance to ends of premaxillaries, owing to the great obliquity of truncation of the end of the rostrum, which brings the profile of nasal orifice into line with that of the forehead; skull highest in parietal region. Palate produced decidedly past the last molars, yet not half-way to ends of pterygoids. Periotic region much as in *Mephitis* proper, but the mastoids rather as in *Spilogale*, projecting more downward than outward. Postorbital processes usually obsolete. Zygomata slightly arched upward. Coronoid process of jaw sloping backward, obtusely falcate, with convex anterior and concave posterior margin, the apex nearly overtopping condyle. Angle of the mandible strongly exflected. Of large size, extremely stout form, and somewhat Badger-like appearance. Snout strongly produced, depressed. Nostrils inferior. Tail short and little bushy (for this subfamily). Soles very broad, entirely naked. Coloration massed in large areas. South, Middle, and (scarcely) North American . . . . *Genus* CONEPATUS.§

---

* *Etym.*—Lat. *mephitis*, a foul or noxious exhalation.

† *Etym.*—Greek σπίλος, a spot; γαλη, a kind of Weasel.

‡ The anterior lower premolar is said to be sometimes wanting.

§ A barbarous word, like many other of J. E. Gray's genera, derived from Conepatl or Conepate, the name of the animal in the vernacular (probably Mexican) of countries it inhabits.

Note on fossil North American species of *Mephitis*.

Mephitis frontata, *Coues.*
  Mephitis frontata, *Coues,* Bull. U. S. Geol. and Geog. Surv. Terr. 2d ser. no. 1, 1875, 7, with woodcut.

From the bone-caves of Pennsylvania. Post-pliocene.

SPECIFIC CHARACTERS.—Skull extremely high in the middle; the profile of the upper outline very rapidly descending in a nearly straight line from this point to the occiput and muzzle. Greatest depth of skull without jaw little less than half its length. Zygoma highly arched; the bone in front compressed vertically instead of laterally.

This species is founded on a skull, No. 2232 of the Smithsonian Museum, obtained by Prof. Baird in the bone-caves of Pennsylvania. The animal was a true *Mephitis*, closely related to *M. mephitica*, if really different. Though the frontal region is always tumid in *Mephitis*, there is seen in the recent species nothing like the protuberance and angulation of the vertex of *M. frontata*. The prominence is also decidedly more posterior; it is something over and above the general tumidity of the interorbital region of recent *Mephitis*; the shape is rather as in *Gulo*, but even the profile of the latter is here exaggerated. The prominence appears to be mainly due to enlargement of the frontal sinuses, as may be seen in this specimen, in which the outer tablet of the skull is abraded in places, exposing the interior. With this general elevation is associated a notably higher arch of the zygoma, and the malar is slenderer than in recent species at its anterior portion, where it is curiously narrowed vertically instead of being laminar throughout. None of these characters obtain in any of the numerous recent skulls examined, notwithstanding the great variability of the latter. The animal was of the size of the common species. The skull in general bulk is intermediate between various specimens of that of *M. mephitica*.

Mr. J. A. Allen[*] takes exception to the specific validity of the species in the following terms :—

" . . . . . Dr. Coues has ventured to describe a 'new species' (*M. frontata*), based on a fossil skull from one of the bone-caves of Pennsylvania, as it seems to me, unadvisedly. The specimen, though that of a very aged individual, is scarcely larger [. . . .] than the average of specimens from the Eastern States, its chief difference from the average skull con-

---

[*] Bull. U. S. Geol. and Geog. Surv. Terr. vol. ii, no. 4, 1876, p. 333.

sisting in an abnormal tumidity of the frontal region, arising evidently from disease. It is a feature by no means confined to the present example, but is merely an extreme enlargement of the sinuses of the frontal region often seen in specimens of the existing animal, evidently resulting from disease. In No. 917 (Albany, N. Y.), No. 8099 (Fort Cobb, Ind. T.), No. 1878 (Calcasieu Pass, La.), and No. 1620 (Indianola, Tex.), the same tendency is strongly marked, which, in some of these specimens, had they attained equal age, must have resulted in a malformation nearly or quite as great as is seen in the fossil skull in question.

"In this connection, I may add that a pretty careful examination of the fossil remains of *Carnivora*, collected by Professor Baird many years since from the bone-caves of Pennsylvania (of which this fossil skull of the Skunk forms a part), has failed to show any of them to be specifically different from the species now or recently living in the same region. Many of them are remains of individuals of large size, but not exceeding the dimensions of the specimens of the recent animal from the same or contiguous regions. These remains include, among others, the following species:—*Lynx rufus, Urocyon virginianus, Mustela pennanti, Mustela americana, Putorius vison, Lutra canadensis, Mephitis mephitica* (other specimens than the '*frontata*' skull), *Procyon lotor, Ursus americanus*, etc."

Granting that the probabilities are against the validity of the species, it may be observed that the disease theory is not proven, and that no recent specimens of *Mephitis* have been found to match this one.

This species, so far as I am aware, is the only fossil Skunk described as such; but compare *anteà*, p. 18, on the question of "*Galera*" *perdicida*.

The Genus MEPHITIS. (CUVIER.)

Viverra *sp.*, of some early authors.
< Mephitis, *Curier*, "Leçons d'Anat. i. 1800" (coextensive with the subfamily), and of authors generally.—*Baird*, M. N. A. 1857, 191.
< Chincha, *Less.* Nouv. Tab. R. An. 1842.
> Spilogale, *Gray*, Proc. Zoöl. Soc. 1865, 150. (Type, *S. interrupta* = *M. putorius*.)
> Mephitis, *Gill*, Arrang. Fam. Mamm. 1872, 66 (restricted to subg. *Mephitis* as characterized in this paper).—*Coues*, Bull. U. S. Geol. Surv. 2d ser. i. 1875 (same restriction).

For characters, see a preceding page (p. 192).

The several North American species of *Mephitis* proper (as restricted to exclude *Spilogale*) indicated by authors are reducible to one, possibly divisible into two or three geographical races. There is a second Mexican species, apparently valid, which will be brought into the present connection to complete a review of the genus. A fossil species is also described in the foregoing pages. *Mephitis* proper and *Spilogale* are both confined, as far as known, to North and Middle America, *Conepatus* being the only South American type of *Mephitinæ*, but also extending through Middle America to the Mexican border of the United States.

The Common Skunk.

Mephitis mephitica.

PLATES X, XI.

(*a. mephitica.*)

Viverra mephitica, *Shaw*, Mus. Lever. 1792, 173, no. 4, pl. 6; Gen. Zoöl. i. 1800, 390, pl. 94, middle fig.
Mephitis mephitica, *Bd. M. N. A.* 1857, 195.—*Coop. & Suckl.* N. H. W. T. 1860, 94.—*Hayd.* Trans. Am. Philos. Soc. xii. 1862, 143.—*Samuels*, Ninth Ann. Rep. Mass. Agric. for 1861, 1862, 161.—*Gerr.* Cat. Bones Br. Mus. 1862, 97.—*Allen*, Bull. M.C. Z. i. 1869, 178; ii. 1871, 169 (critical).—*Allen*, Pr. Bost. Soc. xiii. 1869, 183.—*Gilpin*, Proc. and Tr. N. Scotia Inst. ii. 1870, 60.—*Stev.* U. S. Geol. Surv. Terr. for 1870, 1871, 461.—*Parker*, Am. Nat. v. 1871, 246 (anat. of anal glands, &c.).—*Allen*, Bull. Ess. Inst. vi. 1874, 46, 54, 59, 63.—*Allen*, Proc. Bost. Soc. xvii. 1874, p. 38.—*Ames*, Bull. Minn. Acad. Nat. Sci. 1874, 69.—*Coues*, Bull. U. S. Geol. and Geogr. Surv. Terr. 2d ser. no. 1, 1875, 8 (skull and teeth).—*Coues & Yarrow*, Zoöl. Expl. W. 100 Merid. v. 1875, 62.—*Allen*, Bull. U. S. Geol. Surv. vol. ii. no. 4, 1876, 332 (skull).
Mephitis chinga, *Tied.* Zool. i. 1808, 362 (partly).—*Licht.* Darstell. Säug. 1827-34, pl. 45, f. 1 ; Abh. Akad. Wiss. Berl. for 1836, 1838, 280.—*Maxim.* Reise N. A. i. 1839, 250; Arch. f. Naturg. 1861,—; Verz. N. A. Säug. 1862, 42.—*Wagn.* Suppl. Schreb. ii. 1841, 198.—*Schinz*, Syn. i. 1844, 323, no. 13.—*Aud. & Bach.* Q. N. A. i. 1849, 317, pl. 42.—*Giebel*, Säug. 1855, 766.—*Fitzinger*, Naturg. Säug. i. 1861, 315, f. 63.
Mephitis americana *var.* K, *Desm.* Mamm. i. 1820, 186 ("*Mustela*", *lapsu.* Includes all the American Skunks, vars. A—R); Nouv. Dict. xxi. 515 (var. 7).—*J. Sab.* App. Frankl. Journ. 1823, 653.—*Harl.* Fn. Am. 1825, 70.—*Griff.* An. Kingd. v. 1827, 127, no. 358 (partly). *Less.* Man. 1827, 151, no. 406.—*Godm.* Am. Nat. Hist. i. 1831, 213, pl.-f. 1.—*Doughty's* Cab. N. H. ii. 1832, 193, pl. 17.—*Rich.* Zoöl. Beechey's Voy. 1839, 4.—*Emmons*, Rep. Quad. Mass. 1840, 49.—*De Kay*, N. Y. Zoöl. i. 1842, 29, pl. 12, f. 1.—*Wyman*, Pr. Bost. Soc. 1844, 110 (anat.).—*Warren*, Pr. Bost. Soc. iii. 1849, 175 (anat.).—*Thomps.* N. H. Vermont, 1853, 33.—*Woodh.* Sitgr. Rep. 1853, 44.—*Kenn.* Tr. Illinois Agric. Soc. for 1853-4, 1855, 578.—*Beesley*, Geol. Cape May, 1857, 137.—*Billings*, Canad. Nat. and Geol. i. 1857, 360.—*Hall*, Canad. Nat. and Geol. vi. 1861, 296.
Mephitis americana *var.* hudsonica, *Rich.* F. B.-A. i. 1829, 55, no. 19.
Chincha americana, *Less.* Nouv. Tabl. R. A. 1842, 67.
Mephitis chinche, *Fisch.* Syn. 1829, 160 (includes other species; quotes Tiedemann primarily).
Mephitis varians *var.* chinga, *Gray*, P. Z. S. 1865, 148; Cat. Carn. Br. Mus. 1869, —.
Chinche, *Shaw, l. c.—Geoff. & Cuv.* "Hist. Mamm. ii. 1819, —, pl. —(Louisiana)."
Mephitic Weesel, *Shaw*, Mus. Lever.
Ouinesque, *Sag.-Théod.* Hist. Canad. 1636, 748 (ed. of 1866, iii. 680).
Enfan du Diable, *Charlev.* N. France, v, 1744, 196.
Polecat, *Kalm*, Voy. ——, 452.
Skunk, *Forst.* Phil. Trans. lxii. 1772, 374.—*Penn.* Arct. Zoöl. i. 1784, 85, no. 33.—*Hearne*, Journ. ——, 377.
Chinga, *Schinz, l. c.*

The hairs of the tail which are entirely white (all are usually white basally) are somewhat different in texture from the rest, being even coarser and looser. They appear at the end of the tail in a white tuft that seems to have little connection with the general pelage, and may be early deciduous; or, more curiously, they grow irregularly in various places along the tail, in somewhat isolated fascicles. These singular little bundles are also likely to exceed the rest in length, measuring sometimes seven or eight inches in length. Even without taking these into consideration, the bushiness of the tail is sometimes so great that the width when the hairs are extended sideways rather exceeds the total length. The strictly terminal hairs of the tail are ordinarily not so long as some of those along the sides.

Notwithstanding the endless diversity in the extent and details of the white marking, a certain pattern may be indicated as one of reasonable constancy. This is essentially a sharp, narrow, frontal stripe, and a broad nuchal area, from which last proceed obliquely backward a pair of stripes toward or to the tail, continued or not upon this member, and whiteness, to a greater or less extent, of nearly all the hairs of the tail at base, even when this member is blackest and least bushy. I have not found the frontal stripe either wholly wanting (*Conepatus*) or enlarged into a spot (*Spilogale*); but it varies from a mere trace to a long streak continuous with the nuchal area, and, doubtless, sometimes fails altogether. This last is usually a large spot, beginning squarely and broadly on the occiput in a line between the ears. From the back of it, the two oblique stripes may immediately diverge, forming a V, or it may continue for a considerable distance as a single median stripe before forking into two. The nuchal spot may be again entirely disconnected with the dorsal stripes (rare), or may be broken up into a pair of spots; *i. e.*, the dorsal stripe extended separately on to the nape. The dorsal stripes may extend scarcely any distance beyond the nape; *i. e.*, may be represented by only a slight prolongation of a pair of nuchal spots. They may start over the shoulders independently of the white nuchal area. Ordinarily, they reach, widely divergent, more than halfway along the back; again, they are more nearly parallel, and reach to the tail. They may curve toward each other over the flanks, and even meet there, then completely enclosing an oval vertebral area; or may be interrupted to resume again. They may extend along either side of the tail, in such cases ordinarily being broken into the curious isolated fascicles of white hairs already described, but being sometimes continuous, when the tail is mostly white. In the blackest tails seen, there is always more or less white on the bases of the hairs.

The foregoing may indicate the general range of variation in color. Reference to Audubon's figures of this species and his supposed "*macrura*" will give a fair idea of two conditions very nearly extreme. I have never seen an entirely black Skunk, but in some specimens before me the white is reduced to such mere traces that I have no doubt it may occasionally disappear, as is stated by some. One young specimen has the entire upper half of the body pure white, as in the strongest cases of *Conepatus*, except a slight emargination from behind, just at the root of the tail. Fully aware, as I am, of the endless variability, even in individuals belonging to the same litter, I am satisfied that there is nevertheless a tendency, generally well expressed, to increase of white, in a measure according to certain geographical areas. An average in this respect is the rule in the Eastern and Middle States, where we have a fair frontal stripe and nuchal area sending out obliquely stripes which do not reach the tail, this being black, only white at the end or among the roots of the hairs. In Florida and the South Atlantic and Gulf States generally, the white is at a minimum, frontal stripe a mere trace, nuchal spot small or broken in two, and the stripes almost wanting. Throughout the West, and in British America even as far east as Hudson's Bay, prolongation of the lateral stripes to the tail, or on this member to its end, is the rule; and the stripes do not usually at once diverge from the nuchal spot, but more gradually separate from a single vertebral stripe, into which the nuchal spot is prolonged. Associated with such a condition of the white, we find, almost invariably, in the western forms, a much bushier tail, its width across equalling or even exceeding its total length. Such cases as these, in their minor diversities, have furnished the *mesomelas* of Lichtenstein, *varians* of Gray, *occidentalis* of Baird, and "*macroura*" of Audubon. The figure of the last named represents an extreme of white, with length and bushiness of tail, and might readily be mistaken, as it was, for the altogether different *M. macrura* of Lichtenstein.

Independently of the size of the tail, we may observe a general decrease in stature with latitude. Floridan specimens are notably smaller than those from New England, some, apparently full-grown, being little larger than *Spilogale* at its maximum, about thirteen or fourteen inches long.

Moufette d'Amérique, *Less.* &c. &c.
Fiskatta, *Swedish.*
Bête puante, *French.*
Stinkthier, *German.*

(b. *mesomelas.*)

Mephitis mesomelas, *Licht.* Darst. Säug. 1827-34, pl. 55, f. 2; Abh. Ak. Wiss. Berl. for 1836, 1836, 277.—*Maxim.* Reise, i. 1839, 240; Arch. Naturg. xxvii. 1861, 218; Verz. N. A. Säug. 1862, 36.—*Schinz,* Syn. i. 1844, 322, no. 11.—*St. Hil.* Zool. Voy. Vénus, i. 1855, 133, pl. —.—*Bd.* M. N. A. 1857, 199 (after Licht.).
Mephitis mesomeles, *Gerr.* Cat. Bones Br. Mus. 1862, 97.
Mephitis occidentalis, *Bd.* M. N. A. 1857, 194.—*Newb.* P. R. R. Rep. vi. 1857, 44.—*Coop. & Suck.* N. H. W. T. 1860, 116.
Mephitis mephitica *var.* **occidentalis,** *Merriam,* U. S. Geol. Surv. Terr. for 1872, 1873, 662.
Mephitis varians *var. a, Gray,* P. Z. S. 1865, 148; Cat. Carn. Br. Mus. 1869, —.

(c. *varians.*)

Mephitis varians, *Gray,* Mag. N. H. i. 837, 581.—*Gray,* List Mamm. Br. Mus. 1843, 68.—*Bd.* M. N. A. 1857, 193; Mex. B. Surv. ii. pt. ii, 1859, Mamm. 19.—*Gerr.* Cat. Bones Br. Mus. 1862, 97.—*Gray,* P. Z. S. 1865, 148; Cat. Carn. Br. Mus. 1869, —.
Mephitis macroura, *Aud. & Bach.* Q. N. A. iii. 1853, 11, pl. 102.—*Woodh.* Sitgr. Rep. 1853, 44. (But not of Lichtenstein.)

HAB.—Entire temperate North America. North to Hudson's Bay and Great Slave Lake. South into Mexico (Matamoras, Monterey).

SPECIFIC CHARACTERS.—Black or blackish; a frontal streak, nuchal spot, and two dorsal stripes white; tail black, more or less mixed with white or white-tipped. Tail with hairs not as long as head and body; the vertebræ about half this dimension. Length from nose to root of tail over one foot; soles about 2¼ inches.

Description of external characters.[*]

The Skunk is a stoutly built animal, with a small head, low ears, and short limbs, the trunk thick-set and especially large behind, the back naturally arched as well as broad; tail long and very bushy. The head is pointedly conoidal, with a convex frontal profile and sloping occiput; there is little of the breadth and depression characteristic of the Weasels, the regular conoid being nearly expressed. The eye is small, and nearer the nose than ear. The nasal pad is of considerable size, and protuberant, definitely naked for a closely circumscribed area, the outline nearly circular; the face of the muffle is bevelled a little obliquely downward and backward; the nostrils are chiefly lateral, but their anterior extremity is visible from the front. The ears are low, though the pinna is decidedly better developed than in *Conepatus;* the general set of the conch is rather backward than upward, as its anterior extremity is inserted little below the highest point of the brim; the contour

---

[*] From a large series of specimens in the Smithsonian Institution from various portions of North America.

of the free edge is nearly orbicular, with, however, a slight obtuse angulation. The feet are not so broad and flat as in *Conepatus*, yet they show large plantar and palmar surfaces. These are usually naked, except for a varying distance behind; the soles, in particular, being generally hairy for about a third way from the heel. The palms present behind, just in advance of the wrist, a padded prominence, more or less completely divided lengthwise; in advance of this is a crosswise depression; at the bases of the digits is a crescentic padded area, divided more or less evidently in different specimens into three or four smaller pads. This division is sometimes very evident, the lines of impression being deep and sharp; in other cases, little more than a general horseshoe-shaped padded area is recognizable. There is no constancy about this; and the difference which has been claimed between *Mephitis* and *Spilogale* cannot be satisfactorily substantiated. The digits are short—in fact, they are exceeded in length by the longer ones of the claws they bear. Of these, the third and fourth are subequal and longest, the second is little shorter, the fifth reaches hardly half-way along the fourth, and the first scarcely attains the base of the second. The middle three claws are very long, strong, compressed, little curved, acute and fossorial in character; the lateral ones are shorter, stouter for their length, and more curved. The claws of the hind feet are quite different, being all short, stout, and obtuse, and covered with hairs; the middle three are approximatetly equal in length, the fifth is much shorter, and the first falls short of the base of the second. The naked part of the sole presents a general broad flat area behind, succeeded by an irregular depression, and this by the padding at the bases of the toes, which is imperfectly divided into three. The terminal balls of the toes almost immediately succeed, these digits being very short and extensively connected together.

The tail of the Skunk is remarkably bushy, with long harsh coarse hairs, almost like a kind of tow. The hairs are loose and flaccid, their "set" depending in a great measure upon the movements or position of the member. In the bushiest-tailed examples, the hairs fall loosely all around when the tail is elevated, like the plumes of a pompon, as well represented in Audubon's plate of his so-called "*macrura*" (*nec Licht.*). In other cases, the set of the hairs is more stable. No distichous arrangement is recognizable. There is no fine under fur on the tail.

Description of the skull and teeth. (See Plates X, XI.)

The cranium of no animal with which I am acquainted varies more than that of the Skunk, and few exhibit such remarkable differences, independently of age and sex. Some specimens are more than a fourth larger than others, and twice as heavy; and there is a corresponding range of variation in contour. Compared with an ordinary ratio of osteological variability, the discrepancies are almost on a par with those exhibited by the coloration of the animal when set over against the more constant markings of most animals. In the series of twenty or thirty skulls examined, I find that the western ones, and especially those from the Pacific coast, representing *occidentalis* of Baird, are, as a rule, larger and heavier than others, more widened and flattened behind, with stronger and more flaring sagittal and especially occipital crests. But these extremes shade insensibly into an ordinary pattern, and I can draw no dividing line. Tables of measurements would show these variations, though they would scarcely render that realizing sense of the discrepancies that is gained by laying the two extremes side by side. An average cranium, No. 3816, from New York, is selected for description, in the course of which the variations of the whole series will be brought under review.

The greatest zygomatic width is to the length as 1 to 1.55, or slightly less than two-thirds such length. A similar proportion is generally preserved. Viewed from above, the cranium presents a short, tumid, rostral portion, high at the nose, tapering on either side, but with a protuberance indicating the course of the canine tooth in the bone, subtruncate in front, with large subcircular nasal aperture, in this view much foreshortened. The rostrum is about a third of the whole length, if measured from extreme front to anterior root of zygoma; the zygoma, and then the rest of the skull, being respectively another third. In other skulls, the rostrum is shorter than this, and less vaulted. The general convexity of the rostrum continues on to the forehead in the broad, smooth, interorbital space. Supraorbital processes are very slight, being only indicated in a little bulging at the front, where the anterior forks of the sagittal crest come to the brim of the orbit. There is thus scarcely any definition of the orbit from the general temporal fossa. The point of greatest constriction of the skull is considerably behind the supraorbital process, just about half-way from end of rostrum to occiput, and opposite the apex of the mandibular coronoid, when the jaw is closed. It is a gradual pinching-together of the sides of the cranium for some distance, rather than an abrupt constriction at a particular point. It is sometimes unsymmetrical, one side being more emarginate than the other; is sometimes scarcely narrower than the interorbital space, sometimes about three-fourths as much. Back of this point, the skull widens rapidly to the hinder root of the zygoma and mastoid; the latter being the broadest point of the skull proper, separated from the former by an emargination, in which lies the opening of the meatus auditorius, not visible from above. From each mastoid, the skull narrows in an approximately straight line backward and upward for a distance, and then ends with a straight-across contour, more or less emarginate on the median line. This whole posterior boundary, representing the lambdoidal crest, is extremely variable, not only according to age, but fortuitously. In some skulls—those with the broadest back part and most flaring occipital crest—there is a deep emargination in the middle line of the skull, boldly salient angles on either side of this, and a concave outline thence to the mastoid. This occipital flange hides all the parts beneath it. For the rest, the top of the skull shows a sagittal crest (only in very young skulls a raised tablet), well marked in all but young examples, forking anteriorly (at or a little in advance of the point of greatest constriction) to send a curved leg outward to either supraorbital process. Aside from this crest and the occipital one, the general cranial surface is vaulted. The zygomatic arches, viewed above, show the point of widest divergence near their posterior roots, whence they gradually and regularly converge forward with slight curve.

Viewed in profile, the skull shows its highest point at the interorbital space, whence it slopes gradually with a general slight convexity to the muzzle and occipital protuberance. This highest point is generally a little, sometimes decidedly, in advance of the middle of the skull. The frontal profile may acquire a slight concavity, and the opposite one may be slightly sinuous, owing to irregularity of the sagittal crest. The muzzle is cut squarely off, with an obliquity of perhaps 30 degrees from the perpendicular. The zygoma shows but a slight upward arch, and no bevelling or special curve to define the portion of the orbit which it represents. It is laminar, narrowing midway, stoutest near posterior root. The anteorbital foramen* is a short perforation of a thin upper plate of its anterior root; behind, the glenoid fossa presents rather forward than downward. The prominent orifice of the meatus presents laterally between the root of the zygoma and the mastoid, which latter is a protuberant but blunt process immediately behind the meatus. Behind this, there is an emargination, terminated by the prominent downward-projecting paroccipital; back of this, the semicircular outline, foreshortened, of the occipital condyle appears.

The back of the skull is a subtriangular face, flat and perpendicular in general superficies, bounded above by the overhanging sagittal crest; either lateral corner being the prominent paroccipital, between which appear the faces of the oblique condyles, the upper border of the foramen being transverse with a slight curve.

The skull from below shows a broad, flat, palatal surface for about two-fifths of its total length. The palate ends about opposite, or a little back of, the posterior molars. This terminal shelf, representing the emargination between the pterygoids, is always broad and quite transverse; but the edge varies greatly in detail. It is commonly transverse, with a small median, backwardly-projecting point, producing a double emargination. It may be simply a broad curve, or it may present a median nick. The latter case is oftenest observed in specimens from the West, and constituted a chief character upon which *M. occidentalis* rested; but, with a larger series than Prof. Baird examined, it is shown to be wholly fortuitous. The general shape of the palate is triangular; including the teeth, its greatest width behind is about as much as its length; anteriorly, it presents broad but short incisive foramina, scarcely reaching opposite the molars. The depth of the pterygoid emargination is considerably less than the length of the palate. The pterygoids are simply laminar, with strongly hamulate ends. They are usually parallel, but sometimes converge a little posteriorly, making the inclosed space club-shaped. The general surface of the base of the skull behind is quite flat, owing to slight

---

* As a curious but not very infrequent anomaly, this foramen is sometimes divided into several separate canals, through which branches of the facial nerve pass out apart from each other. I have observed the same thing in *Conepatus*.

inflation of the bullæ. These are decidedly convex only at one place, interiorly, elsewhere flat, and outwardly produced to form a tubular meatus. Traces of separation from surrounding parts long persist, at least in front. About the bullæ are seen the following foramina: one in advance, just inside the glenoid fossa; two at the anterior extremity of the bulla; three along its inner border; one more exterior, near the mastoid; one far posterior, in the occipital. The basi-sphenoid suture, early obliterated, is straightly transverse in advance of the middle of the bullæ. The general basilar area is flat, narrowing forward, unmarked, or with merely a slight median ridge. The border of the foramen magnum represents a deep emargination of the posterior border of this area, with the condylar protuberance on either side.

All the bones of the skull finally coössify, excepting, of course, the mandible, and most are joined at a comparatively early age. The periotic and internasal sutures persist the longest; the latter after the nasals are consolidated with the maxillaries, and the former after the basi-spheno-occipital suture is obliterated. When found separate, the nasals are seen to be regularly concave along their exterior border, truncate anteriorly, with a produced antero-lateral corner, and received by a pointed process in a recess of the frontal. The intermaxillary bone forms less than half of the general naso-maxillary suture. The maxillary extends within a short distance of the supraorbital protuberance. The malar is rather small, and fuses early with the rest of the zygomatic arch. The occipital bone is rather late to coössify; the supraoccipital is then seen to represent most of the lambdoidal crest, reaching, on either hand, from the median line half-way to the mastoid process; thence crossing this crest to the paroccipital, whence the suture runs on the floor of the skull along the border of the periotic by the foramen lacerum posterius to the basi-sphenoid; thence straight across the median line.

The lower jaw in *Mephitinæ* is never locked, as far as known, in the glenoid by the clasping of the condyle in the embrace of the fossa, as is the rule, in adult life, in *Meles* and in *Taxidea*, and as sometimes occurs in the Otters (*Lutrinæ*). The ramus of the mandible is stout and nearly straight along the tooth-bearing portion; the symphysis is thick, short, abruptly ascending obliquely forward. Between the ramus proper and the angle of the jaw, the lower border is decidedly emarginate, and the angle itself is scarcely or not at all exflected (cf. *Conepatus*). The angle itself is obtuse, and there is a decided neck in the outline thence to the condyle. The condyle is horizontal, transverse, very narrow, and acute internally; on the outer half, its articular surface looks upward; on the inner half, backward. The coronoid process rises straight and high, nearly uniformly tapering to the apex, a perpendicular from which falls decidedly in advance of the condyle (cf. *Conepatus*). The general muscular impression on its outer face is well marked. It is pointed below, and reaches forward on the ramus to a point underneath the last lower molar (cf. *Conepatus*).

As remarked under the head of *Conepatus*, the dental formula of the genera of *Mephitinæ* does not, in point of fact, differ. The difference is *nil* as between *Mephitis* and *Spilogale*, while in *Conepatus* a supposed lesser number of teeth is only true in the very small size of the abortive, deciduous, or, at any rate, not functionally developed anterior upper premolar. In *Mephitis*, also, the tooth may be very small, or even abortive, on one or both sides of the jaw; it is, however, normally present and readily recognizable.

Selecting an average skull, of middle age, with fully developed, yet little-worn, dentition (for in very old skulls the teeth are so ground down as not to furnish fair characters), we observe the following points:—

The back upper molar is the largest of the grinders, about as long as broad, quadrate, with rounded inner corners, and entirely tuberculous. It is completely divided across lengthwise by a sulcus, on the outer side of which is a narrow portion, much higher than the broad inner portion, and separated from it not only by the groove across the face of the tooth, but by a nick in the hinder border. This elevated outer moiety is oblique on its face from the general level of the dentition; it runs to a point at its fore and hind ends, and has a central, slightly excavated field, with irregular-raised boundary. The flatter inner moiety of the tooth is chiefly occupied by a large antero-internal tubercle, separated by a curved sulcus from a posterior raised margin. The next tooth—back premolar—differs altogether from the same flesh-tooth in the *Mustelinæ*. It is relatively smaller, and has not a prominent isolated antero-internal fang. On the contrary, it is triangular in general outline, the inner corner of the triangle representing the fang of the *Mustelinæ* just named; this is cuspidate, but this whole inner moiety is low and "tuberculous" in comparison with the elevated and truly sectorial character of the rest of the tooth; for, viewed in profile from the outside, the tooth seems wholly sectorial, with two cusps, an anterior, produced, acute one, and a posterior, shorter and obtuse, separated from the other by an acute re-entrance. Taken together, these two external cusps make the trenchant edge of the tooth. The next premolar is immediately and very markedly reduced in size; it is a small, simple, two-rooted, conical, acute cusp, with a slight posterior "heel" and well-marked cingulum on the inner side. The next—anterior—premolar is exactly like the foregoing, but very much smaller still, and single-rooted; it sometimes aborts. In very old skulls, the foregoing descriptions can hardly be verified. The back molar wears down to a perfectly smooth face, with raised inner and outer borders; the flesh-tooth loses its edge and inner cusp, and becomes almost tuberculous throughout; the other premolars become mere stumps. The canines offer no points for remark. Of the upper incisors, the lateral pair is much larger than the rest, though not longer. I fail to appreciate any tangible difference in this respect between *Conepatus* and *Mephitis*. The tips of the teeth all fall in the same line; they are even and regular; the ends are obscurely lobate. These teeth start from the sockets quite obliquely, but soon turn perpendicularly downward, with an appreciable elbow.

In the lower jaw, the back molar, as usual, is small, simple, circular, single-rooted, with a central depression and irregularly raised margin. The next molar is much the largest of the series, and very notably different from the same tooth in *Mustelinæ*. It is fairly sectorial throughout; for the back portion, though lower than the rest, is decidedly of the same character as the other part. This tooth consists of five cusps: a posterior pair, side by side, inner and outer, of equal size and similar shape; a middle pair, side by side, the outer of which is larger and sharper than the inner; and a single anterior cusp. The latter forms, with the exterior middle cusp, the main trenchant edge of the tooth. The interior middle cusp is a higher development of the "heel", more or less prominent on the inner face of the main cusp of the Musteline tooth. The posterior pair of cusps is the low tuberculous part of the tooth in *Mustelinæ*. The first premolar from behind is a simple conical cusp, two-rooted, with evident heels, both before and behind, and a well-marked cingulum. The next tooth is similar, but smaller, with less of a girdle, and scarcely an anterior heel. The anterior premolar is like the last, but smaller still, and single-rooted. I have not seen its abortion. In very old skulls, the two molars become ground almost perfectly flat, and the premolars become stubby cones. The lower canines are shorter, relatively stouter, and more curved than the upper ones; there is usually quite an elbow at the point of greatest curve. The inferior incisors are more nearly of a size than is usual in *Mustelinæ*, and more regular, *i. e.*, none are crowded out of the general plane; but this is a matter of degree only. The outer pair is larger than the rest; viewed from the front, they widen from base to tip, and the apex is emarginate. The next pair sets a little back from the general plane; for, though their faces are generally quite flush with the others, yet their greater thickness causes them to protrude behind. All the under incisors are approximately of one length. The cutting edge of the outer pair is oblique; of the others, horizontal. The cutting edge of the outer pair is nicked, as already said, and the front faces of the rest are marked by a sulcus ending in a slight bilobation of their cutting edges.

Variation in the skull with special reference to geographical distribution.

Having already called attention to this matter in a general way, I cannot do better than continue the subject with Mr. J. A. Allen's tables of measurements and critical comment, which set forth the subject in more precise detail:[*]—

"The twenty-nine skulls of this species of which measurements are given below show a wide range of variation in size, and a decided decrease southward. The localities embrace such distant points as California and the Atlantic seaboard on the one hand, and Maine and Texas on the other; but, with one or two exceptions, the specimens from any single locality are unsatisfactorily few. The specimens range in length from 2.60 to 3.50, and in width from 1.60 to 2.25! Yet there is not a specimen included in the series that is not so old as to have all the cranial sutures obliterated. A portion of the difference is doubtless sexual, but the specimens, unfortunately, have not the sex indicated. Ten of the specimens may be considered as western, coming mainly from Utah and California; ten others are from

---

[*] Bull. U. S. Geol. and Geog. Surv. Terr. vol. ii. no. 4, 1876, pp. 332–334.

Maine and Massachusetts, and one from Northeastern New York; three are from Pennsylvania; and of the remaining five, four are from Texas, and one from Louisiana. The western series of ten average 3.10 in length and 1.95 in width, ranging in length from 2.85 to 3.50 and in width from 1.70 to 2.25. The New England series of ten average 2.88 in length and 1.72 in width, ranging in length from 2.70 to 3.25 and in width from 1.53 to 1.85. The single New York specimen scarcely varies from the average of the New England series, while the Pennsylvania specimens fall a little below. The five southern specimens average 2.73 in length, or a little below the New England series, ranging in length from 2.60 to 2.90.*

"It thus appears that the western specimens are decidedly the largest of all, and that the northern are somewhat larger than the southern, the specimens compared being of corresponding ages, though of unknown sex, but doubtless comparable in this respect also.

"The difference in size amounts to above one-fourth the size of the largest specimen and above one-third the size of the smallest. Between the western and southern series, the average difference amounts to one-third of the average size of the larger series! The western series includes the so-called *Mephitis occidentalis* of Baird, based on California specimens, and whose chief difference is merely that of larger size; yet the four specimens from Ogden, Utah (Coll. Mus. Comp. Zoöl.), considerably excelled in size the three from California. The southern series represents the so-called *M. varians* of Gray and Baird.

"The unsatisfactory character of the several species of North American Skunks of the *mephitica* group, and the wide range of color-variation among individuals from the same locality, I have previously had occasion to notice,† and a re-examination of the subject confirms the conclusions then announced, which, I am happy to find, have recently received the support of Dr. Coues, who has lately made a study of this group.‡ As Dr. Coues has remarked, and as the subjoined measurements show, few species of animals vary so much in size and in cranial characters as the present, independently even of sex and age. Some

---

* "The range in width is not fairly indicated, owing to two of the smaller specimens being imperfect."

† "See Bull. Mus. Comp. Zoöl. vol. i. pp. 178–181, Oct. 1869."

‡ "Bull. U. S. Geol. and Geog. Surv. of the Territories, vol. i. no. 1, pp. 7–15, 1875."

specimens are not only more than one-fourth larger than others, but 'there is a corresponding range of variation in contour. Compared with an ordinary ratio of osteological variability,' says Dr. Coues, 'the discrepancies are almost on a par with those exhibited by the coloration of the animal when set over against the more constant markings of most animals.'

"*Measurements of twenty-nine skulls of* MEPHITIS MEPHITICA."

Catalogue number.	Locality.	Sex.	Length.	Width.	Remarks.
2617	Petaluma, Cal	......	3.30	2.07	
3271	......do............	......	3.08	2.04	
2434	Port Townsend, Oreg	......	2.93	1.70	
4195	Fort Crook, Cal	......	2.85	......	
417	Ogden, Utah	......	3.12	1.87	
419	......do............	......	3.50	2.25	Very old.
416	......do............	......	3.10	1.90	
418	......do............	......	2.98	1.85	
10008	Wyoming Territory	......	3.15	2.05	
3327	Fort Laramie	......	2.96	1.78	
575	Upton, Me	......	3.25	......	Very old.
580	......do............	......	3.00	1.85	
577	......do............	......	2.87	1.75	
574	......do............	......	2.85	1.73	
583	Norway, Me	......	2.90	1.75	
578	......do............	......	2.70	1.70	
569	......do............	......	2.87	1.78	
567	Massachusetts	......	2.70	1.53	
568	......do............	......	2.75	......	
576	......do............	......	2.72	1.70	
3816	Essex County, New York	......	2.88	1.78	
2232	Bone-caves, Pennsylvania	......	2.90	......	Fossil; *M. frontata* Coues.
610	Carlisle, Pa	......	2.87	......	Imperfect.
4833	Chester County, Pennsylvania	......	2.60	1.65	
1620	Indianola, Tex	......	2.80	1.78	
1004	Eagle Pass, Tex	......	2.60	......	Imperfect.
1113	......do............	......	2.68	1.60	
1395	Matamoras, Tex	......	2.90	1.90	
1878	Calcasieu, La	......	2.68	......	Imperfect."

Anatomy and physiology of the anal glands and properties of the secretion.

The almost insuperable repugnance which the Skunk naturally excites has always been an obstacle to the investigation of its peculiar defensive organs. Until quite recently, when M. Chatin minutely examined the anal glands of *Conepatus mapurito*, no adequate account of any species had been rendered, though these parts in *M. mephitica* had long since been briefly noticed. The first, and for a long time the only accurate, record, was that given by Dr. Jeffries Wyman in the first volume of the Boston Natural History Society's Proceedings (1844, p. 110). This indicated, though briefly, the general structure of the parts which obtains throughout the family, as far as known.

The organ is a true anal gland, without connection with the genito-urinary system, nor yet of a special character; being upon the same plan as other anal glands throughout *Mustelidæ*, though more muscular, with more capacious reservoir, and more abundant secretion. It consists of a strong central capsule, enveloped in muscular tissue, and by the same connected with a bone of the region, the reservoir of a fluid secreted by several small glandular bodies by which it is surrounded, and which is voided by voluntary muscular effort through an orifice on top of a nipple-like eminence, situated on each side of the anus, just within its verge, partially concealed when not in use by a fold of integument. The organ is paired with a fellow on the opposite side. Dr. Wyman's original remarks may be here transcribed:—"The anal pouches", he writes, "consist of two glandular sacs of an oval shape, about three-quarters of an inch in diameter, covered with a muscular envelope, and opening into the rectum, quite near to the anus, by two papillæ. These last, when not protruded, are surrounded by a fold of mucous membrane, and very nearly concealed by it. The fluid is ejected by the contractions of the muscular covering. A small band passes from each sac to the ischium, which rotates these bodies on themselves, and serves to bring their orifices to the anus. The fluid is a peculiar secretion like that of the Civet, and not the urine, as is commonly thought. The common opinion, that the animal scatters it with its tail is erroneous. The fluid is limited in quantity; and, having been discharged, the animal is harmless until the sacs are again filled by gradual secretion."

This account was shortly supplemented in the same publication (vol. iii. p. 175) by a notice from Dr. J. M. Warren, which adds further particulars, though not strictly of an anatomical character. The passages are transcribed as part of the history of the species:—

"Dr. J. M. Warren exhibited, preserved in alcohol, the glands which secrete the acrid fluid which furnishes a means of defence to the American Skunk, *Mephitis Americana*. These glands are situated on either side of the intestine, at the root of the tail, just within the anus, and are about an inch in diameter. When the animal is pursued, the lower part of the intestine is prolapsed through the anus, the tail is elevated over the back, and by the contraction of the muscles of the anus the acrid fluid is ejected in two streams to the distance of six or eight feet.

"Dr. Warren also exhibited to the Society a living specimen of *Mephitis Americana*, which had been deprived of its power of annoyance by a surgical operation. The animal was first made partially insensible by enclosing him in a barrel in which was placed some chloric ether. As he became stupefied, a sponge containing the anæsthetic agent was placed over the nostrils and kept there until entire insensibility was produced. Dr. Warren then cut down, on the outside of the intestine, upon the ducts of the glands and divided them, suffering the glands to remain *in situ*. The animal recovered, being entirely deprived of his means of annoyance by the adhesive inflammation following the operation."

Here the matter rested (so far as I am acquainted with the record) until 1871, when Dr. J. S. Parker published an account of a dissection in the American Naturalist, as above quoted. Besides being not quite accurate in effect, though the observer really recognized the condition of the parts, the account is too diffuse to justify transcription as a whole; yet it is particularly noteworthy as giving the first and probably the only account to date of the physical properties of the fluid itself:—" . . . . I dipped the point of my scalpel in the yellow fluid, put the tenth or twentieth of a drop of it on a glass, covered it with another strip of glass, and placed it under a power of forty diameters in my microscope. The appearance was peculiar. It looked like molten gold, or like quicksilver of the finest golden color. Pressure on the strips of glass made it flow like globules of melted gold. By a power of sixty diameters the same color still appeared, but seemed as if it would by a higher power resolve itself into globules, with some peculiar markings. . . . . To the eye, the peculiar and odoriferous secretion of this animal is of a pale bright or glistening yellow, with specks floating in it. By the microscope it looks like a clear fluid, as water with masses of gold in it, and the specks like bubbles of air, covered with gold, or rather bags of air in golden sacks. The air I take to be the gas nascent from the golden fluid. Had I known that my interest in the dissection would have rendered me so forgetful of the pungent surroundings, I would have had chemical reagents to test the substance so easily obtainable.

"Another thing was a matter of interest. If I correctly made out the capsule of fluid, the commonly called 'glands' are the muscular tunic enveloping and capable of compressing the reservoir, and their sole use is to eject the liquid. The teat-like projections have one large orifice for a distant jet of the substance, and also a strainer, with numerous holes—like the holes in the cones in the human kidney—for a near but diffusive jetting of the matter [?]. The substance is secreted by small glands, dark in color, and of small calibre, connected with the capsule by narrow ducts."

We gather from these accounts that, as already intimated, the secretory apparatus of this species is essentially the same as that of *Conepatus*, described at length by M. Chatin. It is, of course, no longer necessary to refute the vulgar notions once prevalent, that the secretion was that of the kidneys, whisked about by the bushy tail. There remains little to be said on this subject. The fluid is altogether peculiar and indescribable in odor, pungent, penetrating, and persistent to a degree, perhaps, without parallel, outside this subfamily, in the animal kingdom, though probably not more subtilely diffusive than some other analogous emanations. It has been called "garlicky", but this is a mild term. The distance to which the substance, in liquid form, can be ejected, is, in the nature of the case, difficult to ascertain with precision, and doubtless varies with the vigor of the animal and amount of accumulation in the reservoir. But there is no doubt that the squirt reaches several (authors say from four to fourteen) feet, while the *aura* is readily perceptible at distances to be best expressed in fractions of the mile. The appearance of the animal during the act of emission is unmistakable, as I have observed on several occasions. The zigzag course, with mincing steps, by which it leisurely recedes from a pursuer, is arrested for a moment, when the hinder parts are raised and the tail elevated over the back, so that the long hairs, heretofore trailing in one direction, fall in a tuft on all sides, and the sense of smell immediately indicates what has taken place. The discharge is ordinarily invisible in the daytime, but several observers attest a certain phosphorescence, which renders the fluid luminous by night. This is doubtless true, though I have not verified it by actual observation. Statements to the effect that emission is impossible when the animal is held suspended by the tail are, in the nature of the case, not likely to be often proven by experiment. Nor have I found that instantaneous death is always a sure preventive of escape of effluvium. A Skunk which I shot with my pistol, held within a foot of its head, the bullet traversing the whole body from the forehead to the groin, was too offensive to be skinned, though it died without a perceptible struggle, and had certainly not opened its reservoir up to the moment when shot. Nevertheless, there is abundant evidence that life may be taken in such manner that the flesh is eatable, with due care in the preparation of the carcase; and the meat is said to form a regular part of the food of some savage tribes and semi-civilized people. I have seen it stated that emission does not take place when the animal is captured in a deadfall in such way that the small of the back is broken by the falling weight. The "staying" qualities of the effluvium are certainly wonderful; some of the accounts seem incredible, yet they are well attested. Audubon says that at a place where a Skunk had been killed in autumn, the scent was still tolerably strong after the snows had thawed away the following spring. The same author adds that the odor is stronger by night and in damp weather than under the opposite circumstances; and, in speaking of tainted clothes, he continues:—" Washing and exposure to the atmosphere certainly weaken the scent, but the wearer of clothes that have been thus infected, should he accidentally stand near the fire in a close room, may chance to be mortified by being reminded that he is not altogether free from the consequences of an unpleasant hunting excursion." The persistence of the scent in museum specimens depends altogether upon circumstances. Some specimens, in which the fluid had apparently not been discharged at death, and in which care had been taken in the preparation, come directly into our hands with little or no scent; in others, those probably in which the pelage had become impregnated, or in which the fluid had escaped upon surrounding parts, retain their characteristic odor for many years, whether immersed in alcohol, or dried and buried in tobacco-leaves, insect-powder, and other vegetable aromatics. I have also noticed that the scent may be drawn out of seemingly odorless specimens, after several years' keeping, by placing them in the sun. But in proof of the possibility of absolute freedom from scent may be instanced the use, especially of late years, of Skunk furs as wearing apparel, immunity being gained by processes similar to those used by furriers in purifying the pelts of other *Mustelidæ*, as well as of Wolves, Foxes, &c. The enduring and mortifying consequences of actual contact of the fluid with the person or the clothing, as well as of its dissemination in dwellings and outhouses, can hardly be exaggerated, but require no further comment, as these matters have furnished standing accounts since the history of the species began.

It seems, however, that the disgusting qualities of the substance have been given undue prominence, to neglect of a much more important and serious matter. The danger to the eyesight, should the acrid and pungent fluid actually fall upon the eyes, should not be forgotten. Dogs are not seldom permanently blinded by the discharge, and there are authentic cases in which human beings have lost their sight in the same way. Sir John Richardson alludes, on the authority of Mr. Graham, to the cases of " several " Indians who had lost their eyesight in consequence of inflammation resulting from this cause.

The effect upon dogs is described by Audubon and Bachman:—" The instant ", they say, " a dog has received a discharge of this kind on his nose and eyes he appears half distracted, plunging his nose into the earth, rubbing the sides of his face on the leaves and grass, and rolling in every direction. We have known several dogs, from the eyes of which the swelling and inflammation caused by it did not disappear for a week."

These authors also speak of the nauseating qualities of the effluvium. " I have known a dead Skunk", says Sir John, " thrown over the stockades of a trading post, produce instant nausea in several women in a house with closed doors upwards of a hundred yards distant." " We recollect an instance," write Audubon and Bachman, " when sickness of the stomach and vomiting were occasioned, in several persons residing in Saratoga County, N. Y., in consequence of one of this species having been killed under the floor of their residence during the night."

The fluid has been put to medicinal use in the treatment of asthma. One invalid is said to have been greatly benefited by the use of a drop three times a day; but he was soon obliged to discontinue the use of the remedy, owing to the intolerably offensive character which all his secretions acquired. The story is told[*] of an asthmatic clergyman who procured the glands of a Skunk, which he kept tightly corked in a smelling-bottle, to be applied to his nose when his symptoms appeared. He believed he had discovered a specific for his distressing malady, and rejoiced thereat ; but on one occasion he uncorked his bottle in the pulpit, and drove his congregation out of church. In both these cases, like many others, it is a question of individual preference as between the remedy and the disease.

---

[*] By Audubon and Bachman, Quad. N. A. i. 323.

The supposed connection between the suppression of the secretion and the possibility of inoculating hydrophobia is treated beyond under head of "rabies mephitica".

There is one point connected with the varying offensiveness of the substance which has received little attention. It is certain that if its penetration were correspondent with actual quantity of the substance present, no dissection of the parts of a vigorous animal would be reasonably practicable. But the fluid, like other highly odoriferous substances, is perceptible in degree according to its diffusion in the air by minute division of particles. This is well illustrated under the annoying and too frequent circumstance of a Skunk taking up its abode beneath dwelling-houses for the winter, which season is passed in a state of incomplete hibernation in some latitudes. At irregular intervals, the animal arouses, and, to judge from the effluvium, empties its distended pouches; but the stench, when thus caused, soon ceases, as is not the case when it is spirted under irritation or in self-defence.

Chloride of lime has been recommended as the most effectual disinfectant, and there are doubtless other agents which, by chemically decomposing the substance, deprive it of its offensive properties. The professional "earth treatment", of late extensively employed in hospital practice, was long anticipated in this connection, it being a common custom to bury clothes in ground to rid them of the scent. There is also said to be a belief among trappers that the odor may be dispelled by packing the clothes for a few days in fresh hemlock boughs.

The physiological *rôle* of this special secretion is obvious. Its relation to the perpetuation of the species, though overshadowed by its exaggeration into a powerfully effective means of preservation of the individual, is evidently the same as in other species of *Mustelidæ*, each one of which has its own emanation to bring the sexes together, not only by simply indicating their whereabouts, but by serving as a positive attraction. In the case of the Skunk, it would seem that the strong scent has actually tended to result in a more gregarious mode of life than is usual in this family of mammals; and it is certain, at any rate, that the occupancy by one animal of a permanent winter abode serves to attract others to the same retreat. Burrows are sometimes found to contain as many as a dozen individuals, not members of one family, but various adult animals drawn together. One other effect of the possession of such unique powers is seen not so much in mode of life as in the actual disposition of the creature. Its heedless familiarity, its temerity in pushing into places which other animals instinctively avoid as dangerous, and its indisposition to seek safety by hasty retreat, are evident results of its confidence in the extraordinary means of defence with which it is provided. In speculating upon the development of this anal armature to a degree which renders it subservient to purposes for which the glands of other *Mustelinæ*, though of similar character, are manifestly inadequate, it may not be amiss to recall how defenceless the Skunk would otherwise be in comparison with its allies. A tardy terrestrial animal of no great strength or spirit, lacking the sagacity and prowess of the Wolverene, the scansorial ability of the Martens, the agility, small size, and tenuity of body of the Weasels, the swimming and diving powers of the Otters, and even much of the eminent fossorial capacity of its nearest relations, the Badgers—lacking all these qualities, which in their several exhibitions conduce to the safety of the respective species, it is evident that additional means of self-protection were required; while the abundance of the animal in most parts of the country, and its audacity in the face of danger, show that its confidence in the singular means of defence it possesses is not misplaced.

Geographical distribution and habits of the Skunk.

Leaving now that portion of the subject which is properly most prominent in the history of the species of this subfamily, we may turn to other matters. Skunks are common in most portions of temperate North America, and very abundant in some districts. I am not aware that any qualification of the broad statement of their general distribution in this country is required; for the animals seem to be independent of those matters of physical geography, such as mountain or valley, woodland or prairie, which impose restrictions upon the distribution of many quadrupeds. Skunks, moreover, are obviously less affected by the settlement of a country than the more defenceless, wary, and instinctively secretive carnivores, which are sure to be thinned out and gradually forced away by the progress of civilization. In some parts of the West, indeed, I have found Skunks more numerous in the vicinity of the sparse settlements than they are in regions still primitive; they seem to be actually attracted to man's abodes, like some other quadrupeds and not a few birds, which are more abundant in "clearings" than in the depths of the forest or in the loneliness of unreclaimed prairie. I was struck with this circumstance during my recent travels in Colorado, where Skunks were a never-failing nuisance about the ranches, though I never saw or smelled one, to my present recollection, in the uninhabited mountains of that State. Their entire absence, however, is not to be predicated on this score, but simply their relatively lesser numbers; and I have rarely found Skunks more numerous in the West than they were in the entirely unsettled stretches of country in Montana northwest of Fort Benton, and thence to the region of the Saskatchewan. Richardson notes their frequency in this latter portion of the country, and fixes the northern limit of the species at about 56° or 57° North latitude. In the opposite direction, the habitat of the Skunk overlaps that of the Conepate, reaching into Mexico; but exactly how far remains to be ascertained. It is probably replaced, southerly in Mexico, by the closely allied though apparently distinct *M. macrura* of Lichtenstein, treated on a following page. A recent local writer on the quadrupeds of one of our States noted that out of the large number of Skunks attributed to North America only one, the present species, was found in his locality, humorously adding that *one*, however, was generally considered sufficient. Throughout British America, and most of the northern tier of States, New England, the Middle States, and some of the Southern States, the present is the only species of the subfamily certainly known to occur; in most parts of the West, and some of the South, it is associated with the smaller species, *Spilogale putorius;* while the extreme Southwest may rejoice in the possession of all three of the United States species.

The Skunk yields a handsome fur, lately become fashionable, under the euphemism of "Alaska sable"—for our elegant dames would surely not deck themselves in obscene Skunk skins if they were not permitted to call the rose by some other name. Pelts to the number of a thousand or more have annually passed through the hands of the Hudson's Bay Company; and this kind of "sable" is one of the staples of American furriers, many thousands being yearly exported to Europe. The black furs are the most valuable, ranging in price, according to quality, up to $1 apiece for prime; the "half-stripe" and the white bring much less. The trapping of the animal seems to be an easier matter than the subsequent disposition of the prize; for the Skunk is far from cunning, and no special skill is required for its capture. A variety of traps are used with success; the deadfall is particularly recommended, since, if properly constructed, it causes the death of the animal without emission of the fluid.* Audubon and Bachman's statement that the fur "is

---

* Gibson's "Complete American Trapper", pp. 198, 282-3, 286.
The following on the subject of trapping Skunks was contributed by C. L. Whitman, of Weston, Vt., to Forest and Stream of February 17, 1876:—

"I am often asked by friends and brother trappers how I manage to rid my fox traps of skunks without being defiled by their odor. For the benefit of the uninitiated I will state that if there are any skunks living in the vicinity where fox traps are set they are sure to be taken, and till all are thus disposed of there is little chance of capturing foxes. When there is reason to suppose the presence of many skunks, it is best to set the traps early, in order to get them out of the way at once; setting in a manner not to take the fox—that is, less skillfully. To the fox trapper this animal is a pest and annoyance, for where the trap is made fast—as in dirt trapping is desirable—he will in a brief time with teeth and claws greatly impair, if not wholly ruin a good setting-place. Sometimes he frees himself by self-amputation; in such case it is good riddance. They seldom get in a second time, as in their weak and mutilated condition they fall an easy prey to the fox, who is fond of their flesh; so much so that he will sometimes gnaw off the leg by which the skunk is held in the trap, and carry off his booty to be eaten at his leisure. Trappers cognizant of the above trait do not fail to use skunk's flesh for bait. Sometimes he is found asleep after a night of ceaseless toil to get free, when, if in good position, he may be carefully approached from the leeward, and by stepping upon his tail, at the same time dealing a smart blow upon the head with a club, he is easily and safely dispatched. But this seldom occurs, and the attempt to dispatch him when on the alert with clubs or stones, is to risk and often receive defilement. Firearms are out of the question, as a good trapper is chary of their use on his range.

"My favorite method of dealing with them is as follows: With a tough annealed No. 15 or 16 iron wire I form a slip noose about five inches in diameter on one end, and a standing loop of two inches on the other, and a space of five inches between. The loop is attached to the smaller end of a light, stiff pole of eight or ten feet in length. With this firmly grasped in both hands I slowly and carefully approach, and slip the noose over his head, and with a quick jerk backwards and upwards lift him as high as the chain of the trap will allow, and thus hold him until he is strangled. The butt end of the pole may be brought to the ground and there held by a foot, the hands moved further in advance for greater ease. When taken by a hind leg I at once lower the trap to the ground and release the same with one foot pressed upon the spring; the pole may then be set in a secure position against a rock or other support while the trap is being reset. If the jerk upward has not been adroitly made, the wire may not draw as tight as it ought, in which case a discharge of the pungent odor will usually follow; but in this perpendicular position the discharge descends directly downwards, so that if the attack has been made from the windward, as it ought, there is no danger.

seldom used by the hatters, and never we think by the furriers; and from the disagreeable task of preparing the skin, it is not considered an article of commerce" was wide of the mark, unless it was penned before "Alaska sable" became fashionable.

Like other animals of the present family—like most carnivores, in fact—the Skunk is somewhat nocturnal in habits, chiefly prowling for food in the dark, though often abroad in the daytime. In northern portions of its range, it hibernates to some extent, but its torpidity is very incomplete; it appears, moreover, to be under some necessity of arousing itself, perhaps for the periodical evacuation of its reservoirs. In the South, it ranges freely at all seasons. In instances in which the animal has taken up its abode for the winter about dwelling-houses, its temporary activity, during warm spells of weather, is not likely to be overlooked. This propensity to seek retreats in human habitations is strikingly at variance with the disposition of other Musteline quadrupeds, which instinctively shun man's abodes, except when, in foraging for food, the poultry-yard tempts their appetite and their courage. In travelling in some portions of the West, it *did* seem as if I never could approach a ranch without being aware of the visit, past or present, of some prying Skunk; and the outhouses I entered were almost invariably scented. The Skunk is an occasional robber of poultry and eggs, and is said to be fond of milk. When away from human habitations, the retreats of the Skunk are underground burrows, the hollows of decayed logs and stumps, the crevices among rocks—in short, any natural shelter not away from the ground. Audubon and Bachman describe the underground burrows which the Skunk excavates for itself as less difficult to dig out than those of the Fox, generally running near the surface of flat ground for six or eight feet, and ending in a chamber lined with leaves, where may be found during winter from five to fifteen individuals huddled together. Sometimes, these authors add, the burrow divides into two or three galleries. The ani-

---

"The approach is sometimes resented at first, but the gradual arching of the tail gives timely warning, and a careful retreat is necessary for a moment. The second or third attempt is successful. The animal by that time recovers from the alarm, and at most will merely sniff the air in your direction. With this device I have destroyed many hundred during the past thirty years, and do not recollect an instance where I bore any of the odor about me, except I had inadvertently trod upon dirt that was defiled, and now offer it for the consideration of brother trappers."

mals are evidently more gregarious than other *Mustelidæ*, and the numbers which congregate in one burrow are not necessarily members of the same family. They are very prolific, bringing forth in May, it is said, to the number of eight or ten; the period of gestation is probably unknown. Their natural increase is at so high a rate that were they not systematically persecuted, not only for the value of their furs, but on account of their peculiar offensiveness, they would become a serious pest. The reaction of their principal means of self-preservation, in fact, becomes one of the factors in the problem of their undue increase, so nicely are the balances of Nature adjusted.

Skunks are attacked by dogs and other canine quadrupeds, who destroy and devour them in spite of their scent; and some of the larger birds of prey, like the *Bubo virginianus*, or Great Horned Owl, have been observed to capture and eat them. Their own food is of rather an humble nature in comparison with that of other *Mustelidæ* of corresponding size and strength; for they have neither the speed nor the address required to effect the destruction of many animals which the Martens and Weasels, for instance, prey upon. They feed largely upon insects, birds' eggs, such small reptiles as frogs, and small quadrupeds, such as the various species of mice. They are also said to capture rabbits in the burrows into which these timorous beasts sometimes take refuge, though they are manifestly incapable of securing these swift-footed animals in the chase. The depredations committed by the Skunk in the poultry-house have been already alluded to. I recur to the fact to note the way these awkward animals conduct themselves under such circumstances, when their blundering pertinacity and apparent neglect of the most obvious precautions against detection contrast strongly with the stealth, cunning, and sagacity of the Fox, Mink, or Stoat when engaged in similar freebooting. Even after discovery, the Skunk seems to forget the propriety of making off, and generally falls a victim to his lack of wit.

I once tested the speed of a Skunk in a fair race over open prairie. The wind was blowing "half a gale" at my back, and my courage was consequently unchallenged. The animal seemed to be aware of its powerlessness under these circumstances, and, after once or twice vainly discharging its battery, as I saw by its peculiar motions, though the wind carried off the effluvium, made off at its best pace. But I had no difficulty in keeping up with it at an easy jog-trot, scarcely faster than rapid walking, and, after noting its gait and other actions, I shot it dead. The specimen was too offensive to be skinned, however, as some of the fluid had been blown upon its fur. In the course of my various campaigns in the West, I have witnessed not a few ludicrous scenes, and have known the startling cry of "Skunk!" to throw a camp into as great commotion, to all outward appearance, as that other graver, yet not less sudden, warning of Indians. But to recount stories of Skunks would be to go on indefinitely; like the pelt to the furrier, anecdotes to the historian are "staple", and may be read in all the books, such is the facetiousness which this subject seems to inevitably call forth.

History of the species.

The Skunk has figured in literature for more than two centuries, as can be said of comparatively few American animals. The earliest account I have found, one which Richardson also said was the first he had met with, is that given by Gabriel Sagard-Théodat, "Mineur Recollect de la Prouince de Paris", in his History of Canada, 1636. The quaint passage runs as follows:—

"Les enfans du diable, que les Hurons appellent Scangaresse, & le commun des Montagnais Babongi Maniton, ou Ouinesque, est un beste fort puante, de la grandeur d'un chat ou d'un ieune renard, mais elle a la teste un peu moins aigu, & la peau couuerte d'un gros poil rude & enfumé, et sa grosse queuë retroussée de mesme, elle se cache en Hyuer sous la neige, & ne sort point qu'au commencement de la Lunedu mois de Mars, laquelle les Montagnais nomment Ouiniscou pismi, qui signifie la Lune de la Ouinesque. Cet animal, outre qu'il est de fort mauuaise odeur, est tres-malicieux & d'un laid regard, ils iettent aussi (à ce qu'on dit) parmy leurs excremens de petits serpens, longs & deliez, les quels ne viuent neant moins gueres long temps. I'en pensois apporter une peau passée, mais un François passager me l'ayant demandée ie la luy donnay."

From the way in which this passage opens, we may presume or infer that "enfan du diable" was already a recognized name among the French, in spoken at least, if not also written, language. The "devil's own" beast is also mentioned by various other early writers, amongst whom Charlevoix may be cited. It was the "Fiskatta" of Kalm (17..); but the date of the introduction of the term "Skunk" I have not been able to ascertain, nor do I know its meaning. A likeness to the word most suggestive of the animal, and which appears in the German *Stinkthier*, is too obvious to require comment, but the resemblance may be fortuitous. It may be observed that the Cree or Knistenaux word is *seecawk*, which is quite likely the origin of the name, as the sound is not so very different, though the literal discrepancy is great. The American-English name "pole-cat" or "pol-cat", by which the Swedish of Kalm is rendered, and which has long been an appellation of this and other species of Skunks, is simply a transferring of the European-English name of the Fitch, *Putorius fœtidus*, the worst-smelling species of its own continent, to the Western animal, which has the same enviable notoriety. The terms pol-cat or pole-cat and skunk were both used by Lawson about the beginning of the last century. "Polcats or Skunks in America," says he, "are different from those in Europe. They are thicker and of a great many Colours; not all alike, but each differing from another in the particular Colour. They smell like a Fox but ten times stronger. When a Dog encounters them, they void upon him, and he will not be sweet again for a fortnight or more. The Indians love to eat their Flesh, which has no manner of ill smell, when the Bladder is out." "Skunk" was formerly used adjectively, as we see in the "Skunk Weesel" of Pennant, which may be deemed exactly equivalent to the "Mephitic weesel" of Shaw. "Chinche" was a term applied by early French zoölogists to this and other *Mephitinæ*, and in its various forms of *chinche* or *chincha*, *chinge* or *chinga*, was long current. The last-named form, indeed, became with many authors, after Tiedemann, the specific name of the species in binominal nomenclature.

The early history of the *species* in technical nomenclature, as distinguished from that of the animal in non-scientific accounts, is much involved. It may be well to state that authors have gone to opposite extremes in treating of Skunks as species. Some, like Cuvier, "lumped" them all together, whilst others made every streak or spot the basis of a species. We do not find the present species clearly and unequivocally indicated by the founder and earliest supporters of the binomial system; on the contrary, the Linn.-Gmel. accounts, though undoubtedly covering this even then well-known species, are so infiltrated with reference to other species as to be not properly citable in this connection. Linnæus put the Skunks in his genus *Viverra*, transferring this Plinian name of certain Musteline animals to those of the Civet-cat group, and in 1758 named a species *Viverra putorius*. His species at this date was partly based on Kalm's *Fiskatta*, and in so far means the present animal, but the primary reference is to Catesby's Pole-cat, and the description rather suits the *Spilogale*. In 1766, Linnæus made confusion worse confounded by resting his *Viverra putorius* not only upon Catesby and Kalm, as he had done in 1758, but by citing also Hernandez, Ray, Seba, and Brisson, his species being consequently a conglomeration of animals not only specifically but generically distinct from each other, though the drift of his descriptive text is toward the present species.* These accounts, and such as hang upon them, are not properly citable in the present connection. About the end of the last century, Dr. G. Shaw introduced a species, *Viverra mephitica*, which indicates the present animal with sufficient pertinence and exclusiveness, and furnished a specific name, the first tenable one I know of. In consequence, however, of its literal resemblance to the name of the Cuvierian genus *Mephitis*, the term slept until revived by Baird in 1857, when, with those to whom the alteration is not objectionable, the binomial name *Mephitis mephitica* became current.

Shortly afterward, in 1808, Tiedemann introduced a species, *M. chinga*, adapted from the earlier *chinche* as a specific name. This was adopted by Lichtenstein in his special memoirs, by Aubudon and Bachman, and by others. It undoubtedly refers to the present animal, though vitiated to some extent by inapplicable expressions.

Desmarest called all the Skunks *Mephitis† Americana*, having a long array of varieties, from A to R, his var. R being the one which more particularly refers to the present species. In 1829, Fischer rendered the "chinga" of Tiedemann as *chinche*, reverting to the more customary orthography. The same year Richardson introduced a new term, *hudsonica*. Later, nominal species multiplied, not that there were not already names

---

* "Habitat in America septentrionali. Colore variat. Irritatus (cum urina forte) halitum explodit, quo nihil fœtidius; incessu tardus, nec Homines nec Feras metuens; vestes fœtore inquinatæ purgantur sepeliendo per diem. A. Kuhn." (p. 65.) Linnæus's next species, *Viverra zibetha*, the Civet-cat of the Old World, is also tinctured with Skunk, or some other American animal not distantly related.

† Written "Mustela" by an obvious slip.

enough, but apparently in the impossibility of sifting and fixing earlier accounts. *M. varians* was proposed by Gray in 1837 for the southwestern variety, afterward called *macroura* by Audubon and Bachman; and in 1865 Gray had the assurance to set his term over all the prior ones as the specific designation, recounting numerous varieties of the species. *Mephitis mesomelas* of Lichtenstein and *M. occidentalis* of Baird are names of the western strain of ordinary *mephitica*.

Other points in the history of Skunks are reviewed under heads of species to follow.

ADDENDUM TO CHAPTER VII.

ON HYDROPHOBIA FROM SKUNK-BITE, OR THE SO-CALLED "RABIES MEPHITICA".

The importance of this subject induces me to present such facts as have come to my knowledge. Though it has long been known that the bite of the Skunk under certain conditions, like that of various other animals, is capable of inoculating a disease like hydrophobia, it seems that only lately has the subject been thoroughly investigated and adequately presented. This has been done, notably, by two writers, whose respective accounts are here transcribed in full, without further comment.

The points that the Rev. Mr. Hovey makes are these:— That hydrophobia from Skunk-bite is a different species of the disease from *rabies canina;* the term *rabies mephitica* being proposed for it. That *rabies mephitica* is caused by a special hydrophobic virus generated by Skunks. That "possibly there may be a causative connection between inactivity of the anal glands and the generation of malignant virus in the glands of the mouth". That the bite of Skunks in apparently normal state of health (*i. e.*, not rabid in the usual sense of the term) is usually fatal. That "we might go further and seek a solution of the whole dread mystery of hydrophobia in the theory that this dread malady originates with the allied genera of *Mephitis, Putorius,* and *Mustela,* . . . being from them transferred to the *Felidæ* and *Canidæ* and other families of animals". He also suggests that the mephitic secretion might be found to be the natural antidote to the salivary virus.

The article attracted considerable attention, from the novelty of the views put forth, and the intrinsic importance of the subject.

Some months afterward Dr. Janeway replied in an elaborate article, detailing cases and criticising Mr. Hovey's views, coming to the conclusion "that the malady produced by mephitic virus is simply hydrophobia". Following are the two articles in question in full:—

[From Amer. Journ. Sci. and Art, 3d ser. vol. vii. no. 41, art. xliv. pp. 477-483, May, 1874.]

"*Rabies Mephitica*; by Rev. Horace C. Hovey, M. A.

"My subject concerns alike medical science and natural history. For while proving the existence of a new disease, some singular facts will be brought to light about a familiar member of the American Fauna. It is cruel to add aught to the odium already attached to the common skunk (*Mephitis mephitica* Shaw; *M. chinga* Tiedemann). But, clearly, he is as dangerous as he is disagreeable. In a wild state he is by no means the weak, timid, harmless creature commonly described by naturalists; although it is said that, if disarmed of his weapons of offence while young, he may be safely domesticated.

"A peculiar poison is sometimes contained in the saliva of animals belonging to the canine and feline families, the production of which, it has been generally supposed, is limited to them. Other animals, of the same or of different species, may be inoculated with this virus; the result being a mysterious malady, which men have observed from the days of Homer and Aristotle, but which has never been either cured or understood. This frightful disease has been called, from its origin, *Rabies canina*, and from one of its symptoms, *hydrophobia*. Probably it is not communicable by any species but those with which it originates. A few instances have been recorded to the contrary; but they were so imperfectly observed as merely to stimulate us to further investigation. It is stated by the best medical writers (*e. g.*, Watson, Gross, and Aitken), as an undeniable fact, that no instance is known of hydrophobia having been communicated from one human being to another, although many patients, in their spasms, have bitten their attendants. An interesting case, but inconclusive, being the only one of its kind, is reported by M. Guillery, in which an aged man experienced spontaneous hydrophobia (Bulletin of Belgian Academy, No. 8, 1871). In such exceptional instances there may have been previous inoculation, unnoticed or forgotten; for the least particle of this deadly poison will be efficient, and yet it is always tardy in its period of incubation.

"The facts now collated will show, it is thought, one of two things, either that the hydrophobic virus is both generated and communicated by some of the *Mustelidæ* as well as the *Felidæ* and *Canidæ*; or else, that a new disease has been discovered, which generically resembles *Rabies canina*, while differing from it specifically. My judgment favors the latter opinion, decidedly, for reasons to be adduced; and accordingly I may name this new malady, from the animal in whose saliva it is generated, *Rabies Mephitica*.

"The varieties of *Mephitis* are notorious for the singular battery with which they are provided by nature. It consists of two anal glands from which, by the contraction of sub-caudal muscles, an offensive fluid can be discharged in thread-like streams, with such accuracy of aim as to strike any object within fifteen feet. This secretion is either colorless, or of a pale yellow hue. It is phosphorescent. Viewed from a safe distance, its discharge looks like a puff of steam or white smoke. Its odor is far more persistent than that of musk. If too freely inhaled it causes intense nausea, followed by distressing gastric cramp. In minute doses it is said to be a valuable anti-spasmodic. If so, why not experiment with it as a cure for hydrophobic convulsions? It is not known what the effect would be of injecting this fluid beneath the skin. Interesting results might be attained by any one who is willing, in behalf of science, to investigate further in this inviting path! There certainly seems to be some connection between it and the disease under consideration; for, in every instance, the rabid skunk has either exhausted his mephitic battery, or else has lost the projectile force by which it is discharged. Perhaps the secretion is only checked by the feverish state of the system. Possibly there may be a causative connection between this inactivity of the anal glands and the generation of malignant virus in the glands of the mouth.

"An adventure, while on a summer tour amid the Rocky Mountains, first called my attention to the novel class of facts about to be presented. Our camp was invaded by a nocturnal prowler, which proved to be a large coal-black skunk. Anxious to secure his fine silky fur uninjured, I attempted to kill him with small shot, and failed. He made characteristic retaliation; and then, rushing at me with ferocity, he seized the muzzle of my gun between his teeth! Of course the penalty was instant death. An experienced hunter then startled us by saying that the bite of this animal is invariably fatal, and that when in perfect apparent health it is always rabid. He resented our incredulity and confirmed his statement by several instances of dogs and men dying in convulsions shortly after being thus bitten.

"On mentioning this adventure to H. R. Payne, M. D., who had been camping with miners near Cañon City, Col., he said that at night skunks would come into their tent, making a peculiar crying noise, and threatening to attack them. His companions, from Texas and elsewhere, had accounts to give of fatal results following the bite of this animal.

"Since returning to Kansas City, I have had extensive correspondence with hunters, taxidermists, surgeons and others, by which means the particulars have been obtained of forty-one cases of *rabies mephitica*, occurring in Virginia, Michigan, Illinois, Kansas, Missouri, Colorado and Texas. All were fatal except one; that was the case of a farmer, named Fletcher, living near Gainsville, Texas, who was twice bitten by *M. macroura* [of Aud. & Bach.= *M. mephitica* var.—E. C.], yet recovered and is living still. On further inquiry it was found that he was aware of his danger, and used prompt preventive treatment. Another case was alleged to be an exception; that of a dog which was severely bitten in a long fight with a skunk, but whose wounds healed readily and without subsequent disease. It seems, however, that this dog afterward died with mysterious symptoms like those of hydrophobia in some of its less aggravated forms.

"Instead of burdening this article with a mass of circumstantial details, a few cases only will be given best fitted to show the peculiarities of the malady; and those are preferred that are located on the almost uninhabited plains of western Kansas, because there the mephitic weasels would be least liable to be inoculated with canine virus.

"A veteran hunter, Nathaniel Douglas, was hunting buffalo, in June, 1872, fourteen miles north of Park's Fort. While asleep he was bitten on the thumb by a skunk. Fourteen days afterward singular sensations caused him to seek medical advice. But it was too late, and after convulsions lasting for ten hours he died. This case is reported by an eye-witness, Mr. E. S. Love, of Wyandotte, Kansas, who also gives several similar accounts.

"One of the men employed by H. P. Wilson, Esq., of Hayes City, Kansas, was bitten by a skunk at night, while herding cattle on the plains. About ten days afterward he was seized with delirium and fearful convulsions, which followed each other until death brought relief. Mr. Wilson also reports other cases, one of which is very recent. In the summer of 1873, a Swedish girl was bitten by a skunk while going to a neighbor's house. As the wound was slight and readily cured, the affair was hardly thought worthy of remembrance. But on Jan. 24th, 1874, the virus, which had been latent for five months, asserted its power. She was seized with terrible paroxysms. Large doses of morphine were administered, which ended both her agony and her life.

"In October, 1871, a hunter on Walnut Creek, Kansas, was awakened by having his left ear bitten by some animal. Seizing it with his hand, he found it to be a skunk, which after a struggle he killed, but not until his hand was painfully punctured and lacerated. He presented himself for treatment to Dr. J. H. Janeway, army surgeon at Fort Hayes, from whom I have the facts. The wounds in the hands were cauterized, much to the man's disgust, who thought simple dressing sufficient. He refused to have the wound in the ear touched, and went to Fort Harker to consult Dr. R. C. Brewer. Twelve days afterward the latter reported that his patient had died with hydrophobic symptoms.

"Another hunter, in the fall of 1872, applied to Dr. Janeway to be treated for a bite through one of the alæ of the nose. He had been attacked by a skunk, while in camp on the Smoky River, two nights previous. He had been imbibing stimulants freely and was highly excited and nervous. A stick of nitrate of silver was passed through the wound several times. He was kept under treatment for two days, when he left to have a 'madstone' applied. He afterward went home to his ranch, and died in convulsions twenty-one days from the time he was inoculated.

"I give but one more of the cases reported to me by Dr. Janeway. In October, 1871, he was called to see a young man living in a 'dug-out,' a few miles from the fort. He had been bitten by a skunk, seventeen days previous, in the little finger of the left hand. His face was flushed, and he complained that his throat seemed to be turning into bone. On hearing the sound of water poured from a pail into a tin cup, he went into convulsions, that followed each other with rapidity and violence for sixteen hours, terminating in death. This man's dog had also been bitten, and it was suggested that he had better be shut up. He chanced at the time to be in the hog-pen, and he was confined in that enclosure. Ere long he began to gnaw furiously at the rails and posts of the pen and to bite the hogs; until the by-standers, convinced that he was mad, ended the scene by shooting all the animals in the pen.

"It is evidently the opinion of Dr. Janeway that the malady produced by mephitic virus is simply hydrophobia. Should he be correct, then all that is established by these facts would be this, viz: that henceforth the varieties of *Mephitis* must be classed with those animals that spontaneously generate poison in the glands of the mouth and communicate it by salivary inoculation. From this, as a starting-point, we might go further and seek a solution of the whole mystery of hydrophobia in the theory that this dread malady primarily originates with the allied genera of *Mephitis*, *Putorius* and *Mustela*, widely scattered over the earth;* being from them transferred to the *Felidæ* and *Canidæ* and other families of animals. And then, if it could be proved experimentally that the characteristic mephitic secretions contained an antidote for the virus of the saliva, we should have the whole subject arranged very beautifully!

"I am favored by Dr. M. M. Spearer, surgeon in the 6th U. S. Cavalry, with notes from his case-book, of four cases in which persons have died from the bite of the skunk; and he also mentions additional instances reported to him by other observers. He thinks there is a marked difference between the symptoms of their malady and those of hydrophobia. I shall refer to his testimony again, but pause for a moment to notice his final conclusions, from which, original and interesting as they are, I must dissent. He says: 'I regard this virus as being as peculiar to the skunk as the venom of the rattlesnake is to that creature; and not an occasional outbreak of disease as the *œstus veneris* of the wolf or the *rabies canina*.' Singular as this theory may seem, it is not wholly without support. It is remarkable that of all the cases thus far reported to me there is but *one instance of recovery*. It is stated in Watson's Physic (vol. i, p. 615) that of one hundred and fourteen bitten by rabid wolves only sixty-seven died; and of those bitten by rabid dogs the proportion is still less. But mephitic inoculation is sure death. Then again it is to be observed that the only peculiarity noticeable in these biting skunks is the arrest of their effluvium. They approach stealthily, while their victims are asleep, and inflict the deadly wound on some minor member—the thumb, the little finger, the lobe of the ear, one of the alæ of the nose. How different from the fierce assault of a mad dog! How subtle and snake-like! It may be remarked, also, that dogs are generally as cautious and adroit in attacking these odious enemies as they are in seizing venomous snakes. But we must remember, on the other hand, that thousands of skunks are killed annually, partly as pests and partly for the fur trade; and it is incredible that an animal whose ordinary bite is as

---

* "Since forwarding this article for publication, I have obtained an answer to my inquiries made in California through my friend, Dr. J. G. Tidball, respecting the *Mephitis zorilla* [i. e., *M. (Spilogale) putorius*—E. C.]. He described it as a very pretty animal which usually allows itself to be killed without resistance. But he adds that its bite is highly dangerous, causing a fatal disease like hydrophobia.

"I regret that he gives no particulars of actual cases; but his testimony is interesting, as it brings into condemnation a species of *Mephitis* quite different from *M. chinga*."

venomous as that of a rattlesnake, should so seldom resort to that mode of defence, if it be his.

"The resulting disease resembles hydrophobia more than it does the effect of ophidian venom. But here, as observed at the outset, the likeness is only generic, while specifically there are marked differences. These have purposely been kept in the back-ground until now. And in giving a differential diagnosis, I shall avoid repetitious details, and combine facts gathered from many sources with the close and accurate observations which Dr. Shearer has put at my disposal.

"1. The period of incubation is alike in *rabies canina* and *rabies mephitica*. That is, it is indefinite, ranging from ten days to twelve months, with no opportunity meanwhile for subsequent inoculation. But during the incubative period of *R. mephitica*, no perceptible changes take place in the constitution as in hydrophobia. In only one instance was there unusual nervousness, and that might have been due to alcohol. In every case where there was time for it, the wounds healed over smoothly and permanently, and in several instances not even a scar was visible. In no case was there recrudescence of the wound, always seen in hydrophobia. Indeed, there were so few premonitions of any kind that, in most instances, the attending physicians themselves supposed the ailment to be simple and trivial, until the sudden and fearful convulsions came on to baffle all their skill.

"2. Characteristic pustules form in hydrophobia beneath the tongue and near the orifices of the sub-maxillary glands. (See Aitken, Sci. and Pract. Med., vol. i, p. 653.) These were not reported in a single case of *R. mephitica*. Dr. Shearer looked for them carefully in all his cases, but did not find them.

"3. The specific action of hydrophobic virus affects the eighth pair of cranial nerves and their branches, especially the œsophageal branch, the result being great difficulty in swallowing; and the motor nerve of the layx [larnyx—*sic*], causing sighing, catching of the breath, and difficulty in expelling the frothy mucus accumulated in the throat. These invariable accompaniments of *R. canina* are usually wanting in *R. mephitica*; the exceptions being in the case of the Swedish girl, who complained of pain in her chest; and the young man, Dr. Janeway's patient, whose constriction of the throat was decided, as well as his sensitiveness to water. Dr. Shearer's patients had no such trouble. A taxidermist, who has seen four dogs die from *R. mephitica*, in Michigan, says they did not seem to have any fear of water, or other signs which he had supposed were characteristic of *R. canina*. Ordinary hydrophobia, again, is marked by constant hyperæsthesia of the skin, so that the slightest breath of air will precipitate convulsions. But, in *R. mephitica*, fanning the face affords relief, and even cloths dipped in water and laid on the forehead were soothing!

"4. In hydrophobia the perceptions are intensified, so that even the deaf are said to have their hearing restored; the pupils are strongly dilated, imparting to the eyes a wild, glaring expression; the spasms are tonic, *i. e.*, steady and continuous; the pulse is feeble; and delirium is occasionally relieved by lucid intervals. But the symptoms are wholly different in *R. mephitica*: there is oscillation of the pupil; the spasms are clonic, *i. e.*, marked by rapid alternate contraction and relaxation of the muscles; small, but wiry radial pulse and rapid carotids; positive loss of perception and volition throughout, until delirium ends in persistent unconsciousness, simul'aneously with cold perspiration and relaxation of the sphincters.

"5. The mode of death is by asthenia in both forms of rabies; but in *R. canina* the frightful struggles of nature to eliminate the poison are more prolonged than in *R. mephitica;* and in the latter they may, on occasion, be still further abridged by the use of morphine, which has no narcotic effect upon the former, even in the largest doses and injected into the veins!

"I have thus endeavored to describe, and also to explain, these strange and painful phenomena. I must leave the reader to form his own decision, only hoping that some one may be induced to follow this pioneer work in a new path, by further and more able investigations of his own.

"Kansas City, Mo., Feb. 24th, 1874."

[From the New York Medical Record, vol. x. no. 227, pp. 177-180, Mar. 13, 1875.]

"*On Hydrophobia.*—*By John G. Janeway, M. D., Assistant Surgeon, U. S. A.*

"A writer* in the *American Journal of Science and Art*, May, 1874, states that 'it is evidently the opinion of Dr. Janeway that the malady produced by the mephitic virus is simply hydrophobia. Should he be correct, then all that is established by these facts would be this, viz.: that henceforth the varieties mephitis must be classed with those animals that spontaneously generate poison in the glands of the mouth and communicate it by salivary inoculation.'[†] The personal observation of fifteen fatal cases of hydrophobia, produced by the bite of rabid animals, skunks, wolves, and hogs,‡ and the reliable statements of a number of other cases, has fully confirmed me in the opinion above stated, that the malady produced by mephitic virus is simply hydrophobia.

"The following five cases are taken from the fifteen fatal cases that have fallen under my observation:

"CASE I. BITE OF SKUNK.—Was called to visit Wm. P., aged nineteen, a herder, whom I was told by the messenger had been acting strangely all the morning. I found him lying on a bed in a sod-house, dressed, with several of his companions around him. Face flushed, pulse very rapid, the heat of skin intense and dry, eyes brilliant and pupils dilated rather more than natural, extremely restless and frequently catching at his throat; upon questioning, replied that his throat was turning into bone. Had not felt well for two or three days; did not know what was the matter with him. Upon pouring out some water from a pail near by, to administer morphia to him, he went suddenly into convulsions.

"Suspecting hydrophobia immediately, as soon as he regained consciousness I learned that he had been bitten by a skunk, just before daybreak, seventeen days before, in the little finger of the left hand; that the wound was small and soon healed; that for two days preceding my seeing him his

---

* "Rev. Horace C. Hovey, M. A."

† [There is some typographical confusion in the quotation-marks at the opening of Dr. Janeway's article; and Dr. Janeway does not quote Mr. Hovey's literally, leaving it liable to be misunderstood whose opinion is being quoted. I have slightly altered the text in this place, to reproduce the quotation literally from the original.—E. C.]

‡ "Skunks 10, wolves 3, hogs 2."

finger and arm had felt numb. Upon examining the finger, slight redness was observed at the place bitten, tongue slightly furred and somewhat swollen, no so-called 'characteristic pustules' were to be seen. Thirst intense and begged for water, but the sound of dipping the water from the pail threw him immediately into still more terrible convulsions, frequent sighing, and catching his breath. Administer hypodermic injections of morphia without avail. Upon the arrival of chloroform, which I had sent for, its administration gave partial relief for a short time. His endeavors to free himself of the tenacious mucus were terrible, when the incautious upsetting of a pail of water again threw him into convulsions, opisthotonous in character, followed by attempts at biting those holding him, and when consciousness was regained, asking pardon for so doing. Hyperæsthesia existed in a very marked degree in this case. Death came to his relief in about eighteen hours from the time of his first convulsion.

"CASE II. BITE OF SKUNK.—An emigrant from Wisconsin, camped on the north fork of Big Creek, about seven miles from Hays, applied to me in the fall of 1872 for dressing for his hand, which had been bitten between the thumb and index finger of his left hand, the night previously, by a skunk. Cauterized the wound well, and directed him to repeat the cauterization twice a day. Saw nothing of him for twelve days, when I was sent for, and upon arriving at his camp found him in convulsions, which were repeated rapidly. Face flushed, eyes brilliant, pupils rather contracted, skin hot and dry, pulse small and rapid, 120, no so-called 'characteristic pustules' under the tongue. When not in convulsions, mind clear and fully aware of the fate that awaited him. From his wife I learned that after the third day of using the caustic the wound healed and gave him no further trouble; that for three days he had been complaining of some fulness in the head, and a general 'malaise,' neither sick nor well; that the convulsions came on about seven hours previous to my seeing him, suddenly, upon attempting to take a drink from a spring close to their camp; that he would go into convulsions whenever water or tea was offered him, and that the faintest breath of air would cause him the greatest anguish, so that she had to put a blanket up before the door. Death followed in twenty-one hours after seizure.

"CASE III. BITE OF SKUNK.—A hunter, in the latter part of October, 1872, applied to me to be treated for a bite through the right ala of the nose. He had been attacked by a skunk while in camp on the Smoky Hill river two nights previous. Having learned, previous to my seeing him, that skunk-bites would produce hydrophobia, he had imbibed freely, and was decidedly under the influence of liquor when I saw him, evidently nervous about himself, but trying to conceal the fact.

"A stick of nitrate of silver was passed repeatedly through the wound. Actual cautery was proposed, but he would not consent to its use. After being under treatment two days he left and went to Missouri, to have the mad-stone applied; returning from there, he followed his occupation. Twenty-one days after he was bitten he was taken with convulsions, and died about an hour after I got to his ranch, nearly thirty hours after the seizure. From one of his companions I learned that after his return from Missouri he was cheerful and in apparent health up to the day before his seizure, when he complained of pain in his nose and face, headache, chilly, and feeling tired, but had no apprehension concerning himself. The first symptom, the morning the disease developed itself, was a feeling of constriction in the throat, together with dryness, opisthotonos, with decided mania preceding the spasms.

"CASE IV. BITE OF WOLF.—A private of Co. F, Sixth Cavalry, was bitten by a wolf one evening, just after he had come off post, in the lobe of the left ear, in the early part of October, 1873. The wound was freely cauterized with nitrate of silver by the surgeon of the camp. On the 28th of the same month he applied to me for medicine for headache, which was given to him. On the 30th he again applied for medicine, stating that he did not feel sick enough to go on the sick report. Knowing the man's history, I cautiously examined him, and questioned him in such a way as not to excite his fears. I found that the lobe of the ear that had been bitten was quite numb to the touch. No other symptoms presented themselves prominently. There was, however, a general malaise. The day following, the man was in the ranks for muster and inspection. Observing him, I saw at once that something was wrong, and upon reporting his case he was ordered to his quarters, by the commanding officer. Fifteen minutes later I was sent for to see him, and found him in convulsions, which the orderly informed me came on upon his attempts to take a drink of water. He was at once removed to the hospital. He suffered from cold, he told me, whilst being conveyed there. Examination revealed alternately contracting and dilating pupils; skin very hot; temperature 102°, 102.5°, 100°, by three examinations, with the thermometer in the axilla; pulse 120-125, alternating in volume before and after a spasm, but constantly rapid. Tongue somewhat swollen and indented by the teeth on the edge; thick, whitish fur; no so-called 'characteristic pustules' under the tongue. Thirst intense; no irritability or sensation in the wound of the ear; constriction of the pharynx; increasing violent attempts to relieve himself of the thick and tenacious saliva; sound produced resembling more the bark of a wolf than any sound ever heard. Complete inability to swallow any liquid, the attempt ending in a convulsion. Mental faculties perfect when not in spasm; fully aware that death must end the scene. Towards the close the convulsions were longer and of greater strength, with frequent furious attempts to bite his attendants, for which he would beg their pardon time and again. Death took place suddenly in thirty hours.

"CASE V. BITE OF DOG.—A man, aged about 46, attached to a hay-camp, applied to me in August, 1873, to dress his hand, which had been terribly lacerated by a favorite hound that day. He stated that his dog had been acting rather strangely for several days, but that up to that time had always come to him when called, and had appeared as affectionate as ever; that a strange dog had appeared in camp, and that his dog had attacked it furiously; he attempted to separate them, when his dog turned and bit him through the hand, his teeth passing completely through from side to side; that immediately after biting him he (dog) had run off a short distance and laid down in a pool of water. Cauterizing the wound freely I directed him to report at the hospital next morning, when the eschar was removed and his hand was again cauterized. The following day he called at the hospital and stated that he had shot his dog, and was satisfied that he was mad, and that he was going that day to Missouri to have a mad-stone applied. He remained there a week, and then returned and rejoined the hay-camp. On the twenty-fourth day after he was bitten, I was sent for to visit him at the hay-camp, on the Smoky Hill river, lying in a wagon-bed, and was saluted with, 'Doctor, that dog has killed me; I know that I have got the hydrophobia, and that I shall die.' His face flushed; skin hot; pulse very rapid and small, 125; tongue furred, brownish, swollen; complained that his throat was turning into bone, and that he could not swallow; if he saw any liquid, thought he would like to drink a bucketful of water just once. On attempting to give him some morphia in solution the convulsions were ushered in. He had been well up to the morning he sent for me. The first symptom he noticed was the feeling of constriction in his throat, and he noticed a slight increase of redness in the wounds on his hand, though there was no pain. Had seen several cases of hydrophobia, and at the earnest solicitation of his wife had sent for me. Left him powders, of twenty grains each, hydrate chloral, to be given in moist sugar every three hours, and promised to see him next morning. I saw him the following morning, and found him decidedly worse; convulsions more frequent and stronger; pulse smaller and extremely rapid; tongue more swollen; no so-called 'characteristic pustules' to be found after careful search; eyes brilliant, with rather a contracted pupil; great difficulty of swallowing, though he was able to sup up a little water through some straw from a covered cup; had considerable sleep from the chloral, but his stomach had rejected the last dose, and he was unable to take any more; mental faculties clear, could tell the approach of a convulsion, and begged his wife and attendants to take care; much increase of the thick tenacious saliva, and greater difficulty in freeing himself of it. No alteration in appearance of wound. The convulsions became more frequent, stronger, and longer in duration. He insisted upon being chained down to the wagon bed to prevent his injuring any one. Chloroform was left, with directions as to use. The day following I found him barely alive, unconscious, with frequent feeble spasms. Death ended the terrible scene after thirty-seven hours of sufferings. In this case there was no marked hyperæsthesia of the skin complained of.

"Neither can I agree with the writer of the paper mentioned above, that mephitic inoculation is sure death. For the result of one case of bite from a rabid skunk, which will be detailed more fully hereafter, the report of eight others (six hunters and two soldiers) that were bitten, and also from having in my possession two dogs, one a setter and the other a black-and-tan, which have been repeatedly bitten in encounters with these animals and have as yet never evinced any symptom of the disease, will not permit me to concur with him. That more cases, proportionally, may result fatally from the bite of this animal, than from the bite of rabid dogs or wolves, is probably, if not actually, the case; still, there are obvious reasons for it to be so. An animal nocturnal in its habits, generally timid, but armed with a powerful battery to resist any injury or affront; one that will not attempt to bite in defence until the secretion provided for it by nature is exhausted, loses that secretion by the disease. It is a well-authenticated fact that rabid skunks are entirely free from the odor so characteristic of these animals, which could not occur if the secretion was not exhausted, and forgetting its normal timidity will attack any person or animal he may come in contact with, biting the most exposed art of the body, the alæ of the nose, the lobe of the ear, the thumb, or one of the fingers, and passes on. Here is probably the reason these bites are more fatal than those of other animals—always in a vascular part not protected by clothing, which prevents by wiping away the poisonous saliva in the fierce attacks of the mad dog or wolf, and thus saves the life of the one bitten. At a frontier post* this was well illustrated. A mad wolf suddenly sprang upon the officer of the day, who was making his round, and bit him on the arm, through his clothing; passing on, he bit a sentinel on post in the wrist, between the sleeve of his coat and glove, and then sprang upon a woman who was nursing a child near by, and bit her on the shoulder through a thick woollen shawl. All the cases were treated the same. The officer and the woman escaped the dread disease, but the soldier died of hydrophobia. A recent writer † says in reference to bites of rabid dogs: 'The documents of investigation furnish indications full of interest in regard to the more or less innocuousness of bites, according to the different parts of the body upon which they were inflicted. If we compare the fatal with the harmless bites made upon the same region, we find that out of thirty-two cases where the face was bitten, twenty-nine proved fatal, which gives these wounds a mortality of ninety per cent. Out of seventy-three cases, in which the wounds were upon the hands, they have been fatal in only forty-six cases, harmless in twenty-seven, giving an average mortality of sixty-three per cent. In comparing wounds of the arms and legs with those of the face and hands, the ratio is inverted; twenty-eight wounds upon the arms were followed by only eight fatal terminations, and twenty-four bites upon the lower limbs gave only seven fatal cases; seventeen remained harmless, showing a mortality of twenty-eight to twenty-nine per cent., and an innocuousness of seventy to twenty-one per cent., and, lastly, the ratio mortality for wounds upon the body is shown as follows: Out of nineteen bitten, twelve cases were fatal and seven bites proved harmless.'

"These facts are confirmatory of those afforded by other statistics, demonstrating also that rabid wounds upon uncovered or unprotected parts, such as the face and hands, are much more readily contagious than those of the arms and legs, which the teeth of the animal cannot reach without passing through a portion of the clothing, which wipes off the virulent moisture from the teeth. It is true the consequences of bites upon the body seem to conflict with this statement: but we must remember that generally these wounds are more severe, and among them some are uncovered parts, such as the neck and chest, and that, when a man is attacked by a rabid animal and bitten upon the body, he is also bitten upon his hands, which are his material means of defence. Another reason for the apparent large proportion of fatal cases from skunk bite is, that it is only since 1871 that these cases have been collected, or that the fact of hydrophobia existing in and following the bites of these animals has been generally known, and only those cases proving fatal have been reported, the non-fatal cases, from the trivial character of the wound, not being considered of sufficient importance to report.

"A CASE OF SKUNK BITE *not* FATAL.

"W., a young man, twenty-two years old, born in Missouri, commonly known

* "Fort Larned, Kansas."
† "H. Bouley, Gen. Inspector Vet. Schools of France, etc., etc."

by the soubriquet 'Pike County,' driving a team for a party of emigrants for Colorado, was bitten at night, in the early part of May, 1874, upon left cheek, by a skunk, whilst camped at Park's Fort,* Kansas. A companion, who was bitten by the same animal, freely cauterized the wound. Early the next day he presented himself at the hospital for treatment. Removing the eschar I cauterized it again freely with caustic, and directed that he take $\frac{1}{16}$th grain of strychnia every three hours during the day, with vegetable tonics and full diet, the wound to be cauterized morning and night, and a poultice to be applied one hour before retouching to remove the eschar and promote suppuration. No characteristic symptoms being produced by the strychnia on the fourth day, it was increased to $\frac{1}{12}$th grain dose, given as before. Suppuration was fairly set up in the wound and continued; four days after, strychnia increased to grain $\frac{1}{8}$th, and continued at that for four days without any symptoms of its toxic effects. The dose was then increased to grain $\frac{1}{4}$th, and continued for six days without the patient being conscious of any jerkings, though the night nurse and some of the patients stated that he jerked somewhat more than natural when asleep. Suppuration of the wound continued free under the caustic and poultices; the dose of strychnia was then increased to grain $\frac{1}{2}$, and I watched him very carefully, for the slightest appearance of the effect of the medicine, for six days. On the last day I detected some slight involuntary twitching of the muscles of the face, and reduced the dose. Two days after reducing he remarked that he guessed that he was safe from hydrophobia, as the strychnia had not killed him. The wound was allowed to heal up, which it did rapidly, and a few days after he left the hospital, and I saw him three months after perfectly well.

"The above case shows either, first, that the man was not inoculated by the virus when bitten; second, a wonderful tolerance for the drug if he was not so inoculated; or, third, that acting primarily as a tonic to the nerve elements it enabled them to resist the invasion of the disease, and together with the frequent cauterization and free suppuration, to eliminate the poison from the system. (That the strychnia used was a good article was proved by the effect of a small dose upon an obnoxious cur of medium size.) I am inclined to the latter, for that the animal causing the wound was undoubtedly rabid is proved by the fact that the companion who was bitten by the same animal, in the camp, on the same evening, was reported to have died from hydrophobia about ten days after being bitten, and should another case present, would adopt the same treatment and push the drug until its characteristic effects upon the system presented.

"Rabies Mephitica, like Rabies Canina, is evidently epidemical, no cases of it having been reported previous to 1870 in this region.

"The period of incubation is alike in Rabies Canina and Rabies Mephitica (so called), that is, it is indefinite, ranging from ten days to ninety days, no opportunity in the meanwhile being afforded for subsequent inoculation of hydrophobia. Statistics show that the manifestations of the disease have been most numerous during the first sixty days, and that after a bite from a rabid animal the probabilities of escape increase considerably when sixty days have passed and no symptoms of the disease have shown themselves, and that after the ninety days entire immunity is almost certain. Still, I am aware that cases are reported of a longer period of incubation. These

---

* "Park's Fort, K. P. R. W."

are exceptional, and when reported to extend beyond the fourth month it may be questioned whether the patient has not been unconsciously inoculated by the caresses of a pet dog, suffering from the disease unsuspected, from tetanus, or, as Baron Larrey* remarked, when commentating upon Dr. Fereol's case of hydrophobia with two years and a half incubation: 'For my part I should be disposed to regard his case not as an example of rabies, with an incubation of two years and a half, but as one of cerebral hydrophobia or symptomatic of acute delirium, provoked or aggravated by the coincidence of the bite of a dog presumed to be mad.' In all the cases from the bite of a skunk the prodromic stage of the disease was more or less marked, though none of them amounting to acute melancholy. An indefinite feeling of dread and a general malaise—the most prominent symptoms, together, in most cases, with pain or numbness at the seat of the wound, were present from one to three days. To most of these unfortunates the fearful result of the trivial wound they had received was unknown, and unaware of their perilous condition were not incessantly tormented with sad forebodings or dread of the onset of the malady.

"2. The characteristic pustules which the writer of Rabies Mephitica lays stress upon were not found in any of the cases of hydrophobia produced either by the bite of the skunk, wolf, or dog. Niemeyer† states that 'the assertions of Marochetti, who claims that during the incubation stage vesicles form beneath the tongue, and that by destroying these vesicles the outbreak of the disease can be averted, have not been substantiated.'

"3. That the invariable accompaniments of Rabies Canina were not wanting in the cases of R. Mephitica. The specific action of the poison was made manifest first by the œsophageal branch of the eighth pair, giving rise to the characteristic symptom of the disease, or to the extreme difficulty of swallowing, especially of fluids; then the frequent catching of breath noticed in all cases, showing that the recurrent nerve was also affected; later brilliant eye, and the sense of touch becomes painfully excited, hyperæsthesia existing in a marked degree, with the exception of the case reported of R. Canina, all of which point to some lesion of the central and spinal nerves. That the brain itself, and especially the region of the medulla oblongata becomes affected by the terrible convulsions and delirium in the more advanced stage of the malady. The spasms in all the cases were unlike those of tetanus, less continuous, remittent, and often intermittent. In none of the cases produced by the skunk bite was there any loss of perception. In no case that I saw did morphia have any effect in abridging the fearful struggles; death either ended with convulsions, or exhausted by the terrible exertions a sudden calm took place, and, as if nature gave up the conflict, died without a groan."

---

* "London *Medical Times and Gazette*, Aug. 8, 1874, p. 159."
† "Niemeyer, Pract. of Med."

CHAPTER VIII.

MEPHITINÆ—Continued: SKUNKS.

The genus *Mephitis*, continued—*Mephitis macrura*, the Long-tailed Mexican Skunk—Synonymy—Habitat—Specific characters—Description—The subgenus *Spilogale*—*Mephitis (Spilogale) putorius*, the Little Striped Skunk—Synonymy—Habitat—Specific characters—Description of external characters—Description of the skull and teeth—History of the species—The genus *Conepatus*—*Conepatus mapurito*, the White-backed Skunk—Synonymy—Habitat—Specific characters—Description of external characters—Description of the skull and teeth—Description of the anal glands—Geographical distribution and habits.

THE length of the foregoing chapter having rendered a division of the parts of the work relating to *Mephitinæ* advisable, I continue directly with an account of the other species of the genus *Mephitis*, and of the genus *Conepatus*.

Long-tailed Mexican Skunk.

Mephitis macrura.

Mephitis macroura, *Licht.* Darst. Säug. 1827-34, pl. 46, "f. 1, 2"; Abh. Ak. Wiss. Berl. 1836 (1839), 277.—*Wagner*, Suppl. Schreber, ii. 1841, 196.—*Schinz*, Syn. i. 1844, 323, no. 12.—*Baird*, M. N. A. 1857, 200.—*Tomes*, P. Z. S. 1861, 280.—*Gerr.* Cat. Bones Br. Mus. 1862, 97. [Not macroura of Aud. & Bach., nor of Woodhouse.]
Mephitis mexicana, *Gray*, Mag. N. H. i. 1837, 581; P. Z. S. 1865, 149; Cat. Carn. Br. Mus. 1869, —.
? Mephitis edulis, *Berlandier*, MSS. ined.*
Quid Mephitis longicaudata, *Tomes*, P. Z. S. 1861, 280 (Guatemala) ?

---

* The animal referred to by Berlandier is probably this species. "Smaller than the Polecat. Length of head and body 13 inches; tail 11; black; a white frontal line; another on the nape, dividing into two lateral ones, afterward converging near the root of the tail, on which they unite; tail white-tipped. Inhabits most of Mexico. I have found it about San Fernando de Bexar, and in eastern interior States, where it is improperly called *Zorillo*. It may be tamed; is rather nocturnal; hunts various small animals; is slow and heavy in its movements, and bites forcibly. The fluid is highly phosphorescent by night. The natives are fond of its flesh; they kill it, taking care not to irritate it, remove the anal glands entire, cut off the head and feet, singe off the hair, and broil the flesh. I overcame my repugnance on one occasion and tasted the meat, which I found not disagreeable; it resembled young pork." (Freely translated with abridgment from the original MSS.)

Hab.—Mexico. (Not known to occur in the United States.)

Specific Characters.—Tail very long, the vertebræ alone nearly as long as the head and body; tail with hairs not shorter than the head and body. A broad undivided white dorsal area (as in *Conepatus*), with lateral stripe and frontal streak (in the specimen examined).

*Description.**

The specimen which I refer without hesitation to this species is considerably smaller than *M. mephitica*, in fact little exceeding *Spilogale putorius*, with a tail (including hairs) longer than the head and body, and other characters indicating specific validity.

I have no doubt that this specimen represents Lichtenstein's animal, described as above cited, from Mexico. The *M. vittata* of the same author, *op. cit.* pl. 47 (also Abhand. Akad. Wiss. Berlin for 1836, 1838, 278; Wagner, Suppl. Schreb. ii. 1841; Baird, M. N. A. 1857, 200), from Oaxaca, is based upon certain slight peculiarities of color, and does not seem to be specifically distinguishable. But I have seen no specimens corresponding with Lichtenstein's descriptions, and consequently do not venture to commit myself in the matter.

The general physiognomy is that of true *Mephitis*, the snout very acutely pointed but not produced, and closely furred to the small, rounded, definitely naked nasal pad. The nostrils are antero-lateral. The ears are much as in *M. mephitica*—perhaps rather smaller. The fore claws seem to be remarkably long, slender, and curved; the outer reaches more than half-way to the end of the fourth, while the first barely attains the base of the second. There may be some peculiarity in the tuberculation of the soles, but this cannot well be made out in the dried specimen. The palms are perfectly naked; the soles the same, excepting a little space on the side of the heel. The tail-vertebræ appear just to about equal in length the head and body, which is not the case in any other species of the family I have seen. The tail is much less bushy than usual in *Mephitis* proper, but seems to have been in this specimen somewhat worn away.

In coloration, this species curiously combines the broad white dorsal area of *Conepatus* with lateral dorsal stripes and frontal streak of *Mephitis mephitica*. The white dorsal area begins squarely on the nape and continues uninterrupted to the tail, but is only pure white anteriorly, being elsewhere of a grayish

---

* From No. 8566, Mus. Smiths., Orizaba, Mexico, *Botteri*.

cast, and between the shoulders a small irregular black spot appears, leading to the supposition that the white dorsal area in this species is liable to the same variations that are known to exist in other speci s. In addition to this general white area, a slight white lateral stripe starts independently over each shoulder and is continued for a little distance along the sides— in this specimen further on one side than on the other. The frontal streak is short and slight. There is a white area on the breast between the fore legs, prolonged backward as a streak—it is probably not constant, but then I have never seen any white on the under surface of *M. mephitica!* (compare expression in Linnæus's diagnosis of his *Viverra memphitis* of the 10th edition—"*subtus ex albo et nigro variegatus*"). The tail is entirely and intimately mixed black and white—viewed from below, we see chiefly hairs pure white at base and black at end, from above, mixed white and black-and-white hairs, producing a grizzled gray cast, and in direct continuation of the dorsal stripe. (This is exactly as given by Lichtenstein; Gray's description gives the tail as black; the variation is thus seen to be as in the allied species.)

The dimensions of a dried but fairly well-stuffed specimen are as follows: Nose to root of tail about 13 inches (Lichtenstein says 14): tail-vertebræ nearly the same, but rather less (13 inches—*Licht.*), the hairs in this instance under three inches longer (5 inches—*Licht.*). Fore foot 2 inches, of which the longest claw is 0.65; hind foot 2.25.

This species was supposed (but erroneously as far as known) to inhabit the United States by Audubon and Bachman and by Woodhouse, the animal described by these authors being simply the common *M. mephitica* under one of its interminable color-variations.

The Subgenus SPILOGALE. (GRAY.)

The characters of this subgenus having been indicated on a preceding page (p. 192), we may at once proceed to consider the single known species.

The Little Striped Skunk.

Mephitis (Spilogale) putorius, (L.)

PLATES XII, XIII, XIV.

Viverra putorius, *L.* S. N. i. 10th ed. 1758, 44, no. 3; i. 1766, 64, no. 4 (partly. Based primarily upon *Putorius americanus striatus,* Catesb. Car. ii. 1731, 62, pl. 62. Quotes also Kalm, Itin. ii. 378. Includes syns. and descr. of *V. mephitis* of 10th ed. Diagnosis agrees sufficiently with *Spilogale;* general bearing rather upon *Mephitis mephitica*).—*Gm.* S. N. i. 1788, 87, no. 4 (partly).

Spilogale putorius, *Coues,* Bull. U. S. Geol. and Geogr. Surv. Terr. 2d ser. no. 1, 1875, p. 12 (skull and teeth).

Mephitis interrupta, *Raf.* Ann. Nat. 1818, 3, no. 4 ("Louisiana").—*Less.* Man. 1827, 152, no. 411.—*Griff.* An. Kingd. v. 1827, 129, no. 35 r.—*Fisch.* Syn. 1829, 162.—*Licht.* Abh. Akad. Wiss. Berl. 1836 (1838), 283, pl. 2, f. 1.—*Schinz,* Syn. i. 1844, 325, no. 16. (All after *Rafinesque*.)

Spilogale interrupta, *Gray,* P. Z. S. 1865, 150; *Cat.* Carn. Br. Mus. 1869, —.

Mephitis bicolor, *Gray,* Mag. N. H. i. 1837, 581.—*Baird,* M. N. A. 1857, 197.—*Parker,* Am. Nat. iv. 1870, 376; iv. 1871, 761 (Iowa, and probably New York).—*Allen,* Bull. M. C. Z. ii. 1871, 169 (Florida, common).—*Merr.* U. S. Geol. Surv. Terr. 1872, 662 (Idaho).

Mephitis zorilla, *Licht.* Abh Akad. Wiss. Berlin, 1836 (pub. 1838), pl. 2, f. 2 (not of Licht., Darst. pl. 48, f. 2, which is an African species—whether the Zorille of Buff., xiii. 1765, 302, pl. 41 ??).—*Wagn.* Suppl. Schreber, ii. 1841, 199, pl. 123.—*Schinz,* Syn. i. 1844, 325, no. 15.—*Aud. & Bach.* Q. N. A. iii. 1854, 276 (tab. nulla).

Mephitis americana *var. R., Desm.* Mamm. i. 1820, 187 (= *interrupta* Raf.).

Mephitis quaterlinearis, *E. W. Winans,* writing from Williamsport, Kans., in a (Kansas?) newspaper, name unknown, date 1859.*

* My endeavors to complete the reference, and thus place the synonym properly on record, have been unavailing. The newspaper clipping which came into my possession does not include even the name of the paper—nothing but a date, "1859", in *MS.,* which I recognize as that of Prof. Baird, who, however, has no recollection of the source of the clip. The following is the article in full, with typography copied as closely as possible:—

"Mephitis Quaterlinearis.—Win.—Four-striped Skunk.

"DENTAL FORMALA.

"Incisive 6-6; Canine 1-1—1-1; Molar 4-4—5-5 = 34.

" From point of nose to end of tail, vertebrae,	22½ inches.
" From heel to top of shoulder,	4½.
" Length of hair in end of tail,	4½.
" Middle toe nail,	⅜.

" General color, pure black; a spot of white on either side of the head between the eye and ear, another between the eyes, making three on the head. Four parallel lines of white about one fourth of an inch in breadth and three-fourths of an inch apart, have their origon about the posterior part of the head, the two upper originally on either side of the occiput, while the two lower have their commencement behind and at the lower part of each ear, all of which are carried directly backward to the posterior ribs, where the lower lines terminate and the upper curving downward and forward then rather ascend to the hind part of the shoulder where they descend one inch to the elbow joint. A transverse band of white crosses the fore part of the hips with an interruption of one inch at the side of the back. A spot of

??? **Mephitis myotis,** *Fisch.* Syn. 1829, 162 (based on *Bête puante,* Du Pratz, Louis. ii. 97, fig. (not identifiable, but very likely belonging here).
Little Striped Skunk, *Authors.*
Moufette interrompue, *Less. l. c.*
Stinkthier mit unterbrochenen Binden, *Schinz, l. c.*

HAB.—United States, southerly. Carolina, Georgia, Florida. Iowa, Kansas, Wyoming, Idaho, Colorado. Washington Territory. Southwestern States and Territories. Cape St. Lucas. (? New York.)

SPECIFIC CHARACTERS.—Black or blackish, with numerous white stripes and spots, and tail white-tipped. Small; a foot or less long; tail (with Lairs) obviously shorter than the body.

*Description of external characters.**

This animal is the smallest of the American species, as fully indicated by the measurements given beyond. In form, it agrees closely with the common species, excepting in the shortness of the tail, in which there is a decided approach to *Conepatus.* The tail-vertebræ are considerably less than half as long as the head and body, the tail with its hairs being obviously less than the length of the head and body, although this member is full and bushy. The pelage differs from that of *M. mephitica* in being notably finer, softer, and closer. In other respects of external form, the species agrees closely with *M. mephitica*—so closely that further details are not required.

white on the upper part of each thigh; one on either side of the root of the tail; and a tuft of white hair arises from the tip of the tail; nose covered with short hair which is naked. More or less fur is interspersed with the long hair to the extremity of the tail differing in respect from the common American Skunk (Mephitis Chinga); and, also in being less in size and weight and having a finer and denser pelage differently striped and spotted and being of a more slender form. Its habits, so far as they are known, agree with those of the last named species.

"Its geographical distribution is not yet determined the specimens which I have examined were obtained in Kansas and vary in their markings as others of the genus. The above specimen was a male taken on the 16th of Dec., 1852, excessively poor and weighed only one pound three ounces avoirdupois.

"Another male specimen which I examined weighed two pounds; his measurements being rather less than the above; his markings were similar excepting they were finer. The three white spots about the head and the tuft of white hair in the end of the tail do not appear to be subject to any variation. The female differs from the male in being smaller.—After a careful investigation I now venture to introduce this n am mal as here-to-fore being a nondescript.

"EDGAR W. WINANS.

"Williamsport, Shawnee Co., K. T."

* From No. 11899, Mus. Smiths. Inst., Fountain, Colorado, *C. E. Aiken.*

In color, this animal is black or blackish, relieved with white, like the other species. But the number and disposition of the markings are peculiar, affording specific characters in spite of an almost endless diversity in detail of the numerous white spots and stripes by which it is superficially distinguished from all its allies. The fantastic harlequin-like coloration is scarcely duplicated in any two specimens; in fact, the opposite sides of the same specimen show sometimes an appreciably different pattern. The markings are difficult of adequate expression in words that shall cover all their modifications; and those of the same specimen might easily be described in such different ways as to convey an impression of distinct species—as indeed has been done. The following formula, drawn from the most commonly observed state of the markings, probably covers most cases:—

Black. A white spot on forehead between eyes. A white spot on each cheek in front of ear more or less confluent with a white stripe which starts behind the ear. Indefinite white touches on chin and about angle of mouth. Four parallel equidistant white stripes on fore part of body above, beginning opposite the ears; the lower, lateral or external pair of these ending back of the shoulder, the median pair curving around the end of the lateral pair, downward and then forward to the fore leg. A white stripe transversely across the flanks, broken in two by an interruption on the median line of the back. A pair of white spots on the middle of the back just in advance of the last-named stripe. A white spot over each hip. A pair of white spots at base of tail. A white tuft at end of tail.

The notorious inconstancy of the white markings of Skunks, even of those in which the pattern is normally simplest, finds room for exaggeration in the highest degree in this case where the normal markings are numerous and complicated. In some cases, owing to interruption of the usual stripes, I have counted no less than eighteen separate white marks, exclusive of tail-tip and the vague chin-spots. The three head-spots and the four parallel dorsal stripes on the anterior part of the body are the most constant, and may, so far as I have seen, be always traced, though the median pair of stripes are liable to slight interruption. The lateral pair are the firmest of all the markings. There is special liability to a break in these stripes where they begin to curve downward on the side. Complete break here, fusion of the solitary pair of dorsal spots with the transverse flank stripe, and lengthening into a stripe of the hip-spot, result in three vertical crescentic stripes succeeding each other behind the end of the main lateral stripe, that runs from the ear over the shoulder. Interruption of these transverse crescents may give a set of numerous spots, without traceable stripes, on the hinder half of the body; indeed, the markings of this part of the body are wholly indefinite. The lateral spots at the root of the tail often fuse into one. The tail is ordinarily black with definite white tip, but may have white hairs mixed with the black throughout, or be all black or all white. The shoulder-stripe sometimes sends short spurs around toward the throat and breast. The chin and upper throat may be perfectly black, or streaked throughout with white. The part of the ear corresponding to the white markings about it is commonly light-colored; the rest of the ear is black. The naked muffle is dark-colored. The claws are dull horn-color.

The black of this animal is generally quite pure and glossy on every part; but sometimes it has a brownish tinge, especially notable in old museum specimens.

In this connection, the reader will refer, if he is sufficiently interested to do so, to Plate XII, on which is a wood-engraving of a photograph of two skins, showing the complicated markings very clearly.

Description of the skull and teeth.

Numerous specimens before me, labelled "*bicolor*" and "*zorilla*", exhibit surprising variation in size and shape, without, however, warranting presumption that they are not all of the same species. Independently of the usual differences according to age, there is a remarkable range of variation in the width and depression of the skull behind and development of the occipital crest. An average specimen is selected for description, in which the range of variation will be also noted. Comparative expressions used have reference to the skull of *Mephitis mephitica*.

The skull is smaller than that of *mephitica*; excepting one abnormally large example, all are much less in every dimension than the smallest (adult) skulls of *mephitica* which I have seen. Viewed from above, the muzzle appears more tapering, if not also relatively shorter; the angle of obliquity of truncation of the nasal orifice is much the same. Supraorbital processes are small, but well defined, as acute eminences, prolonged from well-defined ridges of bone divaricating from the sagittal crest. This crest is a single and acute ridge in adult skulls; in young ones, it is a tablet of bone, the sides of which separate almost at once from the occipital protuberance. There is little post-orbital constriction of the skulls; the least width there being little, if any, less than the interorbital width. The lateral divergence of the zygomata is much as in the last species; but their upward convexity is usually greater, and the summit of the arch is at its middle. Behind, the skull is notably widened and flattened, almost as in *Taxidea*, the intermastoid diameter being relatively much greater than it is in either *Conepatus* or *Mephitis*; in fact, it is not very much less than the interzygomatic width, in some cases at least. Nevertheless, the mastoid processes are themselves less developed than in *Mephitis* proper, extending little, if any, beyond the orifice of the meatus, instead of flaring widely outward. The occipital crest is strongly developed, and its outline is characteristic in the great convexity of contour on each side and deep median emargination; in other genera, the median emargination is always slight, sometimes *nil*; and the lateral outline from the mastoid to the point where the supraoccipital bones leave the general occipital crest is about straight—if anything, concave.

A notable peculiarity appears in the profile view of the skull. The dorsal outline in *Mephitis mephitica* is strongly convex, with a high point about the middle, and this is carried to an extreme in *M. frontata*; in the present case, the same outline is nearly straight from the ends of the nasals to near the occipital protuberance; in fact, the skull is as flat on top as an Otter's, and flatter than a Badger's. The zygomata are strongly arched upward, with a regular curve throughout, instead of being highest behind; the prominence of the bulla ossea on the floor of the skull is sufficient to bring this part fairly into view from the side, as is scarcely the case in *M. mephitica*; this feature is also due, in part, to an abbreviation of the mastoid process, which is hardly at all produced downward.

On the floor of the skull, the principal feature is the width behind, which, being simply coördinate with the general lateral dilatation already noticed, requires no further comment. The paroccipitals are very small—in fact, mere nibs of bone, hardly able to bear the term "process". There are also strong points in connection with the bullæ auditoriæ and periotic region generally. The bullæ are not only more swollen at the usual point of greatest inflation, but, behind them, the part that reaches between the lateral elements of the occipital bone and the lateral portion of the lambdoidal crest is also turgid, having a general smooth convexity instead of an irregular concavity. The general turgescence is due to the greater development of the mastoid sinuses. The bony palate ends in the same relative position as in *M. mephitica*, and shows the same variation in the character of the edge of this shelf.

The mandible, though, of course, proportionally smaller than in *M. mephitica*, is identical in shape, contrasting equally well with the peculiarities of *Conepatus*, elsewhere mentioned.

The smaller size aside, there is scarcely anything in the dentition of this species calling for comment in comparison with *Mephitis*. The anterior premolar is well developed, and, as far as I can see, the dentition is, in other respects, nearly identical with that of *Mephitis;* the upper sectorial tooth (posterior premolar), however, has the cusp of its inner moiety rather a pointed process of the border itself than a conical cusp, surmounting this inner part.

It should be noted that in one specimen, as an abnormality, the anterior upper premolar has aborted entirely on the right side, though present on the left; while the lower jaw of the same specimen shows an abortive *posterior* premolar on the left side. But, in general, in *Mephitinæ*, abortion or other irregularities of dentition are less frequent than in the *Mustelinæ*, where the smaller teeth are more crowded.

History of the species.

In the case of an animal whose markings are so variable as those of the Striped Skunk, recognition of the species in nature becomes a matter requiring some judgment and experience; and it is not in the least remarkable that compilers of vague and often conflicting descriptions, or of inaccurate figures badly drawn from stuffed specimens, or even prepared from poor descriptions, should have made inextricable confusion. In an attempt to trace the written history of *Mephitis putorius*, it is probably not possible to identify all the names which have been imposed upon it, nor even to fix the date of its first appearance in literature. It is certain, however, that the animal was known to the earlier writers; its characters being clearly traceable in some of the descriptions of the last century, long before the period when Rafinesque and Gray respectively bestowed those names which have become most extensively current. Referring to the above synonymatic list as a *résumé* of the views entertained concerning the record of the species, some points of special pertinence to *M. putorius* may be here noticed.

Le Zorille of Buffon (Hist. Nat. xiii. 1765, pp. 289, 302, pl. 41) is a starting-point of a number of compilations, as at the hands of Erxleben, Gmelin, Shaw, and others; it does not reappear in Linnæus, who carried his grudge against his French rival so far as to ignore him in the "Systema Naturæ", thereby hurting only the book. It is described from South America, and is to be carefully distinguished from an African species, of an entirely different group, also called *Zorilla*. Descriptions of a *Viverra* or *Mephitis zorilla* agree substantially in points of small size and much variegation with white; and thus, perhaps without exception, bear hard upon the present species, if they may not actually represent it. In many cases, however, the accounts are complicated or negatived by introduction among the synonyms of some names which apparently appertain to *Conepatus*, or to *Mephitis* proper. Whether or not we agree with Prof. Lichtenstein that Buffon's *Zorille* was this species, various indications of *Viverra zorilla* which flow from it cannot be satisfactorily and exclusively located here, and are to be passed over. They are, in effect, as they stand upon the pages, compounds which have no actual existence in nature.

The Pol-cat of Catesby, as above quoted, described with five narrow white lines, is a species which authors have found it difficult to locate, as the Common Skunk, *M. mephitica*, the only one supposed to inhabit Carolina, presents no such character. But since the discovery of the existence, in this portion of the United States, of a *Spilogale*, which is the only species having *several* white lines, the pertinence of Catesby's reference here is evident. Catesby is primarily the basis of *Viverra putorius*, the only species of Skunk in the 12th edition of Linnæus; and Linnæus's diagnosis "*V. fusca lineis quatuor dorsalibus albis parallelis*" is exactly and exclusively pertinent to the present species, which is, moreover, the only animal that presents this character. The four white stripes upon the anterior half of the body are its strong and constant character. It is true that the remainder of Linnæus's account does not agree well, but neither does it agree with any Skunk known to me ("*subtus ex albo et nigro variegatus*", &c.); and he also cites some references that probably belong elsewhere. In adopting the name

S. putorius from Linnæus, as I recently did (*l. s. c.*), I rested upon the exclusive pertinence of his diagnosis, and his quotation of Catesby.

The *Mephitis interrupta* of Rafinesque may or may not have been "a pure figment of his imagination". It probably, however, had some basis, and if his account does not wholly agree with specimens of *Spilogale putorius* examined, it will be remembered that even his elastic imagination would be put to the stretch to describe a spotted and striped Skunk in terms too exaggerated to be met by the reality which this species offers. We may accept his name as undoubtedly belonging here, and in fact we should adopt it, as a more definite appellation than *zorilla*, were it not anticipated by Linnæus, as just shown.

Among earlier accounts, the best description I have seen is that presented by Shaw, page 389, vol. i. of the General Zoölogy, under head of "var." of his Striated Weesel. Shaw refers to some miscellaneous plates of animals published a short time previously by Mr. Catton, among which is a representation of an animal "having only four white bands on the back, and the tail almost entirely white; a patch of white appears below each ear, and a small triangular white spot on the forehead. In the description accompanying the plate the animal is said to have measured twelve inches from nose to tail, and to have been brought from Bengal." The probably erroneous locality aside, the whole account is perfectly, and indeed exclusively, pertinent to *Spilogale putorius*.

In Du Pratz's Louisiana, there is a description of a "bête puante", which certainly conforms to no known species, but which was probably meant to be this one, to judge from the locality and the ascribed size. It is the basis of *Mephitis myotis* Fischer, *l. c.*

In 1837, Dr. J. E. Gray bestowed upon this species the name of *M. bicolor*, by which it has been generally known of late years. About the same time, Lichtenstein adopted the name of *M. zorilla*, after Buffon, in which he was followed by Wagner and Audubon. Lichtenstein's earlier *M. zorilla*, of the Darstellungen neuer Säugethiere, u. s. w., is the entirely different African animal.

The only description I have seen in which *four* white lines are prominently indicated since those of Linnæus and Shaw is an account given in 1859, when a certain *Mephitis quaterlinearis* was formally named and described in a Kansas(?) newspaper, as above quoted. It is singular that upward of a century intervened between these two curiously concordant accounts. Mr. Winans's description is accurate in the minutest particulars; it was evidently taken from a specimen exactly like some of those now before me.

Geographical distribution and habits.

The geographical distribution of this species is much more extensive than has been generally supposed. Thus, Prof. Baird, in 1857, gave its habitat as merely "Southern Texas and California", and the indications of most authors are of a western and southwestern animal. But there is no doubt now of its inhabiting the greater part of the Southern States, and of the United States west of the Mississippi. I have examined specimens from Georgia and Florida, in which last State Mr. Allen considers it common, from various portions of the West, and from Cape St. Lucas. Mr. H. W. Parker, in his notes in the American Naturalist, as above quoted, records the species from Iowa, where at least fifty pelts were obtained one season, near Des Moines, and as probably occurring in the State of New York:—"There is reason to believe that the species may be found even in central New York. Dr. S. J. Parker, of Ithaca, N. Y., has twice seen by the roadside, in that region, a small, many-striped skunk, very different from the common one."

Respecting the habits of the species, I have no information to offer. It is not to be presumed that it differs materially from the common species in this regard. Mr. Maynard has stated that in Florida the animals are domesticated and used like cats, the scent-glands being removed at an early age; they become quite tame and efficient in destroying the mice (*Hesperomys*) that infest the houses.

The different species of Skunks, in fact, seem to be susceptible of ready semi-domestication, in which state they are, like the Fitch or Ferret, useful in destroying vermin, if they do not also make agreeable pets. Writers speak of the removal of the anal glands in early life, to the better adaptation of the animals to human society, and such would appear to be an eminently judicious procedure. For, though Skunks may habitually spare their favors when accustomed to the presence of man, yet I should think that their companionship would give rise to a certain sense of insecurity, unfavorable to peace of mind. To depend upon the good will of so irritable and so formidable a beast, whose temper may be ruffled in a moment, is hazardous—like the enjoyment of a cigar in a powder-magazine.

The Genus CONEPATUS. (GRAY.)

Viverra *sp.*, Gulo *sp.*, of some authors.
< Mephitis of most authors.
< Marputius, *Gray*, Charlesw. Mag. N. H. i. 1837, 581.
= Conepatus, *Gray*, Charlesw. Mag. N. H. i. 1837, 581.
= Thiosmus, *Lichtenstein*, Abh. Akad. Berl. for 1836, 1838.

The very well-marked characters of this genus have already been given (p. 192). The peculiarities of the skull and teeth are correlated with certain modifications of external contour, which give the animal a somewhat Badger-like aspect, though there is no mistaking it for anything but a Skunk. It is the only known representative of the subfamily in South America.

I have not been able to examine any specimens of this genus from other than United States and Mexican localities, representing in strictness the *Mephitis* (*Thiosmus*) *mesoleucus* of Lichtenstein and late American writers, the *leuconotus* of Lichtenstein, and the *M. nasuta* of Bennett. The synonymatic list given beyond must be regarded as somewhat tentative or presumptive, indicating that I see nothing in the *descriptions* of authors forbidding the supposition that the seemingly interminable list of nominal species really refers to more than a single good one. In adopting a name for the "Conepatl", I simply take the oldest one I find. Should there prove to be more than one species of this genus included in the synonyms given, my article is to be held to refer solely to that one which occurs in Mexico and the southwestern portion of the United States, and upon which the descriptive matter herewith given is exclusively based.

Nowhere, perhaps, in the literature of mammalogy have greater confusion and uncertainty prevailed than in that portion which relates to the Skunks, and the history of the Conepatl is certainly not less hopelessly involved than that of other Skunks. Views of authors have oscillated between such extremes as those held by Cuvier and his imitators, for whom a Skunk was a Skunk, and those of other persons for whom an inch of tail or a speck of color was a good specific character. Into the tedious discussion of the names cited below I do not propose to enter, but shall content myself with giving a faithful description of the United States animal.

Gray's barbarous term for this genus, *Conepatus*, is obviously the same as the old Mexican *Conepatl*. Of its meaning I am not certain; but it probably refers to the burrowing of the animal; for, it may be observed, *nepantla* in the Nahuatl language signified a subterranean dwelling.* Gray's other generic name, *Marputius*, is similarly related to *Mapurito*. This word may be compared with *Mephitis* itself through such forms as *Mafutiliqui*, French *Moufette*, &c.

The White-backed Skunk.

Conepatus mapurito.

PLATE XV.

Viverra putorius, *Mutis*, "Act. Holm. xxxii. 1769, 68" (*non* Linn.).
Viverra mapurito, *Gm.* S. N. i. 1788, 88, no. 15 (*ex* Mutis).—*Shaw*, G. Z. i. 1800, 392.—*Turt.* S. N. i. 1806, 53.
Gulo mapurito, *Humb.* "Rec. Obs. Zool. —, i. 350".
Mephitis mapurito, *Less.* Man. 1827, 151, no. 407.—*Fisch.* Syn. 1829, 161.—*Schinz*, Syn. i. 1844, 318, no. 1.—*Licht.* Abh. Ak. Wiss. Berlin, 1836 (1838), 270 (*Thiosmus*).—*Tschudi*, Fn. Peru. 1844-46, 113.—*Giebel*, Säug. 1855, 764.
Conepatus mapurito, *Coues*, Bull. U. S. Geol. and Geogr. Surv. Terr. 2d ser. no. 1, 1875, 14 (skull and teeth).
? **Viverra mephitis**, *Gm.* S. N. i. 1788, —, — ("*Chinche*, Buff., xiii. pl. 39").—? *Turton*, S. N. i. 1806, 53.
Viverra conepatl, *Gm.* S. N. i. 1788, 88, no. 12 (*Conepatl*, Hern., Mex. 332).
Mephitis conepatl, *Fisch.* Syn. 1829, 160.
? **Mephitis chilensis**, "*Geoff.* Cat. Mus." (*Moufette du Chili*, Buff. II. N. Suppl. vii. 233, pl. 57).—*Fr. Cuv.* "Dict. Sc. Nat. xiii, 18—, 126".—" *Griff.* Anim. Kingd. ii. —, —, f. —."—*Less.* Man. 1827, 152, no. 408.—*Fisch.* Syn. 1829, 160.—*Licht.* Abh. Ak. Wiss. Berlin, 1836 (1838), 272 (*Thiosmus*).—*Schinz*, Syn. i. 1844, 319, no. 4.—"*Gray*, P. Z. S. 1848, —, —."—*Gieb.* Säug. 1855, 765.
? **Marputius chilensis**, *Gray*, "Mag. N. II. i. 1837".
? **Conepatus "chinensis"**, *Gerr.* Cat. Bones Br. Mus. 1862, 97 (by typog. err. for *chilensis*).
? **Gulo quitensis**, *Humb.* "Rec. Obs. Zool. i. —, 346 [or] 347" (*Atok, Zorra, Glouton de Quito*).
? **Mephitis quitensis**, *Less.* Man. 1827, 152, no. 410.—*Fisch.* Syn. 1829, 161.—*Licht.* Abh. Akad. Wiss. Berlin, 1836 (1838), 273 (*Thiosmus*).—*Schinz*, Syn. i. 1844, 319, no. 4.
? **Gulo suffocans**, "*Ill.* Verh. Berl. Akad. Wissensch. 1811, 109 (*Yaguaré*, Azara, i. 211. of French transl.—*Chinche, Feuillée*, Journ. Obs. Phys. 1714, 272)" (Brazil and Paraguay).
? **Mephitis suffocans**, *Licht.* Darst. Säug. 1827-34, pl. 48, f. 1; Abh. Akad. Wiss. Berlin, 1836 (1838), p. — (*Thiosmus*).—*Schinz*, Syn. i. 1844, 320, no. 5 (*Thiosmus*).—*Gieb.* Säug. 1855, 765.
? |**Mephitis feuillei**, —, "Zool. de la Bonite, —, —, pk 3, f. 1-3" (*Schinz*).
? **Conepatus humboldtii** et *var.*, *Gray*, Mag. N. II. i. 1837, 581; List Mamm. Br. Mus. 1843, 69.—*Gerrard*, Cat. Bones Br. Mus. 1862, 97.
? **Mephitis humboldtii**, *Blainv.* "Osteog. Mustela, pl. 13, f. — (teeth)".
? **Mephitis patagonica**, *Licht.* Abh. Akad. Wiss. Berl. 1836 (1838), 275 (*Thiosmus*) ("*Yaguaré, Maikel*, Falk. Patag. 128").—*Schinz*, Syn. i. 1844, 320, no. 6 (*Thiosmus*).—"*Burm.* La Plata, ii. —, 409."—*Gieb.* Säug. 1855, 765.
? **Mustela (Lyncodon) patagonica**, *D'Orb.* "Voy. Amér. Mérid."
? **Mephitis amazonica**, *Licht.* Abh. Ak. Wiss. Berl. 1836 (1838), 275 (*Thiosmus*).—*Schinz*, Syn. i. 1844, 321, no. 7.—*Tchudi*, Fn. Peru, 1844-46, 115.
? **Conepatus amazonica**, *Gray*, List Br. Mus. 1843, 69.
? **Mephitis molinæ**, *Licht.* Abh. Akad. Wiss. Berl. 1836 (1838), 272 (*Thiosmus*) ("*Chinche*, Molina, Hist. Nat. Chili, 240") (Chili).—*Schinz*, Syn. 1844, 321, no. 8.

---

* *Fide* Prof. G. Barroeta, of San Luis Potosi, Mexico, to whom I applied personally for this information.

?**Mephitis gumilllæ**, *Licht.* Abh. Ak. Wiss. Berl. 1836 (1838) ("*Moufette mapurita* or *Maju-tiliqui*, Gumilla, Orinoco, ii. 276").—*Schinz*, Syn. i. 1844, 321, no. 10.
Mephitis mesoleuca, *Licht.* Darst. Säug. 1827-34, ——, pl. 44, f. 2; Abh. Akad. Wiss. Berl. 1836 (1838), 271, pl. 1, f. 1 (*Thiosmus*) (Chico, Mexico).—*Wagn.* Suppl. Schreb. ii. 1841, 192, pl. 121 A.—*Schinz*, Syn. i. 1844, 319, no. 3.—*Aud. & Bach.* Q. N. A. ii. 1851, 18, pl. 53.—*Gieb.* Säug. 1855, 764.—*Baird*, M. N. A. 1857, 192 (subg. *Thiosmus*).—*Tomes*, P. Z. S. 1861, 280 (Guatemala).—*Maxim.* Arch. f. Naturg. xxvii, 1861, 212.
Thiosmus mesoleucen, *Less.* Nouv. Tabl. R. A. 1842, 66.
Thiosmus mesoleucos, *Chatin*, Ann. Sci. Nat. 5th ser. xix. 1874, 100, pl. 6, f. 59-63 (anat.).
Mephitis leuconota, *Licht.* Darst. Säug. 1827-34, ——, pl. 44, f. 1; Abh. Akad. Wiss. Berl. 1836 (1838), 271 (*Thiosmus*) (Alvarado, Mexico).—*Schinz*, Syn. i. 1844, 319, no. 2.—*Gieb.* Säug. 1855, 764.—*Baird*, M. N. A. 1857, 200.—*Tomes*, P. Z. S. 1861, 280 (Guatemala).
Mephitis leuconota intermedia, *De Sauss*, R. M. Z. 1860, 6.
Mephitis nasuta, *Bennett*, P. Z. S. i. 1833, 39 (California).—*Gray*, Mag. N. H. i. 1837, —.—*Fraser*, Zoöl. Typ. no. 4, pl. —.
Marputius nasuta, *Gray*, Mag. N. H. i. 1837, 581.
Thiosmus nasuta, *Less.* Nouv. Tabl. R. A. 1842, 66.
Conepatus nasutu, *vars.* **nasuta, humboldtii, chilensis** *et* **lichtensteinii**, *Gray*, P. Z. S. 1865, 145, 146, 147; Cat. Carn. Br. Mus. 1869, —.
??"**Mephitis castaneus**, *D'Orbig.* Voy. Amér. Mérid. 21, pl. 12, 13, f. 2."—"*Giebel*, Odontog. 35, pl. 13, f. 7."—*Gieb.* Säug. 1855, 765 (Southern South America).
?"**Mephitis furcata**, *Wagn.* Suppl. Schreb. ii. 1841, 192" (Chili).—*Tsch.* Arch. Naturg. 1842, 242; Fn. Peru. 1844-46, 114 (Peru).
?**Mephitis westermanni**, *Reinh.* Vid. Selsk. Forh. 1856, 270.
?**Mephitis americana**, *vars.* D?, E, F, G, H, I, M?, Q, *Desm.* Mamm. i. 1820, 186-187.'
?**Mephitis americana**, *vars.* a, d?, h, m, n, o, p, s, *Griff.* An. Kingd. v. 1827, 127, no. 358.
?**Maikel**, Patagonian Maikel, *Gray*, *l. c.*

HAB.—Southwestern border of the United States and southward through Mexico and Central and South America.

SPECIFIC CHARACTERS.—Black or blackish, with a white dorsal area sometimes divided by a black vertebral stripe, rarely broken into several portions; tail white, or black and white.

Description of external characters.*

This species differs materially from the North American and Mexican *Mephitis* proper, in many points of external anatomy, as well as the more essential structural peculiarities upon which the genus *Conepatus* primarily rests. These secondary generic characters are the same, as far as known, in all the several slight varieties of the (probably) single species which represents the genus. The general configuration of the body and limbs is much the same as in *Mephitis;* but the physiognomy is wholly peculiar, while the short scrubby tail, almost rudimentary ears, and a particular arrangement of the sole-pads, are all highly characteristic.

The Conepate is the largest of the Skunks, some specimens attaining a length of about two feet, exclusive of the tail; but an ordinary dimension is about a foot and a half from the nose to

---

* From specimens from Texas and Mexico, with uninterrupted white dorsal area and white tail.

the root of the tail. The head is more lengthened and narrowly conical than in *Mephitis*. The facial aspect is highly characteristic in the production and attenuation of the snout and lowness of the ears. The long snout is broad and depressed on top, obliquely truncated in front, with a backward bevelling, so to speak, which brings the nostrils antero-inferior—they are not at all visible from above. The muzzle is almost a little *rétroussé*, and recalls that of a Pig rather than of a Weasel or Badger. It is furthermore peculiar in being entirely bald and callous on top for nearly an inch, this naked part narrowing to a point behind. Underneath, the nasal pad is closely and definitely circumscribed by the line of fur which closely approaches the nostrils. There is no sign, on the front of the snout, of the usual vertical groove, nor of such a division of the hairy part thence to the middle line of the lips. The organ looks as if it were fitted for rummaging among fallen leaves, or even for "rooting" in the ground. The whole muzzle is beset with sparse short bristles, apparently growing irregularly in no determinate direction; the longest moustaches scarcely reach to the eyes. There are similar bristles over the eyes and on the cheeks and chin. The proper pelage of the cheeks and snout is scanty, and it grows upward and forward from the lips.

The external ears are so slightly developed that they have been described as wanting. The pinna is a mere low orbicular rim completing about two-thirds of a circle, lower than the surrounding fur. The entrance of the ear remains broadly open. The eye is considerably nearer to the ear than to the end of the snout. The mouth is wholly far inferior, with short gape; it is nearly an inch from the end of the lips to the extremity of the snout.

The short and rather close-haired (for this group) tail is quite different from the long, full, bushy member in *Mephitis* and *Spilogale*—in fact, it is only superior in these respects to the stumpy tail of *Taxidea*. It is difficult to estimate its relative length accurately, owing to the character of the base; but the vertebræ are certainly less than half the length of the head and body, and with the hairs the whole member is only about half such dimension, more or less. The terminal pencil of hairs is from only about two to four or five inches in length, in different specimens; the width of the hairs in the middle, pressed flat sideways, is from six to ten inches. The tail, especially when white, or the white portion of it when black and white, frequently presents a worn scrubby appearance, as if abraded; and the white tip, when occurring in connection with black, may seem as if imperfectly connected with the remaining portion. There is something very peculiar in the character of the white hairs of the tail of this or other Skunks; it is very coarse, stiff, yet weak and brittle, almost like Antelope hair, to which it has been aptly compared; it seems as if partially devitalized, and readily falls or breaks off. The same character is observable in the white portion of those hairs which are black at the end. In general shape, the tail is rather depressed, or slightly distichous, than uniformly cylindrical or bushy.

The soles of the fore feet are perfectly naked from the wrist-joint, but overhung along the sides with a fringe of long hairs. The tuberculation is not well marked; but at the bases of the digits, in advance of a general broad bare area, two incompletely divided pads are observable. The fore claws are very large, long, strong, compressed, little curved, not excavated underneath, and eminently fossorial. The middle three claws are of approximately the same size, though they are a little graded in length from fourth to second; the fifth is notably smaller, reaching to the middle of the fourth; the first is much shorter and more curved, reaching little beyond the base of the second. The terminal bulbs of all the toes are large.

The soles of the hind feet, like the palms, are perfectly naked from the heel in all the specimens examined, fringed along the side with long hairs. The tuberculation of the soles is better marked than that of the palms, and somewhat peculiar, but not to the extent which Dr. Gray's remarks would indicate. The general broad flat area of the posterior part is divided by a transverse sulcus of variable depth and distinctness from the bulbous part at the base of the toes. This latter is not a continuous pad, as indicated by Dr. Gray—not always at any rate; even in the dried specimens before me it is distinctly divided into three bulbs, much as in *Mephitis*—one at the bases of the three middle toes, and another at the base of each lateral toe. The claws are very much smaller than those of the fore feet, short, stout at base, moderately curved, obtuse, excavated beneath. The third and fourth are subequal and longest, the second is little shorter, the fifth much shorter, and the first shorter still.

The pelage is very coarse and harsh throughout. The peculiarity of the white hairs of the tail has been already indicated.

In coloration, this animal presents greater variations than those of *Mephitis mephitica*, in which the differences are mainly in the extent or restriction of the two normal white stripes. The pattern is essentially a white dorsal area, which may include all or most of the tail. The white, in all the specimens I have seen, begins fairly on the sinciput, in advance of the ears, instead of on the nape, as in *M. mephitica*. This may be the only point of detail that is constant. The white begins squarely in a transverse line, or in a curve, or in a point; it is broad and uninterrupted to the end of the tail, or fails to reach the tail (which then usually only has a white brush at the end), or is divided by a median vertebral stripe of varying width, or, finally, may be interrupted in its continuity.

Viewing the wholly indeterminate character of this white area, and comparing it with corresponding variations in *Mephitis mephitica*, it is easy to account for the extraordinary confusion which prevails in the accounts of this group, by authors who sought to establish species upon the character of the markings. These masses of black and white distracted the attention of all the early authors from the essential generic and specific characters; in fact, it is only about forty years since the true points of distinction were perceived at all, and even subsequently species continued to be made upon a wrong understanding. Some repugnance to handling and closely examining the noisome beasts may not have been entirely inoperative in perpetuating error and confusion; and certainly the group as a whole is not among those best represented in museums, owing to the obviously disagreeable task it becomes to capture and prepare the animals. Once again, the perfect ease with which a fair description will answer to *Conepatus* and *Mephitis* has had its weight in provoking and perpetuating confusion.

Let me illustrate this last point with the following example of fair diagnosis based upon color:—

"Black; back with two broad white stripes meeting on the head; tail end white."

This is a perfectly applicable and exact description of both *Mephitis mephitica* and *Conepatus mapurito* when the latter has the dorsal area divided. Again:—

"Black; back with a single broad white stripe; tail black and white."

This is entirely pertinent to both *Mephitis macrura* and *Conepatus mapurito* when the latter has the dorsal area undivided. I do not know where to look for the parallel to this curious wellspring of error.

Even after the full recognition by Lichtenstein and Gray of the different genera of Skunks, many nominal species endured, upon the basis of coloration alone. In the case of *Conepatus*, these have all latterly been reduced to varieties by Dr. Gray, because, as he very truly says, "the differences in the coloration appear to pass into one another". This is a step in the right direction, but, as it appears to me, does not go quite far enough. The ascribed differences are not of the character to which recognition by name is usually granted; they are apparently not characteristic of particular geographical areas; nor are they accompanied, for all that appears, by any other characters. I see no alternative to regarding them as wholly within the normal range of individual variability of the species.

Nor are the ascribed differences, when sifted of generalities and cleared of mere verbal discrepancies, anything remarkable. I can make nothing more of them than this: the white dorsal area may be entire, or divided by a line of black of varying length or width (giving the "two" stripes of authors); it may stop short of the tail, or go to its end, or may be broken up in its continuity. This is the whole sum of the various accounts I have seen.

In compiling the foregoing extensive synonymy of *Conepatus*, I have not been unmindful of Dr. Gray's judicious caution:— "When we have the power to compare the living animal and the skeleton of each [of his nominal varieties], we may discover that some of them are distinct species, having a peculiar geographical range." This is improbable, yet quite possible; and its prudence impresses me especially, as I have not inspected specimens from South America. But I would urge these points in defence of the synonymy I have prepared: that all the supposed species whose names I have cited rest upon no other basis than that variability which is proven to be merely individual, and that, therefore, they are *ipso facto* synonymous; that should the synonymy be ever shown to embrace more than one species, an attempt to distribute it among two or more species, and to fix upon the proper name for each, would be well-nigh futile, so inextricably blended has it become; that should a second species of *Conepatus* be hereafter established, it should properly receive a new name upon the new basis, as the best means of avoiding further confusion.

It may not be amiss to add, that all the recognized South American references are to *Conepatus* alone, *Mephitis* proper being not known to occur on that continent. This is a simplification of matters which does not hold for Central America and Mexico, where the two genera are found together. Nevertheless, the supposed absence of *Mephitis* proper from South America rests upon negative evidence.

Description of the skull and teeth.

In the following description, reference is had to the same parts of *M. mephitica*, to which all expressions of comparison apply. The account is based mainly upon No. $\frac{1881}{790}$, Mus. Smiths. Inst., from Texas, but several other specimens are examined at the same time.

Viewed from above, the rostrum is notably tapering—decidedly more so than in *M. mephitica*, though the calibre at the base of the zygomata is even greater. The nasal aperture is much less foreshortened in this view. Supraorbital processes are barely, or not at all, recognizable; the prongs of the sagittal crest are faintly indicated or entirely inappreciable. The point of greatest constriction of the skull (about midway between muzzle and occiput) is well marked and abrupt; the skull immediately swelling behind it, forming a decided projection into the temporal fossa, hardly or not seen in *M. mephitica*. The cranial dome is rather higher and fuller. The zygomatic arches are comparatively shorter, more divergent, and more regularly curved. In profile, the differences are more striking. The highest part of the skull is back over the cranial dome, not at the interorbital space; the slope is but slight thence to the occipital protuberance, but is long and regular from the same spot to the incisor teeth; for so great is the obliquity of the nasal orifice that the end of the muzzle comes into this general curve, instead of rising, with slight obliquity, from the teeth to then bend abruptly backward at an angle. None of the specimens, unluckily, are young enough to show the nasal sutures; but I have no doubt that these bones, if not also the neighboring part of the maxillaries and intermaxillaries, will be found to afford good characters. The anteorbital foramen (as in other species, sometimes subdivided into several separate canals) is farther forward and higher up, piercing a thicker zygomatic root, and consequently being rather a tube than a hole. The zygomatic bones are slenderer and less laminar than in *Mephitis*. The arch, as a whole, is shorter and more anterior; in skulls of the same length laid together, the back roots of the arch in *Conepatus* fall in advance of the other when the muzzles are together. Viewed from behind, the occipital surface is much higher and narrower; thus the distance from the bottom of the foramen magnum to the occipital protuberance is greater than the interparoccipital width; in *Mephitis*, it is, if anything, less. Beneath, the palate is seen to end some distance back of a line drawn across behind the molars; the pterygoids and contained interspace are correspondingly shorter than in *Mephitis*, in which the palate ends more nearly opposite the back molars. The edge of the palatal shelf is simply transverse in some specimens, while in others it shows a little median process backward, and we may presume that in other cases it is nicked, for all this variation is now well known to occur in both *Mephitis* and *Spilogale*.

The lower jaw gives excellent characters. The angle of the mandible is strongly exflected and the emargination between this and the condyle is slight. The coronoid process rises with considerable backward obliquity, with a very convex anterior border, and concave posterior one, carrying the apex of the bone backward to a point nearly or directly over the condyle.

A peculiarity of the dentition of *Conepatus* has been unduly exaggerated by some authors, who assign a different dental formula ($pm. \frac{2-2}{3-3}$, as against $pm. \frac{3-3}{3-3}$ in *Mephitis*). But the supposed wanting anterior premolar is often present; though it is always minute, probably never functionally developed, and deciduous or abortive on one or both sides. I see this small tooth plainly in two skulls before me, but do not find it in a third; in which last there is instead an unusual diastema between the canine and the nearest premolar. This point disposed of, nothing in the dentition of *Conepatus* calls for special remark. The detailed account given under the head of *Mephitis mephitica* is here equally applicable.

Vertebræ: C. 7; D. 16; L. 5; S. 3; Cd. 18 (*Gerrard*).

Description of the anal glands.

The anal armature of this species has been investigated by M. Chatin, who has published a thorough description, illustrated with excellent figures, in the *Annales des Sciences Naturelles*, as above quoted. Although the specimen had been preserved in alcohol for several years, it was still extremely offensive, and the preservative fluid was tainted with the same fetor. M. Chatin's account is substantially as follows:—The anus presents 56^{mm} behind the root of the penis, in a large irregularly elliptical depression, crossed by numerous grooves; the surrounding integument forms a sort of flap folded about the anus and excretory pores of the glands. These appear as openings pierced in the centre of two thick, prominent, umbilicated papillæ, situated on each side of the anus, about 9^{mm} from the middle line. Moderate pressure suffices to bring up to these orifices the dirty brown fetid liquid which has made Skunks famous. Removal of the cutaneous fold which partially covers these nipples shows that they are situated in a kind of recess rising about 5^{mm} above the floor of this small pouch; the calibre of the pore which opens at the summit is sufficient to admit a probe about a millimetre in diameter. Dissection of the perinæum brings to view the whole secretory apparatus, the size of which is so considerable that it is surprising Cuvier had nothing to say on the subject in his chapter upon anal glands and those of neighboring parts. Stannius and Siebold were also silent, while Owen confined himself to mere mention of the anal glands of the Skunks.

The glandular mass is nearly trapezoidal; it begins 33^{mm} from the prostate, on a level with the origin of the corpora cavernosa, the roots of which thus extend upon the posterior or prostatic portion of the gland; superiorly it lies in relation with the urethra, which passes over it. It is enveloped in a thick muscular tunic, the origin of which may be readily determined. It is well known that the ischio-cavernous muscle (ischio-penial of German authors) passes obliquely downward and inward to the root of the corpus cavernosum, where it ends in tendinous or muscular fibres. "But sometimes", says Leyh, "there are found below this muscle some isolated muscular fibres which appear to have no function." They may be so regarded in domestic animals; but, in the present case, nature furnishes another instance of her economy in giving them no inconsiderable part to play. These same fascicles form a large part, but not the whole, of the muscular envelope of the gland; the bulbo-cavernous muscle is equally concerned in the formation of the tunic; nor can we entirely separate from it the prostatic muscles which are blended with it, and which cover the upper portion of the urethra in those animals that, like the present species and the Dog, have no Cowper's glands. The muscular coat, about 3^{mm} thick, is composed of two layers readily distinguished by the direction of their fibres; those of the superficial layer being transverse, that is to say, perpendicular to the median or antero-posterior axis of the gland, while the fibres of the deep layer run in the opposite direction, parallel with the same axis.

Below these muscular layers, that is to say, within the general muscular envelope, is found the follicular or glandular portion proper of the organ; it is not regularly distributed around the central reservoir, as in most *Carnivora*, but occupies only a limited portion of the surface of this receptacle. The follicles are rather large, and of a reddish-brown color; their numerous well-developed culs-de-sac measure on an average $0^{mm}.55$ in diameter, and are variously rounded, ovoidal, club-shaped, &c.

The reservoir, which is of great size, is covered with a thick whitish tunic composed of dense laminated tissue and elastic fibres, the presence of which is readily determined by means of acetic acid. In the specimen examined, the receptacle was empty, containing only a few dirty brown pellicles, which showed under the microscope nothing but laminated fibres and fine granules. Toward the anterior extremity of this cavity is found the opening of a duct, through which the fluid secreted is conducted to the lateral anal pore, as may be easily ascertained by passing a probe. The surface of the reservoir, marked with numerous folds and furrows, resembles to some extent that of *Herpestes fasciatus*.

It is an established fact, then, that the fetid humor which was long supposed to be urine is the secretion of true anal glands. It is to be regretted that the anatomy of the various species of *Mephitis* is not better known, for it would be interesting to compare them in the details of this structure. Much is still required to complete their history, and it is to be hoped that the missing links may be soon supplied.

Geographical distribution and habits.

The general extra-limital dispersion of the species southward has been already indicated. Lichtenstein's *M. mesoleuca* was procured by Deppe, in 1825, near Chico, Mexico; his *leuconota* was from the Rio Alvarado, Mexico. Bennett's *M. nasuta* came from "California". Audubon and Bachman describe the animal from Texas. The only specimen Baird had seen in 1857 was also from this State (Llano Estacado), beyond which I am not aware that the animal has been actually observed north of the Mexican border. I obtained no evidence of its presence in New Mexico, Arizona, or Southern California during my residences in those regions, and the species may be confined, in the United States, to the valley of the Lower Rio Grande, like various other quadrupeds and birds.

From Audubon's account, it is to be inferred that the animal is not rare in portions of Texas, where the specimen which is figured in his work was procured by his son, John Woodhouse. His notice of its habits is as follows:—

"The *Mephitis mesoleuca* is found on the brown, broomy, sedgy plains, as well as in the woods, and the cultivated districts of Texas and Mexico. Its food consists in part of grubs, beetles, and other insects, and occasionally a small quadruped or bird, the eggs of birds, and in fact everything that this carnivorous but timid animal can appropriate to its sustenance.

"The retreats of this Skunk are hollows in the roots of trees or fallen trunks, cavities under rocks, &c.; and it is like the northern species, easily caught when seen, (if any one has the resolution to venture on the experiment,) as it will not endeavour to escape unless it be very near its hiding place, in which case it will avoid its pursuer by retreating into its burrow, and there remaining for some time motionless, if not annoyed by a dog, or by digging after it.

"The stomach of the specimen from which our drawing was made, contained a number of worms, in some degree resembling the tape-worm at times found in the human subject. Notwithstanding this circumstance, the individual appeared to be healthy and was fat. The rainy season having set in (or at least the weather being invariably stormy for some time) after it was killed, it became necessary to dry its skin in a chimney. When first taken, the white streak along the back was as pure and free from any stain or tinge of darkness or soiled color as new fallen snow. The two glands containing the fetid matter, discharged from time to time by the animal for its defence, somewhat resembled in appearance a soft egg.

"This species apparently takes the place of the common American skunk (*Mephitis chinga*), in the vicinity of the ranchos and plantations of the Mexicans, and it is quite as destructive to poultry, eggs, &c., as its northern relative. We have not ascertained anything about its season of breeding, or the time the female goes with young; we have no doubt, however, that in these characteristics it resembles the other and closely-allied species.

"The long and beautiful tail of the Skunk makes it conspicuous among the thickets or in the musquit [mezquite] bushes of Texas, and it most frequently keeps this part elevated, so that in the high grass or weeds it is first seen by the hunter who may be looking for the animal in such places."

CHAPTER IX.

SUBFAMILY MELINÆ: THE BADGERS.

The genus *Taxidea*—Generic characters and comparison with *Meles*—*Taxidea americana*, the American Badger—Synonymy—Habitat—Specific characters—Description of external characters—Description of the skull and teeth—Geographical variation in the skull—History of the American Badger—Its geographical distribution—Habits—*Taxidea americana* var. *berlandieri*, the Mexican Badger—Synonymy—Habitat—Subspecific characters—General remarks.—ADDENDUM: Description of the perinœal glands of the European Badger, *Meles vulgaris*.

ANIMALS of this subfamily inhabit Europe, Asia, and America. There are four well-marked genera, though the species are so few: the European *Meles*, the Asiatic *Mydaus* and *Arctonyx*, and the American *Taxidea*, long time confounded with *Meles*.

I have already (p. 10) given the characters by which the North American representative of the *Melinæ* is distinguished from our other subfamilies. The expressions used, however, are rather diagnostic of the particular genus *Taxidea* than of the subfamily *Melinæ* at large, the various members of which differ sufficiently to require greater latitude of definition. It being not to my present purpose to consider the *Melinæ* further than as represented by the American genus, we may at once take up the latter.

The Genus TAXIDEA. (WATERH.)

× Ursus, pt., of Schreber.
< Meles, Taxus, of authors referring to the North American Badger.
= Taxidea, *Waterhouse*, Proc. Zoöl. Soc. Lond. vi. 1838, 154; Trans. Zoöl. Soc. Lond. ii. pt. v. 1841, 343.—*Baird*, Mamm. N. A. 1857, 201, and of late authors generally.

GENERIC CHARACTERS.—*Dental formula*: i. $\frac{3-3}{3-3}$, c. $\frac{1-1}{1-1}$, pm. $\frac{3-3}{3-3}$, m. $\frac{1-1}{2-2} = \frac{16}{18} =$ 34.* Back upper molar a right-angled triangle, with hypothenuse postero-

---

* Prof. Baird (M. N. A., 201), after correctly stating the dental formula, makes the total "32" instead of 34, by an obvious slip of the pen. He further states, "In young specimens there is an additional premolar, (first,)

external. Back upper premolar similar in size and shape (though the eminences of the crown very different), but the hypothenuse postero-internal. Back under premolar with two tubercles. Anterior under molar comparatively small, not dilated behind, mostly opposing the back upper premolar (instead of the upper molar as in *Meles*). Cerebral portion of skull depressed-cuneiform, very wide across the flaring occipital crest; the intermastoid diameter nearly equalling the inter-zygomatic; sides of the braincase straightened and strongly convergent anteriorly. Bony palate reach-

which soon disappears; this will add 1-1 premolar to the lower jaw, making 34 in all." But such additional under premolar of the early dentition (which I have not seen, though I have examined skulls with the teeth scarcely cut) would make 36 in all, not "34", the latter being the correct total of the adult formula.

Audubon and Bachman state (Quad. N. Am. i. 361) that "the present species has one tooth less than the latter [*Meles vulgaris*] on each side in the lower jaw", which is certainly not the case, as the dental formula is the same in the two genera. Quoting Waterhouse, Trans. Zool. Soc. ii. pt. v, p. 343, these authors continue:—"'The subgeneric name, TAXIDEA, may be applied to the American Badger, and such species as may be hereafter discovered with incisors $\frac{6}{6}$; canines $\frac{1-1}{1-1}$; false molars $\frac{2-2}{2-2}$ ... molars $\frac{2-2}{2-2}$ ...'" I have not Waterhouse's article at hand to verify the quotation; if his words and figures are correctly quoted, Waterhouse did not give the right formula, for his total is only 32, instead of 34; besides which he reckoned the upper sectorial tooth as a molar, instead of a premolar, as it is. Making this change, but retaining his original numbers, Waterhouse's formula becomes $pm. \frac{3-3}{2-2}, m. \frac{1-1}{2-2}$. But there are five grinding teeth on each side of the lower jaw of *Taxidea*. I have never seen an American Badger's skull with teeth otherwise than as given in the text above.

A peculiarity in dentition of *Meles vulgaris*, which may account for discrepant statements of the dental formula, has been pointed out by Professors Moseley and Lankester (Journ. Anat. and Phys. iii. 1868, 79):—"Mr Flower, in his recent admirable paper on the Dentition of Marsupials, has laid some stress on the fact that, in several diphyodont mammalia, some of the anterior maxillary teeth never have any predecessors, as in the case of the second anterior maxillary teeth of the dog, and the corresponding lateral mandibular teeth, and in the hog also. We are led to believe, from the examination of a fine series of Badgers' skulls in the university museum, that this animal furnishes an additional example. In three skulls, possessing the permanent dentition, we found a small peg-like tooth implanted in the jaw immediately behind the caniniform maxillary, and somewhat internally to the general line of the teeth, and obviously corresponding to the small anterior lateral tooth (præmolar) of the lower series abutting against the large caniniform. We found no trace of this tooth in a young skull with the perfect deciduous dentition, nor in De Blainville's figure of the same. It is described neither by Owen nor De Blainville, and is evidently easily lost, since it has dropped out of one skull, leaving only its alveolus as evidence of its former presence; and in two other skulls no traces of it were to be seen at all. The addition of this tooth makes the dentition of the Badger the same as that of the Glutton."

ing half-way to ends of pterygoids. Bullae auditoriae at a maximum of inflation, impinging behind upon paroccipitals. Condyles of jaw often locked in the glenoid. Coronoid of jaw erect, pointed, its posterior edge angulated by the meeting of two straightish lines. (For further cranial characters see page 269.)

Body extremely stout, squat, and clumsy, owing to great depression; tail short, broad, flattened; pelage loose; coloration diffuse; fore claws extremely large, highly adapted for digging. Habits thoroughly terrestrial and fossorial.

Taxidea is confined to North and Middle America. "This genus," as Prof. Baird* has remarked, "is so strikingly different from *Meles* as to render it a matter of astonishment that the typical species were ever combined." It is represented by a single species, divisible into two geographical races.

The American Badger.

Taxidea americana.

PLATE XVI.

Ursus taxus, *Schreb.* "Säug. iii. 1778, 520, f. 142 B. (After Buffon.)"
Meles taxus *var.* americanus, *Bodd.* Elench. Anim. i. 1784, 136.
Meles americanus, *Zimm.* Penn. Arktische Zool. i. 1787, 74. (Quotes Boddaert.)

---

* Mamm. N. Am. p. 201. From direct comparison of skulls, which I have not made, this author has concisely set forth many leading points of dissimilarity. I quote his article, with some abridgment:—

"The most striking peculiarity of *Taxidea* consists in the great expanse of the occipital region, the width of the occiput being [nearly or about] equal to that of the skull, measured between the outer surfaces of the zygomatic arches. Thus the general shape is that of a depressed wedge, widest behind and truncated anteriorly, instead of being very much widest across the zygomatic arches, as in *Meles*. . . . The occipital crests are well developed in *Taxidea*, the sagittal very moderate. The auditory bullae are very large and convex. The processes of the glenoid cavity are not so well developed as in *Meles*, though occasionally sufficiently developed to lock the condyles of the lower jaw. The coronoid process has its apex pointed instead of rounded or truncated; its posterior margin is formed by two lines, the lower rising nearly perpendicularly a little in advance of the condyle, the other rather longer than the first, making a very obtuse angle with it. The differences in the character of the teeth are equally striking, though their number is the same. The penultimate or sectorial upper molar [last premolar] is very large and triangular; fully equal in size to the last molar, instead of being much smaller; it has likewise a large tubercle on the inner lobe, scarcely observable in *Meles*. The last molar is also triangular, (nearly right-angled,) somewhat resembling half of the quadrilateral tooth of *Meles*. In the lower jaw the last premolar is larger than in *Meles*, and has two tubercles. The penultimate molar is smaller and not dilated behind. The portion of its crown which is applied against the upper sectorial molar [premolar] is larger than that in contact with the last upper molar, instead of being smaller, as in *Meles*."

Taxidea americana, *Baird*, M. N. A. 1857, 202, pl. 39, f. 2.—*Newb.* P. R. R. Rep. vi. 1857, 45 (habits).—*Coop.* N. H. W. T. 1860, 77.—*Suckley, ibid.* 94.—*Suckley & Gibbs, ibid.* 117.—*Hayd.* Trans. Am. Philos. Soc. xii. 1862, 134 (Upper Missouri country).—*Gray*, P. Z. S. 1865, 141; Cat. Carn. Br. Mus. 1869, —.—*Coop.* Am. Nat. ii. 1868, 529 (Montana).—*Ster.* U. S. Geol. Surv. Terr. for 1870, 1871, 461.—*Allen*, Pr. Bost. Soc. N. H. xiii. 1869 (published February, 1870), 183 (Iowa, still numerous); Bull. Ess. Inst. vi. 1874, 46 (Kansas), 54 (Colorado), 59 (Wyoming), 63 (Utah); Pr. Bost. Soc. xvii. 1874. 38.—*Ames*, Bull. Minn. Acad. Nat. Sci. 1874, 69 (Minnesota).—*Coues & Yarrow*, Zoöl. Expl. W. 100 Merid. v. 1875, 63.—*Allen*, Bull. U. S. Geol. and Geogr. Surv. Terr. vol. ii. no. 4, 1875, 330 (skull).

Ursus labradorius, *Gm.* S. N. i. 1788, 102, n. 7.—*Kerr*, S. N. i. 1792, 187.—*Shaw*, G. Z. i. 1800, 469, pl. 106.—*Turt.* S. N. i. 1806, 63.

Meles labradoria, *Meyer*, "Zool. Arch. ii. 1796, 45."—*J. Sab.* App. Franklin's Journ. 1823, 649 (compared with European).—*Harl.* Fn. Amer. 1825, 57.—*Griff.* An. Kingd. v. 1827, 116 ("*labradorica*").—*Less.* Man. i. 1827, 141, no. 372 ("*labradorica*").—*Fisch.* Syn. 1829, 151.—*Rich.* F. B. A. i. 1829, 37, no. 12, pl. 2.—*Godm.* Am. Nat. Hist. i. 1831, 179.—*Rich.* Zoöl. Beechey's Voy. 1839, 4.—*Wagn.* Suppl. Schreb. ii. 1841, 182.—*De Kay*, N. Y. Zoöl. i. 1842, 27.—*Schinz*, Syn. i. 1844, 315 ("*labradorus*").—*Aud. & Bach.* Q. N. A. i. 1849, 360, pl. 47.—*Bd.* Stansbury's Rep. 1852, 311.—*Kenn.* Tr. Illinois Agric. Soc. for 1853–4, 1855, 578.—*Giebel*, Säug. 1855, 761 ("*labradorius*").—*Hall*, Canad. Nat. and Geol. vi. 1861, 294 ("*labradoricus*").—*Maxim.* Arch. Naturg. 1861, —; Verz. Säug. 1862, 33.

Taxus labradoricus, *Say*, Long's Exp. i. 1823, 261, 369.

Taxidea labradoria, *H. Smith*, Nat. Lib. xiii. 1840, 210.—*Gray*, List Mamm. Br. Mus. 1843, 70.—*Baird*, M. N. A. 1857, 745 (expl. of pls.).—*Gerr.* Cat. Bones Br. Mus. 1862, 99.

? Taxidea labradoria, *Waterh.* P. Z. S. vi. 1838, 154; Tr. Z. S. ii. 1841, 343, pl. 59 (may be the other subspecies).

Meles jeffersonii, *Harl.* Fn. Amer. 1825, 309 (based on Lewis and Clarke).

American Badger, *Penn.* Syn. Quad. 1771, 202, no. 143; Hist. Quad. ii. 1781, 15, no. 298 b.—*Erxl.* Syst. i. 1777, 164 (in text).—And of authors generally.

Common Badger, *Penn.* Arct. Zoöl. i. 1784, 71, no. 23 (in part; includes the European).

Blaireau d'Amerique, "*F. Cuv.* Hist. Nat. Mamm."

Blaireau du Labrador, *Less. l. c.*

Amerikanische Duchs, *Schinz, l. c.* (*Dachs*, cf. Martens, Zoöl. G art. xi. 1870, 251, philological).

Braro, *Lewis & Clarke*, Trav. Allen's ed. ii. 1814, 177; Rees's ed. 4to, 471; Rees's ed. 8vo, iii. 40 (also called "badger" *passim* in this work; rendered "blaireau" in the McVickar ed. ii. 349; basis of *Meles jeffersonii*, Harlan).

Prarow, "Gass's Journ. p. 34."--(*Richardson*.)

Brairo or **Lacyoti**, *Gray*, List Mamm. Br. Mus. 1843, 70.

Braibo or **Lacyoti**, *Gerr. l. c.*

Carcajou, *Buff.* "Hist. Nat. Suppl. iii. 242, pl. 49" (cf. Desm., Mamm. i. 1820, 173; Ency. pl. 38, f. 2).

Carcajou ou Blaireau américain, *F. Cuv.* Suppl. Buff. i, 1831, 267.

Carkajou ou Blaireau d'Amerique, *Gervais*, Proc. Verb. Soc. Philom. Paris, 1842, 30.

Brairo et Siffleur, *French Canadians.*

Nannaspachæ-neeskæshew, Mistonusk, Awawteekæoo, *Cree Indians* (Richardson).

HAB.—United States, from Wisconsin, Iowa, and Texas westward. British America, east to Hudson's Bay at least, north to 58°. Replaced near the Mexican border by var. *berlandieri*, which extends into Mexico. Formerly further east (Michigan, Indiana, Illinois). "Ohio, near Toledo, about 20 years since; now extinct" (Edw. Orton, *in epist.*).

SPECIFIC CHARACTERS.—Top of head darker than other upper parts, with a median white stripe; sides of head below the eyes, and its under surface, white, with a dark patch before the ear; limbs blackish; body-coloration above a grizzle of blackish with white, gray, or tawny, or all of these; below uniform whitish, shaded or not with gray or tawny. Length about 24 inches to root of tail; tail 6; head 5½; longest fore claw 1¼.

Description of external characters.*

Form stout, thick-set, indicative of great strength and little agility; body broad depressed; head flattened, conoidal; tail and limbs short; feet broad and flat; fore claws enormous, highly fossorial. Pelage of body and tail long, loose, shaggy, and of coarse texture; shorter and closer on the head and feet. Coloration blended, diffuse, grizzly above; below, uniform; on the head definitely marked in certain areas.

The head is nearly one-fifth of the total length exclusive of the tail; it is conoidal, but depressed very broad across the temples and cheeks, contracting gradually to the prominent snout. It is covered with short, close, coarse hair, only lengthening about the ears. The muzzle is completely furry, excepting the nasal pad itself; this is completely anterior, with a downward-backward obliquity; there is a median vertical furrow; the nostrils, not at all visible from above or laterally, are pyriform, lengthening slitwise at the lower outer corner. The naked pad is black; below it, the upper lip is completely furred across, and the fur elsewhere extends to the very edges of the rather thin lips. The rictus is ample; the canines are visible in life. The eye is remarkably small, and rather high up, a little back of the angle of the mouth. The vibrissæ are sparse and short, the longest scarcely or not attaining the eye; here and there other bristles grow about the eyes and on the chin. The ears are low, rounded, and very broad, with remarkably large external meatus, partly defended by long loose hairs growing in front, completely hairy outside and for some distance inside around the border; but most of the concavity of the conch is naked, with some sparse isolated tufts.

The fore limbs are short, stout, and the fore feet very large, broad, and flat, bearing immense claws. The digits are much abbreviated and consolidated, appearing from above almost entirely grown together, from below as five closely appressed oval pads. They are shorter than the claws they respectively bear; the 2d-5th, are subequal and longer than the 1st or 5th, which are mere claw-bearing bulbs. The back of the hand is hairy to claws, the bases of which are overhung by the longer anterior hairs; the palm shows the following disposition: a crescent of five large closely apposed naked digital bulbs, separated by a profound excavation from a single large irregu-

---

* From numerous specimens in the Smithsonian Museum.

larly shaped palmar pad, either entirely naked or partly overgrown with coarse hair extending crosswise from the inner border over more or less of its extent; this main pad divided by a decided transverse groove from a much smaller postero-exterior one, which is entirely hairy, or partly so, or perfectly naked, in different cases. The claws are all compressed, arched, with rounded ridge and sharp edge underneath, originally acute but generally blunted with use. The three middle ones are subequal in length, much longer than either of the lateral ones, and project still further, owing to the shortness of the lateral digits; they are also stouter than the others. The lateral claws are subequal to each other, and they reach half-way (more or less) to the ends of the middle ones; they are more compressed and not so strong, the inner one especially being thin, sharp, and falcate. As usual in such cases of special developments of parts, the rate of variation in size, both absolute and relative, is high, not permitting more strict statements than the above. Either one of the three larger middle claws may exceed the other two in size, and, of the lateral ones, either may surpass the other. The inner claw, however, apparently preserves its decidedly thin and falcate condition.

The hind feet are much like the fore, on the whole, with, however, a decided reduction in size, and especially in the development of the claws. The foot is about four times as long as broad, of nearly equal width throughout, hairy above and completely so below, more than half-way from the heel to the ends of the toes. Much as in the fore foot, the digits underneath present a crescent of five bald pads, of which the lateral ones, and especially the outer one, are somewhat disconnected; with a deeply excavated interval, these pads are succeeded by a single large bald callosity, heart-shaped in general contour, incompletely divided by several radiating impressions into four, sometimes five parts. These grooves are not constant, nor are the resulting partitions always of the same size and shape. The relative lengths of the digits, and of the claws they bear, are essentially the same as in the fore foot; but all the claws are very much smaller, and the lateral ones scarcely or not attain the base of the 2d and 4th respectively. These hind claws, however, as compared with the fore, differ remarkably in construction, though of much the same shape; they are less compressed, and, instead of being sharp-edged along the median line below, they are deeply excavated underneath—

sometimes so deeply as to be merely a thin shell of horn, the edges of which only unite at the base of the claw.

The short, broad, flattened tail has no sharp distinction from the body at its base, but the body tapers toward it somewhat as in the Porcupine. It is densely covered, in a somewhat distichous manner, with long coarse hair like that of the body; the end is obtusely rounded.

The perinæal region shows, immediately beneath the root of the tail, a large transverse fissure leading into the peculiar subcaudal pouch of the *Melinæ*, and, in advance of this, a large hemispherical protuberance, more or less naked, or covered with a few sparse hairs, and imperfectly divided by a median raphé into lateral oval masses. (The anatomy of the peculiar organs of these parts, as illustrated in the European *Meles taxus*, is given beyond.)

The Badger varies greatly in color, as a fortuitous matter of age, season, or condition of pelage, aside from certain geographical differences, to be shown in the sequel. The variation, however, is mainly in the relative amounts of the blackish tawny-gray and white which produce the general grizzle, the pattern of coloration being well preserved, especially as to the markings of the head. The top of the head is dark brown or blackish, generally increasing in intensity and purity from the nape to the snout, since it is commonly more or less blended with gray or hoary encroaching from behind. This dark area is divided lengthwise by a sharp white or whitish median stripe, which runs from the snout, or from just back of the snout to the nape, where it is gradually lost in the grizzle of that part. I have never known this stripe to be entirely wanting; but it varies much in extent, both laterally and longitudinally. The sides of the extreme muzzle are dark, like the top of the head; from about opposite the canines, the sides of the head and the ears are white, continuous with the white of the chin and throat, but interrupted by a large (generally crescentic) dusky patch in front of the ear. Another dark patch usually shows, though less conspicuously behind the ear. The whole body and tail above are an intimately blended mixture of blackish with white, hoary gray, and tawny, or pale dull fulvous (dilute *helvolus*). The individual hairs are for the greater part of their length of one of the lighter colors above mentioned, then black or blackish for a distance, and finally tipped with hoary gray or whitish. This pale ending of the hairs seems constant, even when there is most tawny in the body of the furs. Owing to the length and coarseness of the pelage, the animal usually presents, when prepared for the museum, a patchy or streaky appearance, the completely blended grizzle being interrupted by the slightest disturbance of the set of the hairs. Beneath, the animal is uniformly as above, *minus* the black or blackish. The feet are dark brown or blackish; the claws are generally light-colored, especially those of the fore feet.

In examining a large suite of specimens from various localities in the United States west of the Mississippi, I find decided expression of a variation dependent upon climatic influences. Specimens from the comparatively fertile and well-watered regions upon the eastern border of the great central plateau are identical in tone with others from the Pacific slopes, and both much more heavily colored than those from the arid intermediate region. In the former, the fulvous or tawny tinge predominates among the lighter colors, mixed with a large amount of nearly pure black. As remarked by Prof. Baird, the resemblance of these specimens to the Woodchuck (*Arctomys monax*) is striking. In all the specimens from the interior dry region, and especially from the Upper Missouri, where the animal is extremely common, there is little if any of the fulvous. At a distance, the animal appears nearly white; the general color is white, soiled with a faint tawny or dirty yellowish-tinged and mixed with but little blackish, the dark part of the individual hairs being less extensive than even the terminal hoary portion, and the area where the black occurs at all being restricted. In these cases, also, the general grizzle encroaches most on the head, and the frontal white stripe reaches farthest along the nape. Under these conditions, the animal very closely resembles in coloration the brindled gray Wolves of the same geographical area.

It is almost needless to add that the gradation between the extremes above noted is unbroken and insensible.

None of the specimens now under consideration show the slightest trace of a vertebral white stripe beyond the nape. Those exhibiting this peculiarity are treated under the next head.

Without alcoholic specimens, or measurements taken in the flesh, I cannot give the dimensions with desirable precision, as all the dried skins before me are more or less distorted. The range of variation in size, though considerable, is nothing unusual. To the figures above given may be added: Nose to eye about 2½ inches; to ear about 5; fore foot from the posterior callosity, and including claws, 3½; hind foot from heel, including claws, 4; Lowest hairs of the back 3 or 4 inches; of the tail, 2 or 3. Height of the ear above the bottom of the meatus 1¼–1¾.

Penis bone 4 inches long, clubbed at one end, compressed, and with shallow sulcus in the continuity; the other end bent nearly at right angle, abruptly and irregularly flattened and grooved.

It is surprising that this animal should ever have been confounded with the *Meles taxus* of Europe, since the decided structural characters upon which the genus *Taxidea* rests are coördinated with readily appreciable superficial distinctions. In the European Badger, the snout is much larger, more protuberant, more extensively naked, and differently shaped, being not very dissimilar to that of a hog in miniature. It is definitely naked on top for some distance, as well as in front for a space below the nostrils; these occupy but a small part of its subcircular front. The fore claws are much smaller and weaker. There are some differences in the details of the pads upon the palms and soles. The general body-color above is not dissimilar; but the under parts are black like the limbs, this color extending on the chin, where our species is white. The head is otherwise white, with a broad black stripe beginning on each side opposite the canines, running back, embracing the eye and ear, and losing itself on the side of the nape. The edge of the ear is white in this otherwise uninterrupted black bar. In the dried specimen before me, the naked part of the snout appears to have been flesh-colored, and the claws are dark.

Description of the skull and teeth.

I have no skull of *Meles* in hand for direct comparison, but this is less to be regretted in view of the numerous striking differences which any accurate and detailed description will show, even without use of strictly antithetical expressions. (See also *anteà*, p. 263, note.)

A striking peculiarity of the adult skull is perceived in a view from above, in the great width behind, the distance across the terminations of the occipital crest being equal to (a little more or less than) the inter-zygomatic width; the lateral outline in general is therefore wedge-shaped. The rostral part of the skull (all that in advance of the zygomatic arches) is about one-third, or rather less, of the total length. The sides of the rostrum are approximately parallel in old skulls, owing in a measure to the swollen tract of rooting of the canines, in young specimens somewhat tapering; the nasal extremity is abruptly narrower than the rest; the obliquity of bevelling of the nasal aperture is about 45°. In old skulls, the nasal and maxillary sutures are obliterated; in young ones, the nasal bones are seen to be narrow, with approximately parallel edges for the anterior half of their length, where they begin to narrow, and extend as slender acute processes very far back—to opposite the middle of the orbits. Their suture with the superior maxillaries is very brief; for the intermaxillaries reach far up, and for nearly all the rest of their extent they are received betwixt long, pointed processes of the frontal. Similarly, the superior maxillary runs up in a recess of the frontal to a point opposite the ends of the nasals. This deep wedging of lateral processes of the frontal between processes of the nasals and maxillaries forms a complete letter **W**, better marked than in any other North American genera of the family, though they all, excepting *Lutra* and *Enhydris*, show an approach to the same character. The anteorbital foramen is large and rather triangular than circular. The orbits are much better defined than in *Mustelinæ* and *Mephitinæ*—not that supraorbital processes are stronger than usual, but because the zygoma sends up a spur to mark the orbital brim below—much as in the others. The approximation of these two (zygomatic and supraorbital) processes completes about two-thirds of a circle. The point of greatest constriction of the skull is a little back of the supraorbital processes, at a point about midway in the whole length of the skull; except in some very old skulls, the constriction is little, if any, greater than that of the interorbital space. The top of the skull is marked with an average sagittal crest, whence forks curve outward to the supraorbital processes. In young specimens, there is little or no trace of these ridges. The occipital crests appear more flaring than they really are, owing to the general breadth of the skull behind; they are in fact only moderately developed in the oldest specimens, excepting at their lateral extremities. From a moderate median emargination, the crests proceed on either hand with a moderate convexity, which suddenly increases at the bend around to the mastoid. The cranial dome considered alone has little inflation; the lateral outlines run nearly straight from the point of greatest constriction to the back root of the zygoma. This wedge shape contrasts with the greater inflation of the cranial dome in most other *Mustelidæ*, notably *Lutra* and *Enhydris*.

The back of the skull has a general triangular shape, with perpendicular flat face and irregular strong muscular impressions. The paroccipital processes are rather short, wide, and blunt; they descend to the level of the lower border of the foramen magnum, which latter is low and broad across. The condyles are short and very broad, their articular surfaces being prolonged toward the paroccipitals.

In profile, the skull shows a single general declivity from near the occiput to the end of the nasals, thence an abrupt bend down to the teeth. This general curve is sometimes a little sinuous, owing to slight depression just behind the orbits, and elevation over them. The posterior outline is truncate, with the occipital crest curving into full view below. The zygomata are very little arched indeed—almost straight; they are stoutly laminar, with a strong superior orbital process anteriorly, and remarkably developed borders of the deep glenoid fossa. Such development of the glenoid, in connection with its peculiar shape (the front border overlaps on the outer half, the hinder on the inner half), is sufficient to ordinarily lock the jaw in old age. But this peculiarity is not so strongly pronounced as it is said to be in *Meles*. The same thing occasionally occurs in old Otter skulls. The orifice of the meatus auditoriûs lies wholly between the border of the glenoid and the well-developed mastoid process.

The floor of the skull, aside from its mere shape, resembles that of the *Mustelinæ* in the two prominent features of far backward extension of the bony palate and great inflation of the bullæ. In the skulls, with moderately inflated bullæ, the palate ends nearly opposite the last molars; in the Otters, with the same extent of palate that *Taxidea* shows, the bullæ are quite flat. The palate reaches considerably more than half-way from the incisors to the foramen magnum; about half-way from the molars to the end of the pterygoids the palate is quite plane; the incisive foramina are very short and broadly oval. The palate ends behind with a simple concave edge, or nearly straight transverse one, indifferently nicked on the median line or with a little median process. The alvelolar borders are approximately parallel—more nearly so than usual in this family. The spread of the zygomata is rather more than three-fifths the total length of the skull; the outward curvature is greatest behind. The great width of the skull behind leaves a very broad basilar space, notwithstanding the size of the bullæ. This space narrows but little as it advances between the pterygoids, and is nearly flat throughout. A curious character is seen in the division of the posterior nares into two by a vertical bony septum running to the very edge of the palate, and thence projecting into view. Skunks and Land Otters have such a septum, but it is not complete to the end of the bony palate. In the Sea Otter, it is represented by a lamina depending from the roof of the nares, but not reaching the palate for about an inch from the end of the latter. In the Martens, Weasels, and Wolverene (*Mustelinæ*), there is nothing whatever of the kind. The pterygoids are simply laminar, with some little irregularity outside, as usual; they are moderately hamulate. The (comparatively) immense inflation of the large bullæ occupies nearly all the extent of the periotic bones; the swelling is immediate all along the interior border; outwardly it subsides in a moderately tubular meatus, and behind it is replaced by a concavity around the foramen.

The mandible is massive; the ramus lower and thicker before than behind; the symphysis long, strong, and early completed; the coronoid low and of peculiar shape. Its apex is obtuse; its front border nearly straight, but its hinder border divided at an abrupt angle into a lower perpendicular part and an upper strongly oblique portion. The lower border of the ramus is a gentle curve along the symphyseal portion, thence a straight line to a considerable angle abreast of the last molar, thence straight again to the proper angle of the jaw, which is small and not at all exflected. In young animals, the same border is more nearly a continuous slight curve from symphysis to the end. The condyle is very broadly transverse; its articular surface is extensive, with a peculiar twist to correspond with the above described formation of the glenoid fossa.

A young animal should be examined with reference to the teeth, as the characters of the molars become much obscured by wear. The back upper molar is neither narrowly transverse as in the Martens and Weasels, nor quadrate as in the Skunks and Otters, but triangular; and in size and shape it is not very dissimilar to the last premolar. Details aside, it is a right-angled triangle of nearly equal base and perpendicular, the right angle being antero-internal, the longest side postero-external. When entirely unworn, it shows six or eight irregularly disposed tubercles, all small, the general surface being quite flat, and there being no notable division, by sulcus or otherwise, into different portions, such as the crown of this tooth presents in most *Mustelidæ*. The lowest part of the tooth is a small circular area posteriorly. This tooth roots by a long fang exteriorly, but is otherwise simply set in an irregular shallow depression. The last premolar is likewise approximately an equilateral right-angled triangle; but in this case the right angle is antero-exterior, the hypothenuse postero-interior. It is well divided into an outer and inner moiety. The former is produced into a large main cusp, with prominent heel on its front base, and a smaller posterior cusp. The low inner moiety shows *two* perfectly distinct conical cusps; one anterior, the larger, with a cingulum around its base, and a smaller posterior elevation directly from the border of the tooth. The middle premolar is a simple conical cusp with a slight heel posteriorly; it is two-rooted. The front premolar is like the last named, but still smaller. I have never seen, in the adult dentition, the small first premolar which is said to occur " in young animals", nor do I observe any trace of such tooth in a young specimen which was just shedding when killed. In this one, the first and last permanent premolars have just displaced the earlier ones, but the middle milk-premolar is still present, with the future one visible below it, about to push it away. The presence of four upper premolars can, therefore, only characterize the milk dentition.*

The upper canines offer no special points. Of the incisors, the outer is very much larger than the rest; indeed, it is hardly more exceeded by the lower canine than the latter is by the upper canine, and its superficial resemblance to a canine is striking. The other incisors are of the same size, regular, with dilated trilobate ends.

The back lower molar, as elsewhere in the family, is small, circular, with a border a little higher before and behind than at the sides. It abuts against the depressed back part of the

---

* The small anterior upper premolar is the most variable tooth in *Mustelidæ*. It persists in the *Lutra, Enhydris, Gulo, Mustela*, and usually for a time in *Conepatus;* it is absent in the adult dentition of *Taxidea, Mephitis, Spilogale, Putorius.*

upper molar. The next molar is large and more complicated than usual. It presents, in front, a large cusp, which, with the outer one of a pair of median cusps, constitutes the trenchant edge of the tooth; the inner cusp of the median pair is little smaller than the other. The lower back part of this tooth, or its tuberculous portion, which abuts against the back upper molar, is seen, when entirely unworn, to present four cusps, three transversely abreast, whereof the middle one is the largest, and a posterior one. These all wear down level in the course of time, and indeed very old skulls show this whole tooth almost flat. The next tooth—last premolar—is a strong conical cusp, with a secondary cusp half-way up its back border, and well-developed posterior heel; the anterior border is straight. The remaining premolars, successively decreasing in size, are like the last, but without the secondary cusp. The lower canines are not peculiar. The lower incisors are smaller than the upper; the exterior pair are little larger than the rest, and obscurely trilobate. The next pair reach backward further than the rest, but all are flush on the front face; the four inner teeth are slightly bilobate.

Geographical variation in the skull.

Like other species of the present family, the Badger has been discussed in this regard by Mr. J. A. Allen.[*] His results are here transcribed:—

"The subjoined measurements of eleven skulls of this species (embracing all at present available) show also a well-marked southward decrease in size. A fuller series would be more satisfactory, but would doubtless only confirm what is here indicated. Six of the specimens are from rather northern localities and five from rather southern localities, the region represented extending from the Upper Missouri southward to the Lower Rio Grande. The specimens composing the two series are of very nearly corresponding ages. The northern series (four from different points on the Upper Missouri, one from Iowa, and one from Oregon) average 5.00 in length and 3.18 in width, the extremes being, in length, 5.22 and 4.92 (4.75 if we include one rather young example), the width ranging from 3.50 to 2.97. The southern series (including two or three from the vicinity of Matamoras, Mexico, and one each from New

---

[*] Bull. U. S. Geol. and Geogr. Surv. vol. ii. no. 4, pp. 330, 331.

Mexico and California) averages 4.62 in length and 2.92 in width, the extremes being, in length, 4.75 and 4.50, and in width, 3.07 and 2.80.

"The skulls, and especially the molar teeth, in the American Badgers, vary considerably in different individuals, as long since pointed out by Professor Baird.* Southern specimens differ from northern ones not only in being smaller, but somewhat in color, so that the *T. berlandieri* of Professor Baird may perhaps be entitled to subspecific rank (*T. americana* subsp. *berlandieri*), though the material at hand indicates that the two forms will be found to thoroughly intergrade. The chief differences in coloration consist in the more reddish-gray tint of the southern form, with a decided tendency to a continuous light dorsal stripe, instead of this stripe being restricted to the head.

"*Measurements of eleven skulls of* TAXIDEA AMERICANA.

Catalogue number.	Locality.	Sex.	Length.	Width.	Remarks.
11505	Upper Missouri		5.22	3.50	
1178	......do......		5.12	3.12	
2148	......do......		4.75	3.07	Rather young.
2078	Quisquaton, Iowa		5.06	......	Imperfect.
1290*	Fort Randall, Dak		4.95	3.25	
2033	Upper Des Chutes, Oreg		4.92	2.97	
4106	Fort Crook, Cal		4.60	3.07	
3767	New Mexico		4.50	2.80	Rather young; *berlandieri*.
1390	Matamoras, Mexico		4.75	2.94	*berlandieri*.
	......do......		4.66	2.85	......do.
4135	Texas		4.57	2.94	......do."

History of the American Badger.

The early history of the American Badger is curiously involved, not only with that of the European species, but with several entirely different animals. The celebrated traveller, Kalm,† speaks of the occurrence of the common Badger in Pennsylvania, where, he adds, it is called "ground hog". But this is a common appellation of the Woodchuck, *Arctomys monax*, to which, doubtless, Kalm's note is to be considered to apply. In 1756, Brisson‡ describes a "*Meles alba*" from New York; but this, it seems, proved to be an albino Raccoon, *Procyon lotor*. "Buffon," (says Sir John Richardson,§) "in the body of his great work, doubts whether the Badger be an inhabitant of the American continent," . . . . "but afterwards, in the first

---

* U. S. and Mex. Bound. Survey, Zoöl. p. 21."   † Trav. i. p. 189.
‡ Règne Anim. p. 255.   § F. B.-A. i. p. 38.

addition to his article on the Glutton, described the skin of a true Badger, which he received, it is said, from Labrador, under the misapplied name of Carcajou." We find the same confusion with a vernacular name of the *Gulo luscus* or Wolverene to continue for many years among French naturalists; thus, in 1842, Gervais still speaks of the "Carkajou, ou Blaireau d'Amérique". I am not able, at the time of present writing, to consult Buffon's work. His figure, given in the supplement of the Histoire Naturelle ("pl. 49"), is stated to have been afterward given by Schreber, in 1778, as plate 142 B of the "Säugthiere". Schreber's work is not just now accessible to me. He is cited for a name, "*Ursus taxus*", as applicable to the American Badger, though quoted as considering our species as distinct from the European.

Pennant, one of the more accurate and reliable among the early writers, is sadly at fault in the present case. After treating of an "American Badger" in his earlier works, he afterward, in the Arctic Zoölogy, as above quoted, united it with the European *Meles*, and, besides thus confounding it with a totally different species, he perpetuates several errors. Thus he quotes Kalm (see last paragraph) for its supposed occurrence in Pennsylvania, and speaks of its being "sometimes found white in America", evidently having Brisson's albino Raccoon in view.

Boddaert, in 1784, seems to have been the first to bestow a technical appellation upon our animal, calling it *Meles taxus* var. *americanus*. Zimmermann, citing Boddaert, adopted the name *Meles americanus* in his German translation of Pennant's Arctic Zoölogy. This name has priority over all others that have come to my knowledge, though it was suffered to rest almost unnoticed until very recently, when, in 1857, it was formally adopted by Prof. Baird, whose example has been generally followed by subsequent writers.

Linnæus (1766) makes no mention whatever of an American species of Badger. Supposing him to have had any knowledge of the animal from Buffon, his unworthy jealousy of the great French naturalist would have led him to studiously ignore the fact, in gratification of his absurd and puerile whim.

The name *labradoria*, or *labradorius*, by which our Badger has been usually known, was imposed by Gmelin in 1788. His *Ursus labradorius* is based primarily upon Schreber's plate 142 B. Other citations given by him are, the "American Badger" of Pennant, Quad. no. 143, and the "Carcajou" of Buffon,

Suppl. pl. 49. The habitat is given as "Labrador to Hudson's Bay". But there is grave reason to doubt that Buffon's animal, which furnished the material for his and Schreber's plate, came from Labrador, as implied in Gmelin's name. This point was brought up by Richardson, who, quoting Buffon's words "qu'il venoit du pays des Esquimaux", adds, "but in fact it may have been brought actually from the banks of the Saskatchewan by some of the Canadian fur hunters." In this uncertainty, it is fortunate that Gmelin's name, most probably objectionable on the score of geographical inapplicability, is also anticipated in point of date. Gmelin's diagnosis is also incorrect, for it seems that his phrase "palmis tetradactylis" arose in the circumstance that Buffon's specimen had accidentally lost one of its fore toes.

Early in the present century, the Badger attracted the attention of Lewis and Clarke, being then as now extremely abundant in the regions traversed by these intrepid explorers. Under the curious name of "braro", the animal is fully described by them in the narrative of their journey, published under the editorship of Paul Allen in 1814, for the first time, and in many subsequent editions. This word "braro" is obviously a corrupt rendering of the French "blaireau", like "brairo", by which name the animal was known to the Canadian voyageurs; the orthography is corrected in the McVickar edition. It is curious to trace the further typographical mangling of this word, originally written wrong by the travellers, being phonetically rendered according to the sound which caught their ears. It is spelled "brarow" or "prarow" in the Journal of Gass, one of their companions; and "braibo" is found in Gerrard's work above cited.* The animal described by Lewis and Clarke furnished Harlan, in 1825, with the basis of a nominal species, *Meles jeffersonii*, considered distinct from *M. labradoria*, which is also given by this author.

In 1823, Thomas Say treated of this species, under the name of *Taxus labradoricus;* and the same year Mr. Sabine gave us a detailed and the first satisfactory account of the actual differences in external characters between the American and European Badgers. His comparison was transcribed by Sir John Richardson, in the Fauna Boreale-Americana, 1829. The latter

---

* The old Mexican name of the southern Badger, said to be rendered "Tlacoyotl" by Fernandez, has suffered as badly, being rendered "Layotl" and "Tlacoyole" by some late writers.

author's article remains to-day one of the best, on the whole, that has appeared, covering as it does the then known ground, with a history, description, account of habits and geographical distribution, a synonymy, and a very characteristic plate, drawn by Landseer. Sir John, however, noted none of the characters by which our animal is generically distinguished from the European *Meles*, the establishment of a genus *Taxidea* being left to Waterhouse, 1838.

An index to the general later history of this species is afforded by the synonymatic list given on a preceding page (p. 263); it is unnecessary to recite the various authors who have contributed to our knowledge of the subject. Audubon and Bachman's article, however, is specially noteworthy as illustrating the habits of the animal in confinement. Portions of it are quoted beyond. Baird's notice of the species enters very fully into the technicalities of the case. J. A. Allen has discussed the variability of the skull, with special reference to geographical distribution.

Geographical distribution.

I am not aware that the Badger has ever been traced northward beyond the limit of its distribution long ago assigned by Richardson, namely, latitude 58° north. "The *Meles labradoria*", says this author, " frequents the sandy plains or prairies which skirt the Rocky Mountains as far north as the banks of the Peace River, and sources of the River of the Mountains, in latitude 58°." The doubts respecting its extension in British America to the Atlantic Ocean have already been expressed. Mr. Donald Gunn, in some inedited MSS. which have come into my possession, temporarily, through the Smithsonian Institution, speaks of the animal in the following terms:—"The Badger, called by the Indians 'Weenusk', inhabits all the woody districts south and west of Hudson's Bay. It hybernates during the long winters, entering its retreat early in October and remaining under ground until the middle of May. It is not often met with on the shores of the Bay, but is found at a distance of thirty to fifty miles. It does not appear to inhabit the woody districts east of Lake Winipeg, but is found in open places to the westward of that lake, and is occasionally met with along the river of the same name." It is well known to abound in the region of the Saskatchewan, and the British territory in general lying immediately north of Dakota and Montana Territories.

Audubon and Bachman, writing in 1851, state that they were not able to trace the Badger within a less distance from the Atlantic than the neighborhood of Fort Union (which stood at the southeastern corner of the Territory of Montana as at present bounded). But there is abundant evidence that the species formerly occurred far east of the Mississippi; and even now its range extends to that river. One of the States along the Mississippi has in fact acquired a cant name from this animal, being known by the soubriquet of the "Badger State". In 1858, Prof. Baird gives the *habitat* as "Iowa and Wisconsin to the Pacific coast, and from Arkansas to 49° N. lat. (To 58° N. lat., Rich.)" The animal formerly extended eastward in the United States to Ohio at least. A letter addressed by Mr. Edward Orton, not long since, informs me of its occurrence near Toledo in that State, about twenty years previously, and of its extinction there. Mr. Robert Kennicott, in 1853-54, has the species among the mammals of Illinois; while in Iowa, writes Mr. Allen in 1869, "the species is probably nearly as numerous as formerly." The eastward range in the United States to Ohio, Indiana, Illinois, Michigan, Iowa, and Minnesota, as well as the oblique trend in British America to Hudson's Bay, thus makes the distribution of the animal more or less closely coincident with that of some of the Spermophiles; these animals, with the Badger and Kit Fox, being highly characteristic species of the central treeless region of the United States, where they occur in countless multitudes.

To the southward, the range of the typical Badger cannot be precisely given, for the reason that there the characters of the animal melt insensibly into those of the Mexican subspecies *berlandieri*. The change becomes marked in Texas, New Mexico, Arizona, and Southern and Lower California. I have personally traced the typical form into Colorado, and it is said by Drs. Coues and Yarrow to be very common throughout Western Utah and Eastern Nevada, but less abundant in New Mexico and Arizona. To the extraordinary abundance of the animal in the Upper Missouri country I shall again refer in speaking of its habits. Dr. J. S. Newberry has indicated its abundance in Eastern California, Utah, and Oregon. Mr. George Gibbs says that the Badger, called by the Yakima Indians *Weehthla*, was not seen by him west of the Cascade Mountains of Washington Territory, though very common on the dry barren plains on the Yakima River, and also on the timberless mountains between the Yakima and the Columbia. In certain sections of that Territory, as for instance the Simcoe Valley, the species is represented by Dr. Suckley as so abundant that riding becomes dangerous from the number of the burrows. The writer last mentioned adds a paragraph on the general east and west distribution of the species, as follows:—

"Found sparingly in the eastern portion of Minnesota; being more abundant near the Missouri. From thence, after entering Nebraska [*i. e.*, the present Territories of Dakota and Montana], it extends almost all the way to the dividing ridge of the Cascade Mountains, near the Pacific coast. Farther west it does not go, at least north of the Columbia. I have seen it in the St. Mary's Valley, at the western base of the main chain of the Rocky mountains, and as far south in Oregon as the vicinity of Fort Boisé on the Snake or Lewis river. They are most abundant (north of Utah) in the vicinity of Powder river, Oregon, and the Yakima, one of the northern tributaries of the Columbia."

Habits.

The Badger is one of the most secret animals of this country— one whose habits and whose whole nature tend to screen it from observation so thoroughly that much of our knowledge is a result of reasonable inference rather than a matter of actual experience, while some of the most important points respecting its economy remain to be ascertained with precision. As will have been gathered from what has preceded, it lives altogether in holes in the ground, for the excavation of which its whole structure is adapted. Other animals are as decidedly fossorial as the Badger, and like it live underground, but the Badger, unlike its usual associates, the Prairie-dogs (*Cynomys*) and other Spermophiles, does not continually appear in view; rather, it leads a life almost as completely subterranean as that of the Gophers (*Geomys* and *Thomomys*), or even of the Mole itself. In the colder latitudes, moreover, it hibernates during a considerable portion of the year. I have travelled for days and weeks in regions where Badgers abounded, and where their innumerable burrows offered the principal obstacle to progress on horseback or by wheeled conveyance, yet the number of Badgers I have actually seen alive, in a state of nature, might be told off on the fingers of either hand. Most of the individuals I have laid eyes on were in sight but a few moments, as they hurried into the nearest hole. On one occasion, however, a Badger, crouching at the mouth of its burrow in fancied security, allowed me to approach and kill it with a shot; but I should add that this imprudent individual was but half-grown, and probably had never seen a man before.

I have found Badgers in countless numbers nearly throughout the region of the Upper Missouri River and its tributaries. I do not see how they could well be more numerous anywhere. In some favorite stretches of sandy, sterile soil, their burrows are *everywhere*, together with those of Kit Foxes, Prairie-dogs, and Spermophiles, and, as already said, these holes are a source of annoyance and even danger to the traveller. In ordinary journeying, one has to keep constant lookout lest his horse suddenly goes down under him, with a fore leg deep in a Badger-hole; and part of the training of the western horse is to make him look out for and avoid these pitfalls. In the Buffalo country particularly, Badgers live in extraordinary numbers, attracted and retained by the surety of abundant food-supply; and there are places where the chase of the Buffalo on horseback is absolutely impracticable, except at a risk to life or limb which few are willing to run.

The burrows of the Badger are known from those of the Prairie-dog and other Spermophiles by their greater dimensions; besides, they differ from the former in never being built up around the entrance into the regular mound or circular buttress which usually surmounts the well-kept domicile of the *Cynomys*. From the holes of Kit Foxes and Coyotés, they are not distinguishable with any certainty; in fact, it is probable that these animals frequently or almost habitually occupy deserted burrows of the Badger, remodelled, if need be, to suit their convenience. But it must not be supposed that all of the innumerable Badger-diggings are the residences of these animals. The Badger, too slow of foot to capture the nimble Rodents which form its principal food, perpetually seeks them in their own retreats; and it is the work of a few minutes for this vigorous miner to so far enlarge their burrows that it can enter and reach the deepest recesses. In places where the Badgers and Spermophiles most abound, the continual excavation of the soil by these animals fairly undermines and honeycombs the ground. The Badgers, though not migratory, are sometimes attracted or focussed as it were, from a large area in some particular spot which temporarily offers special attraction in the way of a food-supply. Thus, I have in mind a place on the Mouse River, Dakota, where there had been not long before a grand battue of Buffalo by the Indians, and where the number of Badger-holes, then deserted, exceeded anything I had before seen or have since witnessed.

The abundance of the Badgers might be expected, in view of the fact that they have very few enemies. The animals are stout and determined enough to stand off Wolves and Foxes; they seldom venture far from their secure retreats; and in fact I know of no indigenous mammal which habitually preys upon or otherwise destroys them. A Badger ensconced in its hole would be a formidable antagonist which few animals would care to molest. Their immunity from danger, partly the result of their physical prowess, partly secured by the practically impregnable nature of their resorts, together with the abundance of food and the ease with which it is secured, tends to the firm perpetuation and continual increase of the species in all unsettled portions of the country. Man is the principal enemy of the Badger, destroying thousands annually for his convenience or luxury.

Besides the Spermophiles, Arvicolas, and other small quadrupeds which furnish its staple diet, the Badger is said to prey upon a variety of humbler animals, even insects and snails, and to eat birds' eggs. As to the last named, there is a large supply on the western prairies, where many kinds of small birds, in great multitudes, nest upon the ground. Mr. W. H. Gibson refers to an especial fondness of the Badger for the stores of wild bees; the honey, wax, and grubs being alike devoured.* Audubon has figured the Badger with a Shore Lark (*Eremophila alpestris*) in its grasp. Mr. J. A. Allen speaks of finding

---

* I am not sure, however, that the actual reference is not to the European Badger, whose apivorous habit has long been known. "Buffon states that it digs up wasps' nests for the sake of the honey; — a fact which has received an interesting confirmation from the observation of a correspondent of Loudon's Magazine of Natural History, who seems, however, to attribute the destruction of these nests to the fondness of the Badger for the larvæ of the wasp, as he says that the combs were found scattered about, but none were left that contained the maggots. This predilection of the Badger for honey offers a striking analogy to several others of the group, particularly to its Oriental relation the Ratel, *Mellivora Capensis*, which is known to live principally upon it."—(BELL'S *British Quadrupeds*, ed. of 1837, pp. 123–4.)

the bones and wool of lambs in its burrows, though the animal is not generally regarded as injurious by the farmers.

The Badger has been called a "timid" animal. So it is, in the sense that it avoids rather than confronts impending danger; but this is simply the instinctive prudence and discretion of a creature which prefers the absolute immunity of its subterranean resorts to the chances of unequal combat in which it is at disadvantage. Certainly, no lack of courage, determination, and physical endurance is seen when the creature, captured or cut off from its retreat, is brought to bay. Its pluck is then as conspicuous as its really formidable strength. The cruel sport of "Badger-baiting" is sometimes indulged in the West; and if the animal be given a barrel or similar retreat in which it is secure from attack in the rear, it may prove more than a match for a strong dog. Indeed, the fighting qualities of the Badger, and stubborn resistance it offers at whatever unfair odds, have supplied our language with a word of peculiar significance: to "badger" is to beset on all sides and harass and worry. The stout, thick-set, and depressed shape of the animal is greatly in its favor, combining with the long loose hair to prevent a dog from reaching vulnerable parts, and to embarrass it in attempting to take hold; the snap of the jaws inflicts a serious wound; and, finally, the tenacity of life is at a high rate.

A sketch of this animal, from the pen of Dr. J. S. Newberry, gives corroborative evidence of the Badger's powers of self-defence; I transcribe the passages at some length, as they afford other items in the natural history of the species:—

"In traversing the arid surfaces of the sage plains of eastern California, Utah, and Oregon, there is, perhaps, no one thing which the traveller may be more sure of seeing every day of his journey than the burrow of a badger; and, after cursing the country, and the folly which led him to cross these barren, hot, and dusty surfaces, there is nothing he will more certainly do, whether on foot or mounted, than tumble into one of these same badger holes, and yet the chances are more than equal that he never sees a living badger on which to revenge himself; for the badger is a shy and timid animal, and the country he inhabits is so open, it rarely happens that he is surprised at a distance from his burrow. During our march of several hundred miles through the country inhabited by the badger this did occur, however, on one or two occasions, and gave rise to some ludicrous scenes. The badger, though far from formidable, is too well provided with teeth to be handled without gloves; and knowing that his only safety when attacked is in plunging to the bottom of his burrow, his pig-headed pertinacity in endeavoring to reach it is such, that an unarmed man finds it difficult to stop him.

"Mr. Anderson, who gave me most efficient aid in collecting, came one day suddenly upon a badger at some distance from his hole; of course he made for it with all possible speed, which, it should be said, is not so great but that a man could easily overtake one. Mr. Anderson at first endeavored to trample him under his horse's feet, but, though he ran over him several times, the badger avoided the hoofs and received no injury. As we had not then obtained a specimen, he was particularly anxious to secure this one, so he drove his horse before him, and brought him to bay. He then jumped off, hoping, by means of kicks and his sheath-knife, to dispatch him; but the badger, instead of retreating, came at him open-mouthed, and with such a show of ferocity that he was fain to let him pass, trusting to find a club to kill him with; but in that region clubs do not 'grow on every bush,' for most of the bushes are sage bushes, and before he found any sort of stick the badger had reached his hole. Two days after I became indebted to him for a fine specimen, which a long rifleshot had dropped at the entrance of his burrow. Another, while leisurely following an old trail, apparently on a journey, was overtaken and killed by some of our soldiers. Seeing, perhaps the hopelessness of the attempt, he made no effort to escape, but a vigorous defence, and was only dispatched with some difficulty."

Sir John Richardson narrates an incident which further illustrates the prowess of this stubborn, sullen customer. "The strength of its fore-feet and claws is so great," says he, "that one which had insinuated only its head and shoulders into a hole, resisted the utmost endeavors of two stout young men who endeavored to drag it out by the hind legs and tail, until one of them fired the contents of his fowling-piece into its body." This is quite a match for the stories told of the Armadillo itself. "Early in the spring, however," the author continues, "when they first begin to stir abroad, they may easily be caught by pouring water into their holes; for the ground being frozen at that period, the water does not escape through the sand, but soon fills the hole, and its tenant is obliged to come out."

The author of the "Complete American Trapper" also refers to this method of taking Badgers, and adds others:—"Although his general appearance would not indicate it, he is a sly and cunning animal and not easily captured in a trap of any kind. He has been known to set at defiance all the traps that were set for him, and to devour the baits without suffering from his audacity. He will sometimes overturn a trap and spring it from the under side, before attempting to remove the bait. Although not quite as crafty as the fox, it is necessary to use much of the same caution in trapping the badger, as a bare trap seldom wins more than a look of contempt from the wary animal. The usual mode of catching the creature is to set the trap, size No. 3 [the so-called fox-trap, with springs at each end], at the mouth of its burrow, carefully covering it with loose earth and securing it by a chain to a stake. Any of the methods used in trapping the fox will be found to work admirably. The dead-fall or garrote will also do good service. Bait with a rat, mouse, or with whatever else the animal is specially fond, and scent with oil of anise or musk. In early spring, while the ground is still hard, badgers are easily captured by flooding."

The reproduction of the species does not appear to be fully known. I have no personal information on this score, beyond the fact that I once secured a still ungrown animal in Colorado during the latter part of August. The writer last quoted says that the nest is made in the burrow (as indeed is unquestionable), and that the young are three or four in number. Richardson, referring to the hibernation of the Badger in British America from November to April, states that, like Bears, the animals do not seem to lose much flesh during the winter, for they are observed to be very fat when they come abroad in the spring; and adds that, as they pair at once, they soon become lean. The periods of gestation and lactation are probably unknown.

The habits and manners of the Badger in confinement, to which we will next turn attention, have been attentively studied by Audubon and Bachman, who have given an interesting account, here transcribed in full:—

"During our stay at Fort Union, on the Upper Missouri River, in the summer of 1843, we purchased a living Badger from a squaw, who had brought it from some distance to the fort for sale; it having been caught by another squaw at a place nearly two hundred and fifty miles away, among the Crow Indians. It was first placed in our common room, but was found to be so very mischievous, pulling about and tearing to pieces every article within its reach, trying to dig up the stones of the hearth, &c., that we had it removed into an adjoining apartment. It was regularly fed morning and evening on raw meat, either the flesh of animals procured by our hunters, or small birds shot during our researches through the adjacent country. It drank a good deal of water, and was rather cleanly in its habits. In the course of a few days it managed to dig a hole under the hearth and fire-place nearly large and deep enough to conceal its body, and we were obliged to drag it out by main force whenever we wished to examine it. It was provoked at the near approach of any one, and growled continuously at all intruders. It was not, however, very vicious, and would suffer one or two of our companions to handle and play with it at times.

"At that period this Badger was about five months old, and was nearly as large as a full grown wood-chuck or ground-hog, (*Arctomys monax*.) Its fur was of the usual colour of summer pelage, and it was quite a pretty looking animal. We concluded to bring it to New York alive, if possible, and succeeded in doing so after much trouble, it having nearly made its escape more than once. On one occasion, when our boat was made fast to the shore for the night, and we were about to make our 'camp,' the Badger gnawed his way out of the box in which he was confined, and began to range over the batteau; we rose as speedily as possible, and striking a light, commenced a chase after it with the aid of one of the hands, and caught it by casting a buffalo robe over it. The cage next day was wired, and bits of tin put in places where the wooden bars had been gnawed through, so that the animal could not again easily get out of its prison. After having become accustomed to the box, the Badger became quite playful and took exercise by rolling himself rapidly from one end to the other, and then back again with a reversed movement, continuing this amusement sometimes for an hour or two.

"On arriving at our residence in New York, we had a large box, tinned on the inside, let into the ground about two feet and a half and filled to the same depth with earth. The Badger was put into it, and in a few minutes made a hole, in which he seemed quite at home, and where he passed most of his time during the winter, although he always came out to take his food and water, and did not appear at all sluggish or inclined to hibernate even when the weather was so cold as to make it necessary to pour hot water into the pan that was placed within his cage, to enable him to drink, as cold water would have frozen immediately, and in fact the pan generally had a stratum of ice on the bottom which the hot water dissolved when poured in at feeding-time.

"Our Badger was fed regularly, and soon grew very fat; its coat changed completely, became woolly and a buff-brown color, and the fur by the month of February had become indeed the most effectual protection against cold that can well be imagined.

"We had an opportunity in Charleston of observing almost daily for a fortnight, the habits of a Badger in a menagerie; he was rather gentle, and would suffer himself to be played with and fondled by his keeper, but did not appear as well pleased with strangers; he occasionally growled at us, and would not suffer us to examine him without the presence and aid of his keeper.

"In running, his fore feet crossed each other, and his body nearly touched the ground. The heel did not press on the ground like that of the bear, but was only slightly elevated above it. He resembled the Maryland marmot in running, and progressed with about the same speed. We have never seen any animal that could exceed him in digging. He would fall to work with his strong feet and long nails, and in a minute bury himself in the earth, and would very soon advance to the end of a chain ten feet in length. In digging, the hind, as well as the fore feet, were at work, the latter for the purpose of excavating, and the former, (like paddles,) for expelling the earth out of the hole, and nothing seemed to delight him more than burrowing in the ground; he seemed never to become weary of this kind of amusement; and when he had advanced to the length of his chain he would return and commence a fresh gallery near the mouth of the first hole; thus he would be occupied for hours, and it was necessary to drag him away by main force. He lived on good terms with the raccoon, gray fox, prairie wolf, and a dozen other species of animals. He was said to be active and playful at night, but he seemed rather dull during the day, usually lying rolled up like a ball, with his head under his body for hours at a time.

"The Badger did not refuse bread, but preferred meat, making two meals during the day, and eating about half a pound at each.

"We occasionally saw him assuming rather an interesting attitude, raising the fore part of his body from the earth, drawing his feet along his sides, sitting up in the manner of the marmot, and turning his head in all directions to make observations."

The assuming of this attitude may have been a result of confinement, as I have not observed it when I have seen the animal in a state of nature, nor does it appear to have been noticed by others. The Badger, above all our other animals, is notable for its *flatness;* even when running it looks broad and flat, and the belly seems to sweep the ground during its rather slow, heavy, and awkward progress. Seen when crouching in fancied security or hoping to escape observation (and it will sometimes remain long motionless in this posture, permitting near approach), the animal might easily be mistaken for a stone or clod of earth; the very hairs lie flat, as if "parted in the middle", and form a fringe along either side, projecting, as one writer has remarked, like the shell of a turtle or the eaves of a house. The peculiar pattern of coloration is then displayed to best advantage. Under anger or irritation, the animal bristles up its hair, and appears much larger than it really is.

The flesh of the Badger, like that of the Skunk, is eatable, and doubtless often eaten by savage tribes, though not to be recommended to a cultivated palate. The specimens I have skinned, even the young one before mentioned in this article, emitted during the process such rank and foul odor as to be simply disgusting. The Badger yields a valuable and at times fashionable fur, used for robes, and for muffs, tippets, and trimmings. Thousands of shaving-brushes are said to be annually made from the long hairs, which are also extensively used in the manufacture of artists' materials, one of which is a "badger-blender". In 1873, the London sales of Badger skins by the Hudson's Bay Company were 2,700, at prices varying from one to seven shillings, averaging 1$s.$ 6$d.$ The leading American journal of the fur trade in 1876 quoted Badger skins at $1 for prime, 50 cents for "seconds", and 10 cents for "thirds". The colors of the Badger pelt, though not striking, are pleasing, being an intimate and harmonious blending of gray, tawny, black, and white, the colors ringed in alternation gray, tawny, black, and white, the colors ringed in alternation on individual hairs. The gray predominates, the general "tone" or effect being a grizzled gray, which has given rise to the well known adage, "as gray as a Badger".

The Mexican Badger.

Taxidea americana berlandieri.

Meles labradoria, *Bennett*, P. Z. S. 1833, 42 ("California"). Vertebral stripe continuous).- *Rich.* Zool. Beechey's Voy. 1839, p. 9*.
? Taxidea labradoria, *Waterh.* P. Z. S. 1838, 154.
Taxidea berlandieri, *Baird*, M. N. A. 1857, 205, pl. 39, f. 1.—*Bd.* Mex. B. Surv. ii. pt. ii. 1859, Mamm. 21.
Taxidea americana *var.* californica, *Gray*, P. Z. S. 1865, 141 (from Bennett); Cat. Carn. Br. Mus. 1869, p. —.
Taxidea americana *var.* berlandieri, *Gray*, P. Z. S. 1865, 141 (from Baird); Cat. Carn. Br. Mus. 1869, p. —.
Taxidea americana *subsp.* berlandieri, *Allen*, Bull. U. S. Geol. and Geog. Surv. Terr. vol. ii. no. 4, 1876, 331.
Meles tlacoyoté, *Berl.* MSS. ined.
Tlalcoyotl, *Nahuatl.*
Tlacoyotl, "*Fernandez.*"
Tlacoyote, *Mex. Vulg.*
Texon *or* Tejon, *Mex.* (cf. *Taxus, Tasso, Taisson*).

HAB.—Southwestern border of the United States and southward. Llano Estacado, Texas, *Pope;* Canton. Burgwyn, N. M., *Irwin;* Cape St. Lucas, *Xantus.* "Interior and Eastern States of Mexico, especially Nuevo Leone and Tamaulipas.—(*Berlandier*, MSS.)

SUBSPECIFIC CHARACTERS.—Similar to *T. americana*, but with a white dorsal stripe, sometimes interrupted, from nose to tail.

General remarks.

The extreme manifestation of this form of *Taxidea* which I have seen is exemplified in a specimen in the Smithsonian Museum, collected at Cape St. Lucas by Mr. John Xantus. Here the white frontal stripe is remarkably broad, nearly equalling in width the dark part of the head, and it continues uninterrupted thence to the tail as a sharp white vertebral line. This is a conspicuous character, and, were it constant, there would need be no hesitation in recognizing a second species, even in default of correlated difference from *T. americana*. But it is not constant; on the contrary, other specimens show various degrees of interruption of the white dorsal line. Thus, the one from Texas noticed by Prof. Baird in the works above cited shows a prolongation of the white frontal line past the nape to a point opposite the shoulders, its interruption there for about three inches, and its reappearance for about four inches at the middle of the back. I fail to appreciate any other decided peculiarities of this form, though it may average rather smaller, and somewhat more heavily colored, owing to its southern habitat. Certain cranial characters noted by Prof. Baird, according to the material then in hand, are negatived in the later examination I have made of much larger series of specimens.

This is clearly the animal referred to by Bennett, as above, as a Californian variety, with darker ground-color, and a white line showing in several places along the back, or continuous to the tail. In the United States it has only been noticed, to my knowledge, in the localities already indicated; but there is no doubt that this form, more or less decidedly pronounced, ranges over the intermediate ground. I find it noticed at considerable length, with an unmistakable description, under the name of *Meles tlacoyoté*, in Dr. Berlandier's manuscripts, where it is considered to be the Tlacoyotl of Hernandez. The fore claws are described as "blackish"; otherwise the account agrees exactly with the specimens before me named *berlandieri* by Baird. Dr. Berlandier was evidently familiar with the animal, which he represents as common in Northeastern Mexico, and gives several biographical notices—nothing, however, to indicate any differences of moment in its habits as compared with those of *T. americana*. The following are his measurements of a female in the flesh:—Nose to end of tail 24 inches; head 5; tail 5½; whole fore leg 6; hind leg 5½; stature at shoulders 7.

ADDENDUM TO CHAPTER IX.

Description of the perinæal glands of the European Badger, Meles vulgaris.

The American Badger has not, so far as I am aware, been examined anatomically with reference to the peculiar organs of the perinæum and neighboring parts. These, however, have been studied in the European species by M. Chatin, whose results may be here reproduced in substance, in default of information respecting our own species, as it is improbable that any material difference in these respects subsists between the two. It does not appear that the *Meles vulgaris* itself had been sufficiently studied prior to M. Chatin's investigations. The Badger is found to be peculiar* in the presence not only of anal glands of an ordinary character, but also of another, per-

---

* But is the anatomy of *Mydaus* and *Arctonyx* known?

fectly distinct gland, the secretion of which is emptied into a pocket back of the anus, just beneath the root of the tail.

1. *Anal glands.*—The parts present, near the termination of the rectum, as appendages at its sides, two oval slightly recurved masses, 20^{mm} long and about half as broad in the middle. Upon the slightest pressure, a liquid gushes from the two excretory pores, which open at the sides of the anus at the bottom of well-marked recesses. This substance is very viscid, of a rosy-yellow color, and extremely fetid; it is almost entirely soluble in sulphuric ether, and contains numerous fatty particles and epithelial remains. The two glands are embedded in adipose tissue, and entirely covered with a muscular tunic arising from the anal muscles, especially the retractor. The secretory portion is of the same general character as in allied species; the tissue enveloping the culs-de-sac is principally of laminated fibres, strengthened, however, by elastic ones; the diameter of the culs-de-sac is from 0.04 to 0.08^{mm}. On longitudinal section of the gland, the centre is seen to be occupied by a large reservoir lined with a delicate brownish membrane, much as in other Carnivores. The product of secretion is turned by a small opening into a rather narrow duct, leading to the external orifice already indicated.

2. *Glands of the subcaudal pouch.*—In the Badger, as in the Porcin and domesticated Carnivores, the rectum is attached to the sacrum and first coccygeal bones by a strong muscular band, which, in the present species, leaves the rectum at a point 25^{mm} from the insertion of the anal glands, at an angle of about 60° from the axis of the intestine, and proceeds to its insertion upon the sacro-coccygeal bones. At a point in front of the insertion of this muscle is found a deeply bilobate mass, apparently formed of two separate glands, each of oval shape, and apposed along a flat internal face, the exterior surface being convex and lobate; but the organ is really single, as it has but one receptacle for the product of all the follicular portion of the apparatus. This is a new glandular organ peculiar to the coccygeal region. Each moiety measures 24^{mm} in depth with an average breadth of 11^{mm}; for the rest, they are embedded in abundant adipose tissue, and the surface is whitish and papillate. The secretory portion is about 2^{mm} thick; it is composed of follicles, each comprising a large number of culs-de-sac, of an average diameter of 0.06^{mm}, lined with polyhedral epithelial cells. The product of secretion is received in an extremely large central sac, the surface of which is furnished throughout with numerous short, stiff, brownish hairs. This sac is distended with a yellowish fetid substance mixed with numerous hairs, like the viverreum of the Civet. This central reservoir may be regarded as the beginning of the pocket which opens beneath the tail; it communicates freely, and, in fact, is part of one and the same cavity.

In the female, it is observed that the end of the genital organs and the rectum form, by their union, a kind of cloacal vestibule, in front of which is the clitoris, with the urethra immediately below. Beneath this sexual portion of the vestibule is the orifice by which the rectum opens externally, after receiving at its sides the excretory ducts of the anal glands. Finally, back of these parts, is a broad transverse depression,—the subcaudal pouch; it measures 23^{mm} across. The anal glands are 18^{mm} long and about 9^{mm} broad; they are club-shaped, and quite similar to those of the male. To the subcaudal pouch there is also attached a secretory apparatus like that already described, but smaller; its internal structure is the same.

Briefly, then, the Badger possesses perinæal glands, remarkably peculiar not only in their formation, but in their general relations, since the true anal glands are supplemented with another special glandular mass. This latter is not situated between the genitalia and the anus, as in *Viverra*, but between the tail and the anus, so that it is behind the latter and not in front of it, like the scent-bags of the Civets. We cannot, however, deny their analogy, as seen in the central cavity, clothed with hairs, and the bilobation of the gland; but they differ in situation, in the nature of the product, and to a certain extent in histological structure, thus warranting, from the present standpoint, recognition of two quite distinct types.

CHAPTER X.

Subfamily LUTRINÆ: The Otters.

General considerations—The genus *Lutra*—Generic characters and remarks—The North American Otter, *Lutra canadensis*—Synonymy—Habitat—Specific characters—Description of external characters—Description of the skull and teeth—Variation in the skull—History of the species—Geographical distribution—Habits of Otters—Extinct species of North American Otter.

IN the Otters, we encounter a fourth decided modification of the family characters in adaptation to a highly aquatic mode of life. Among the true *Mustelinæ*, indeed, we found some aquatic species, like the Mink, *Putorius vison;* but in none of the foregoing subfamilies is the structure modified to any great extent with reference to natatorial abilities. The short, broad, fully webbed feet of the Otters, the cylindrical body, the stout tapering tail, and very turgid blunt muzzle result in an unmistakable physiognomy, as characteristic of the *Lutrinæ* as are the more important structural modifications of the skull and teeth.

The *Lutrinæ* have been defined as *Mustelidæ* with the number of teeth equal in both jaws. This expression, however, is equally applicable to the *Enhydrinæ*, or Sea Otter, in which, very curiously, lack of one pair of under incisors brings about the same adjustment of total teeth of the two jaws, though the grinders are unequal in number.

The *Lutrinæ* as here limited to the exclusion of *Enhydris* may be recognized as the only *Mustelidæ* in which the number of grinding teeth (molars and premolars together) is the same in both jaws, the formula being $pm. \frac{4-4}{3-3}; m. \frac{1-1}{2-2} = \frac{10}{10}$. The total of the teeth is $\frac{18}{18} = 36$. The upper molar is large and quadrate in shape.

After throwing out the very different genus *Enhydris* as type of a separate subfamily, the *Lutrinæ* still include a number of well-marked genera. Of these, *Lutra* is the principal genus, with the greater number of species and most general distribution, occurring in both hemispheres. Nearly allied genera, by some considered only as subgenera, are based chiefly upon modifications of the claws, which, in some of the Old World Otters, are small, rudimentary, or even wanting, as in *Leptonyx* and *Aonyx*. The most remarkable genus is the South American *Pterura* or *Pteronura*, peculiar in the lateral dilatation of the tail.

Lutra itself is the only North American genus of the subfamily, our species having been unnecessarily, if not unwarrantably, distinguished by generic name from the European type of *Lutra*.

The Genus LUTRA. (Linn.)

GENERIC CHARACTERS.—*Dental formula:* i. $\frac{3-3}{3-3}$; c. $\frac{1-1}{1-1}$; pm. $\frac{4-4}{3-3}$; m. $\frac{1-1}{2-2} = \frac{1}{1}\frac{8}{8} = 36$. Teeth of ordinary carnivorous pattern. Molar of upper jaw large, quadrate. Back upper premolar with a large internal shelf, making the contour of the whole crown triangular. Skull much depressed and flattened on top, the dorsal outline more or less nearly straight and horizontal; rostrum extremely short, bringing the fore ends of the nasals nearly opposite the anterior root of the zygoma, the sides of the rostrum erect, the top flat. Cerebral portion of the skull swollen backward, with strongly convex lateral outline. Postorbital processes variable (well developed in some species, as in the North American, wanting in others). Anteorbital foramen very large, bounded above by a slender bridge of bone. Posterior nares thrown into one conduit.* Palate extending far back of molars. Pterygoids strongly hamulate. Body stout, but lengthened and cylindrical; muzzle very obtuse; ears very small. Feet short, broad, naked, or partly hairy on the palm and sole, the digits full-webbed. Tail long, tapering, cylindrico-depressed, but without special lateral dilatation. Claws, though small, well formed. Pelage without striking color-contrasts.

Many of the foregoing expressions are applicable to the subfamily at large, as well as to the present genus. Particular points of *Lutra* proper are the presence of perfect claws, in comparison with their absence or rudimentary condition in some other genera, and the lack of special dilatation of the tail.

The uniformity of coloration and the great individual variability in size throughout this genus render the determination of the species difficult. The points which I have found most available in specific diagnosis, when cranial and dental characters fail, are, the size and shape of the nose-pad and the con-

---

* In some species, there is an incomplete septum extending further back than in *Mustelinæ*, but never, so far as I know, to the very end of the bony palate, as in *Mephitinæ* and *Melinæ*.

dition of furriness or nakedness of the soles, together with the special tuberculation of the latter. The various American species may readily be determined by attending to these particulars.

There is but a single well-determined North American species of this genus. This is so distinct from the European, with which it was long confounded, that I am tempted to place it in a different subgenus, grounded on various cranial peculiarities that might be enumerated, and only refrain from so doing in my ignorance of what intermediate forms of crania other species may present to connect the extremes seen in *L. vulgaris* and *L. canadensis*. Other American species agree closely with *L. canadensis* in cranial characters; and it is not improbable that the species of this hemisphere may all be subgenerically different from those of the Old World. I shall, however, consider them as simply *Lutra*.

Besides *L. canadensis*, moreover, there is a perfectly distinct Mexican species, *Lutra californica* of Gray (not of Baird), which is said, and I believe correctly, to extend into the United States along the Pacific side. I think it will be found, as already supposed by some, to be very extensively dispersed along the western shores of North, Middle and South America. It appears to be as distinct from the *L. brasiliensis* as it is from *L. canadensis*, and I have no doubt will ultimately be established as a second good species of Otter of the United States, though under a name long prior to that imposed by Gray. But as I have seen only Mexican skins of this animal, I cannot now introduce it to our fauna. The point is discussed beyond.

The North American Otter.

Lutra canadensis.

Plate XVII.

Mustela canadensis, *Turton*, S. N. i. 1806, 57 (not *Mustela canadensis*, *id. ibid.* 59, which is *M. pennanti*, the Pekan. Not of Schreber nor of Erxleben nor of authors).
Lutra canadensis, "*F. Cuv.* Dict. Sc. Nat. xxvii, 1823, 242.—*Is. Geoff.* Dict. Class. ix. 520."—*J. Sab.* App. Frankl. Journ. 1823, 653.—*Less.* Man. 1827, 154, no. 414.—*Griff.* An. Kingd. v. 1827, 130, no. 362.—*Fisch.* Syn. 1829, 225.—*Rich.* F. B.-A. i. 1829, 57, no. 20.—*Emmons*, "Rep. Quad. Mass. 1838, 25"; Rep. Quad. Mass. 1840, 46.—*Rich.* Zoöl. Voy. Beechey, 1839, 4.—*Maxim.* Reise N. Am. i. 1839, 211; Arch. Naturg. 1861, 236; Verz. N. A. Säug. 1862, 60, pl. 8, f. 6 (os penis).—*De Kay*, N. Y. Zoöl. i. 1842, 39, pl. 3, f. 1, pl. 33. f. 1, 2, 3 (skull).—*Linsley*, Am. Journ. Sci. xliii. 1842, —.—*Schinz*, Syn. i. 1844, 349, no. 5.—*Aud. & Bach.* Q. N. A. ii. 1851, 2, pl. 51.—*Woodh.* Sitgreaves's Rep. 1853, 44.—*Kenn.* Tr. Illinois Agric. Soc. for 1853-4, 1855, 578.—*Giebel*, Säug. 1855, 789.—*Beesley*, Geol. Cape May, 1857, 137.—*Bd.* M. N. A. 1857, 184, pl. 38, f. *a, b, c, d. e.*—*Billings*, Canad. Nat. and Geol. i. 1857, 235.—*Samuels*, Ninth Ann. Rep. Mass. Agric.

for 1861, 1862, 160.—*Hayd.* Tr. Amer. Phil. Soc. xii. 1862, 143. *Hall,* Canad. Nat. and Geol. vi. 1861, 297.—*Ross,* Canad. Nat. and Geol. vi. 1861, 35.—*Barnston,* Canad. Nat. and Geol. viii. 1863, ——, f. —.—*Gerr.* Cat. Bones Br. Mus. 1862, 101.—*Allen,* Pr. Bost. Soc. xiii. 1869, 183.—*Bull.* M. C. Z. i. 1869, 178; ii. 1871, 169 (Florida).—*Gilpin,* Proc. and Tr. N. Scotia Inst. ii. 1870, 60.—*All.* Bull. Ess. Inst. vi. 1874, 46, 63 (Kansas and Utah).—*Ames,* Bull. Minn. Acad. Nat. Sci. 1874, 69.—*Coues & Yarrow,* Zoöl. Expl. W. 100 Merid. v. 1875, 63.—*Allen,* Bull. U. S. Geol. and Geog. Surv. Terr. vol. ii. no. 4, 1876, 331 (skull).

Lutra canadensis *var., Aud. & Bach.* Q. N. A. iii. 1853, 97, pl. 122 (figure of Gray's type of *Lataxina mollis*).

Latax canadensis, *Gray,* P. Z. S. 1865, 133; Cat. Carn. Br. Mus. 1869, —.

Lutra vulgaris *var.* **canadensis,** *Wagn.* Suppl. Schreber, ii. 1841, 256.

Mustela hudsonica, "*Lacépède*".

Lutra hudsonica, *F. Cuv.* Suppl. Buffon, i. 1831, 194.

? Lutra gracilis, *Oken,* Lehrb. Naturg. Th. iii. Abth. ii. 1816, 986 ("Staatenland, Insel an Amerika bei New-York").

Lutra brasiliensis, *Desm.* Mamm. i. 1820, 188 (in part).—*Harl.* Fn. Amer. 1825, 71 (in part).—*Godm.* Am. N. H. i. 1831, 222, pl. —, f. 2 (in part).—*Thomps.* N. H. Vermont, 1853, 33.

Lutra lataxina, *F. Cuv.* "Dict. Sc. Nat. xxvii. 1823, 242"; Suppl. Buffon, i. 1831, 203.—"*Is. Geoff.* Dict. Class. ix. 520."—*Griff.* An. Kingd. v. 1827, 131, no. 364.—*Less.* Man. 1827, 154, no. 416.—*Fisch.* Syn. 1829, 226, no. 4.—*De Kay,* N. Y. Zoöl. i. 1842, 41.—*Schinz,* Syn. i. 1844, 350.

Latax lataxina, *Gray,* Ann. Mag. N. H. i. 1837, 119.

Lataxina mollis, *Gray,* List Mamm. Br. Mus. 1843, 70 (type figured by And. & Bach. *l. c.*).

Lutra americana, *Wyman,* Pr. Bost. Soc. ii. 1847, 249 (on articulation of mandible).

Lutra californica, *Bd.* M. N. A. 1857, 187.—*Newb.* P. R. R. Rep. vi. 1857, 42.—*Coop. & Suck.* N. H. W. T. 1860, 115. (Probably not of Gray.)

Lutra destructor, *Burnst.* Canad. Nat. and Geol. viii. 1863, 147, f. — (Lake Superior).

Loutre du Canada, *Buff.* "Hist. Nat. xiii. ——, 322, 326, pl. 44 (4to ed.); ed. Pillot. xv. p. 80",—*Fr. Cuv. l. c.*

Loutre de la Caroline, *F. Cuv. l. c.*

Loutre d'Amérique, *Cuv.* (in part; unites the Brazilian species).

Common Otter, *Pennant,* Arct. Zoöl. i. 1784, 86, no. 34 (in part; unites the European).

Land Otter, *Warden,* United States, i. 1819, 206.

American Otter, *Godman, l. c., Baird, l. c.,* and of authors.

Canada Otter, *Sabine, Rich. l. c., Aud. & Bach. l. c.,* and of authors.

(**Otter,** see *Martens,* Zool. Gart. xi. 1870, 279; philological.)

Neekeek, *Cree Indians.*

HAB.—North America at large, being rather sparingly distributed over most of the waters of the continent. Said to occur in Central America (Costa Rica, *v. Frantzius*).

SPECIFIC CHARACTERS.*—Orbits well defined by prominent conical post-orbital processes, the distance between the tips of which is one-half or more of the intermastoid width of the skull. Inner depressed moiety of posterior upper premolar as large and nearly as long as the main outer moiety; general dentition strong. Naked nasal pad large (upward of an inch long or broad in full-grown individuals), extending back above the nostrils in a ∧-shaped outline, reaching below the nostrils with a straight transverse border, which sometimes sends a slight spur part way down the median line of the lip. Palms hairy between the digits, isolating the individual bald digital bulbs, and having an isolated patch or carpal peninsula of hair posteriorly. Soles hairy between the digits, isolating the individual digital bulbs, much encroached upon by hair from behind, and having three or four peculiar small circular elevated callosities arranged around the posterior

---

* Drawn up with special reference to antithesis with *L. vulgaris* of Europe.

border of the main bald plantar surface. (Form, stature, and coloration not diagnostic.) Finally attaining a total length of four feet or more; liver-brown, with purplish gloss, paler on the under surface of the head, throat, and breast.

Description of external characters.*

This Otter shares the well-known form common to most species of the genus—the massive columnar body, without constriction of neck, small globose head, small eyes and ears, long taper tail, short stout limbs, and broad webbed feet, with close-set glossy fur and abundant woolly under-fur. Externally, the special form of the nose-pad and the state of furring of the palms and soles are the chief, if not the sole, characters distinguishing the species from several of its congeners.

The nose-pad is remarkably well developed—almost as much so as in *Enhydris*—perfectly bald, and in adult life tessellated by subdivision into very numerous small flat-topped papillæ. In general shape, it is an equilateral pentagon, with one side inferior, horizontal, and straight across, the next side on either hand irregular, owing to the shape of the nasal apertures, the two remaining sides coming together obliquely above to a median acute angle, high above a line drawn across the tops of the nostrils. It somewhat resembles the ace of spades.† The lower horizontal border is below a line drawn across the bottom of the nostrils; it sometimes sends down a small naked spur vertically towards the tip, sometimes not; either of the borders not occupied by the nostrils may be a little convex or a little concave, or sigmoidal. (In *Lutra vulgaris*, the nose-pad is very small, and entirely confined between the nostrils. In a common species of Mexico, said also to inhabit California, and in fact to extend from Chili to Kamtschatka, the nose-pad is considerably more developed than in *L. vulgaris*, yet much less so than in *canadensis;* the upper outline is deeply double-concave, like ᴗᴗ, and the lower outline, which does not reach below the nostrils, is concave, like ᴧ. In the Saricovienne, *Lutra brasilien-*

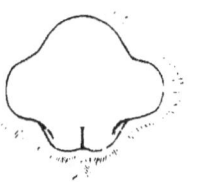

Nose-pad of *L. canadensis*.
Nat. size.

---

* From various specimens in the National Museum from different portions of North America.

† The figure, copied from Baird, is perhaps rather too near an ace-of-clubs shape; according to the dried specimens from which I drew my text, the top of the figure should be more pointed, and the lines thence rather less curved.

sis, with which ours used to be confounded, the nose-pad is described as divided by a line of hair coming down from above.)

The upper border of the nostrils is in *L. canadensis* represented by a prominent overhanging bulb. The whiskers are short, stout, stiff bristles, arranged in numerous series; others equally long and stiff grow from the sides of the chin near the angle of the mouth, and in front of the ears; others again spring over the eyes, and at the point of the chin. The eyes are small, far forward, nearer to the muzzle than to the ear. The ears are comparatively minute, with a thin, obtusely pointed conch, about as long as the surrounded fur, though they project somewhat, since the hairs lie flat; the entrance of the meatus is completely occluded with fur.

The tail is about half as long (more or less) as the head and body; regularly tapering from base to tip, elliptical in transverse section.

The short fore limb is succeeded by a stout wrist and broad flat hand. The fingers are very short, and when divaricated their tips describe nearly a semicircle around the centre of the palm. The toes are almost completely webbed by membranes reaching out to about the middle of the conspicuous digital bulbs—the median digit is a little freer than the rest, the lateral ones most completely united. The hand is entirely hairy above; below, the bulbs of the digits are perfectly bald, but the connecting membranes are more or less completely hairy, separating the naked bulbs from each other and from the main palmar surface. After this hairy membranous surface comes the single large palmar pad, naked for the most part, but having posteriorly a scant patch of hair, either isolated or connected by a hairy isthmus with the fur upon the wrist. In life, this main pad has no decided subdivision, though it sometimes shows certain lines of impression which in the dried state may be exaggerated into partitions. All the bald parts of the palm and the digital bulbs are tessellated with minute papillæ.

The soles, in general, resemble the palms in the webbing of the toes by a hairy membrane, and encroachment from behind of hairs upon the main plantar pad; but the shape of the hind foot is quite different. The 4th digit is much elongated, the 3d a little shorter, the 2d and 1st rapidly much graduated, with the 5th intermediate between the 3d and 2d. The terminal bulbs of the toes are naked and papillate, and completely isolated, by the hairiness of the intervening membrane, both from each other and from the main plantar pad. This last is perfectly naked and papillate for a broadly crescentic space, there being a central furry projection from behind. But the most remarkable feature, peculiar to this species, as far as is known, is the presence of three or four small, definite, circular, elevated papillæ, arranged along the posterior border of the naked space. I do not understand these singular structures, the appearance of which almost forces the presumption that they are the excretory pores of glandular organs beneath the integument.

The claws are similar on both fore and hind feet. They are short, stout, compressed, much arched, rapidly contracted from the thick base to an acute point. Those in front are rather larger, sharper, and more arched than the hinder ones.

The fur of the Otter is of great beauty, very thick, close, short, and shining, an exaggeration, in correspondence with the completely aquatic habits of the animal, of that of the Mink or Muskrat. The longer hairs are stout and glistening; the very copious under fur is lanuginous and lustreless. The sheen is only visible in its perfection when the pelt is viewed with the lay of the hairs; from the other direction the color is plain. As in most other species, the color is a rich dark liver-brown, or deep chestnut-brown above, more or less blackish or with a purplish gloss; paler below, especially anteriorly, on the under parts and sides of the head, the throat, and breast. This paleness is very variable, from a slight lightening of the general tone to a pale dull brownish or grayish, or even muddy white. The change is insensible, and there are no special markings anywhere. The roots of the hairs, even on the darkest parts of the pelage, are quite light brown, or often even dingy white, but the fur is so close that this does not appreciably affect the tone of the surface. The top of the tail is ordinarily the darkest part of the animal. The whiskers are partly colorless, partly brown. The nasal pad, palms, and soles are dark-colored.

Beneath the root of the tail are two glandular eminences.

Few animals vary more in stature than the Otter. Some individuals are, in round terms, twice as large and heavy as others apparently equally mature, and, at any rate, capable of reproduction. An average total length of full-grown individuals is 4 to $4\frac{1}{2}$ feet; some specimens, however, touching 5 feet, while others fall short of the first-named dimension. The species appears to grow several years after puberty. Nose to root of tail 3 feet; tail $1\frac{1}{2}$ feet; nose to eye $1\frac{3}{4}$ to 2 inches; nose to ear 3½ to 4 inches; ear less than an inch high, and about as broad; fore foot from wrist 3½ inches; hind foot 4 inches; girth of body about 1½ feet; stature a foot or less; weight ordinarily 20 to 25 pounds. I have recognized no particular sexual differences, though the female may, as usual in this family, average smaller than the male.

Comparison with allied species. (See plates XVII, XVIII.)

The differences between the present species and *L. vulgaris* of Europe are decided and unmistakable, in fact much stronger than those usually subsisting in this intricate group, where recognition of species is rendered difficult by similarity in form and color. Some of the characters of *L. vulgaris* have been already noted. It is a much smaller animal; the nasal pad is reduced to a small bald spot strictly confined betwixt the nostrils; and there is no hair on the soles or palms. The cranial characters are still stronger. A great many details of difference that might be adduced may be summed in the statement that the skull of *L. vulgaris* is less massive, narrower for its length, and with weaker dentition. The prominent peculiarities are these: There are no decided postorbital processes defining the orbit above. The postorbital constriction is great, the skull being at this point less than one-fourth as broad as it is across the mastoids, instead of nearly one-third such measurement, as in *L. canadensis*. The zygomatic width is contained one and four-fifth times in the total length, instead of only about one and two-fifths, as in *L. canadensis*. The rostrum is decidedly more produced and narrower, and the nasal bones are of a correspondingly different shape. The inner spur of the posterior upper premolar in *L. vulgaris* is a semicircle, only about half as long as the tooth; in *L. canadensis*, the same part of this tooth is developed along the whole inner border of the main moiety. There are other minor dental peculiarities. (Compare Plates XVII and XVIII.)

A skull of the Mexican Otter above mentioned as entirely distinct from *L. canadensis*, and which is probably the species named *L. californica* by Gray (but certainly not the one so called by Baird), is of the same general character as that of *L. canadensis*, in fact presents no very strong points of difference. The inner part of the back upper premolar, however, is rather triangular than quadrate, lacking the bulge of the posterior part, conspicuous in *L. canadensis*, which causes the part to be closely apposed to the succeeding molar, whereas in this Mexican example there is a decided angular interval at the same place. The point is difficult of expression on paper, but is readily perceived when the specimens are laid together. The whole skull is rather broad and massive.

The same characters of skull and teeth are witnessed in a large series of Otters' skulls before me from the coast of Alaska and some of the adjacent islands. The skulls, unfortunately, are unaccompanied by skins; but they lead me to suspect that they may be those of an animal the same as the Mexican species. This would accord with the ascribed range of the species (from Chili to Kamtschatka); but the point cannot be determined until skins are examined from this region, as the skulls alone do not furnish grounds for separation. The Mexican animal is entirely distinct, as already noted, in the smaller and differently shaped nose-pad, perfectly naked palms and soles, and absence of the peculiar callosities seen on the latter in *L. canadensis*. If the ascribed range of this Otter prove to hold, we have, in North America, a second perfectly good species, the characters and supposed synonymy of which are presented in the accompanying foot-note.*

---

* "**Lutra felina.** *Molina*, Hist. Nat. Chili."
"**Lutra chilensis,** *Bennett*, Proc. Zoöl. S. li, 1832, p. 1."—*Tschudi*, Fn. Peru. 1844–46, 113.
"**Lutra platensis,** *Waterh.* Zoöl. Voy. Beagle, i. 22.—*D'Orbig.* Voy. Amér. Mérid."
Lutra californica, *Gray*, Mag. Nat. Hist. i. 1835, 580, *nec Baird*.
Nutria felina, *Gray*, P. Z. S. 1865, 128.

Description (No. 9425, Mus. Smiths. Inst. ♂, Jan. 15, 1869, Santa Efigenia, Tehuantepec, *F. Sumichrast*).—Of about the size and with much the general appearance of *L. canadensis*. Tail very long, at least ⅔ the head and body. Feet notably small. Nose-pad contracted, broader than deep, not deeper than the length of nostril, the lower border lightly concave, the upper border strongly doubly concave, with a central pointed projection upward, and similar acute produced lateral upper corners. Soles and palms naked, the palmar pad divided into a posterior circular part and a larger anterior portion, the latter subdivided by several lines of impression. Soles without peculiar circular callosities, with several well-marked subdivisions by lines of impression. Color above a lighter and more chocolate-brown than in *L. canadensis*; below, decidedly paler and grayer brown than usual in *L. canadensis*, becoming dingy whitish anteriorly. Estimated length 2¼ feet to root of tail; tail about 1¼ feet; nose to ear perhaps 3½ inches; fore foot, measured from beginning of the naked part, only 2¼ inches; hind foot about the same. Additional specimens from the same locality, with others from Orizaba and Central Guatemala, appear to be identical. A skin from Buenos Ayres is not materially different, though the upper outline of the nose-pad is less decidedly double-concave above; the size is less, the coloration lighter, and of a peculiar yellowish-brown on the under parts; the whole pelage is remarkably harsh and hispid.

Description of the skull and teeth. (See Plate XVII.)

The lateral view of the skull shows its most characteristic shape in its general depression, the flatness on top, and the shortness of the blunt muzzle. In the adult, the dorsal profile, from occiput to end of nasals, approximates to a straight line; in younger examples, the frontal outline is also about straight, but the cranial portion arches a little, and then curves down to the occiput. The profile of the nasal orifice is sinuous, convex above, concave below. The rostrum is only about one-fifth of the total length of the skull. The anteorbital foramen is widely open, obliquely elliptical in shape, and only bridged over by a slender process of the root of the zygoma; its obliquity of position is so great that, viewed from above, it presents within the orbit instead of before it. The orbit is small, subcircular, and well defined, not only by an acute malar process of the zygoma, but by a very prominent supraorbital process, these two together completing more than two-thirds of the circumference of a circle. The zygoma is not very strong; it is moderately arched upward, with quite an abrupt rise near the middle, rather than a regular general curve. The glenoid appears rather far forward on its posterior root. The orifice of the meatus auditorius is small, and high up in a deep recess between the glenoid and mastoid; the latter is notably large and prominent.

Viewed from above, the skull displays the great brevity and obtuseness of the rostrum already mentioned. This seems to be due, in a measure at least, to the anterior position of the orbits, and the forward encroachment of the broad flat anterior roots of the zygoma; other topographical points are less different from an ordinary Musteline type. The sides of the rostrum are about parallel, its width is fully equal to its length. Just in front of the orbital brim, at its upper corner, is seen a well-marked depression. The interorbital area is a broad elevated tablet, perfectly smooth and flat, bounded behind by the forks of the sagittal crest, proceeding in curved lines from the termination of the sagittal crest to the supraorbital processes. All the surface of the skull behind these is roughened by muscular impressions. Supraorbital processes are much more largely developed than usual, acute, directly transverse. The skull is very narrow just behind these, the point of greatest constriction being decidedly in advance of the middle of the skull. From this point backward, the skull bulges considerably, with a general ovoidal contour. The occipital crest is moderately developed; the line of contour it represents is emarginate on the middle line, then strongly convex on each side, thence about straight to its termination at the mastoids. There is rarely, if ever, even in the oldest skulls, a decided sagittal crest, the median line being in fact rather a groove, at least behind; in front, however, there is a slight raised line. In young animals, there are several parallel grooves and striæ along the median line.

The occipital face of the skull is in general flat, with various muscular irregularities, curving around laterally to the mastoid region. It is bounded above by the occipital crest, the general contour of which, in this view, resembles the dorsal profile of a military chapeau. The condyles are large, and the region around the foramen magnum is prominent; it descends far below the level of the slight obtuse paroccipitals, the apices of which fall on the level of a line drawn from the mastoid to the middle of the foramen magnum. The articular surfaces of the condyles are obliquely oval, with no outward prolongation, but, on the contrary, an extension toward the median line till they nearly meet each other beneath the foramen magnum. This aperture, in general outline, is transversely elliptical, broader than deep, with a strong emargination posteriorly.

The zygomatic width of the skull, best viewed from below, is seen to be not much less than three-fourths the total length (2.90 × 4.20 inches for example); the intermastoid width is about three-fifths the length. The zygomata are widest apart behind, thence approximately moderately in a nearly straight line. The alveolar borders of the palate are about parallel anteriorly, and, though divergent behind, this is mostly due to the size of the back teeth themselves, the general palatal margins inside the teeth being parallel. The palate extends far back of the last molars (not so far, however, as in *Taxidea*), ending about half-way to the ends of the pterygoids. The incisive foramina, very short and broadly oval, are directly between the canines. The emargination between the pterygoids is broad, and ends with a rounded outline (with a median process or median emargination indifferently); these bones are laminar, smooth inside, thickened with various muscular ridges outside, end in long hamular processes. The glenoid fossa is deep; it develops a broad overlapping shelf at its inner back corner, and a similar but slighter one at the outer anterior corner, together frequently sufficient, in old individuals, to lock the jaw. The posterior nares are only separated by a vertical median septum for a short distance; they débouch together at the edge of the bony palate as a single orifice, as in *Mustelidæ* generally, but not as in *Taxidea* (*q. v.*). The bullæ osseæ are flattish, about as in *Mephitinæ*, strongly contrasting in this respect with *Mustelinæ* and *Melinæ*. They are most vaulted at the antero-internal angle; exteriorly they are produced into a long slender tubular meatus. The basilar space betwixt these periotic bones is very broad, with its sides little convergent anteriorly. The foramen lacerum posterius sometimes appears as several distinct circular foramina through which the cranial nerves respectively emerge separately, a state I have not noticed elsewhere in the family, though it may occur; it is analogous to the division of the anteorbital foramen frequently seen in the Skunks. This state of the lacerate fissure is usually unsymmetrical; that is, it is not alike on both sides of the same skull.

The bones of the skull are early confluent in *Lutra*. Thus even the nasal sutures, usually among the most persistent in *Mustelidæ*, are obliterated at an age when the skull is still thin and papery. In a very young specimen in which the bones are still mostly distinct, I observe the following disposition of the sutures: The nasals are received behind in a shallow semicircular recess of the frontal; their sides are approximately parallel; the intermaxillary and maxillary form each about half of the rest of the nasal boundary. The maxillary ends about opposite the middle of the orbit; there is but a beginning of wedging of a process of the frontal between the nasal and superior maxillary (cf. *Taxidea*). The coronal suture is ex- extremely irregular; it lies altogether back of a line drawn across the apices of the coronoid processes when the jaw is *in situ*. Nearly all the dome of the cranium is parietal, the squamosal forming only a low irregular border along the side, not a fourth of an inch above the root of the zygoma, though it occupies much surface beneath the skull. Owing to the width of the glenoid, it is entirely separate from the sphenoid. The occipital crest is also chiefly parietal, as the lambdoidal suture passes across it to gain the back aspect of the skull at a point nearer to the median line than to the mastoid. The latter is a sizable element, wedged between the parietal and squamosal above, periotic below and in front, and a small piece of the occipital behind; it is already partly confluent with the periotic. The basilar suture is distinct, directly transverse, near the anterior end of the bullæ. Similarly, the spheno-vomerine suture is open; it appears back of the end of the palate. The pterygoids are already lost in the sphenoid, but the pterygo palatal suture is evident, opposite the spheno-vomerine. The contour of the palatine bones may be traced all around, though their palatal plates are fused with each other. The maxillo-palatine suture is opposite the anterior portion of the last premolar. The palatal plate extends far backward, as already twice indicated in noticing other points, and its orbital portion curves over into the temporal fossa, though it forms but an insignificant portion of the orbit proper, being only prolonged by a slight process fairly into the orbit. The orbito-sphenoid remains instructively distinct from all surroundings, bounded above and in front by the frontal, behind and partly below by the alisphenoid, for the rest below by the palatal. The lachrymal is similarly distinct, except anteriorly. The malar is seen to form most of the zygomatic arch; though the pointed process of the squamosal overlies it nearly half-way, its bevelled posterior extremity reaches almost to the glenoid fossa; its anterior extremity runs along on top the maxillary to the lachrymal, forming an upper layer of the bridge over the anteorbital foramen. The palatal plates of the intermaxillaries extend in a V past the canines to a point on the median line opposite the second premolar; the incisive foramina are not pierced entirely in these bones, their posterior periphery being completed by a nick in the corresponding border of the palatal plate of the superior maxillary.

Returning to the adult skulls for examination of the mandible, we find that this bone has a stout thick ramus, with long slanting symphysis, an irregular continuous curve from incisors to angle, with a slight emargination just in advance of the latter, and a rather low broad obtuse coronoid, the front border of which is nearly straight and vertical, the posterior border curving forward with quite an elbow to the apex. The condyle is wide across, but narrow in the other direction; it slants oblique both to the horizontal and vertical plane, its inner end being both higher and further back than the other. There is a deep notch between the condyle and angle of the jaw, which last is not exflected. The muscular impression on the outside of the jaw is, as usual, well marked; it ends below in a rounded outline beneath the last molar.

For the dentition, a young subject is preferably selected, in which the teeth are fully formed but entirely unworn. The back upper molar is quadrate in general contour, as in *Mephitinæ* (cf. *Melinæ*, *Mustelinæ*), but rather lozenge-shaped, the inner posterior and outer anterior corners being less than a right angle, while the opposite ones are obtuse. All the corners are rounded off. The tooth is, if anything, a little smaller than the next one. Its face presents an exterior, narrow, longitudinal, raised portion, in the closed jaw wholly external to the anterior lower molar. The exterior moiety is divided across by a sulcus; its inner border is deeper and more trenchant than the outer; its front part is also deeper than its back part. The rest of the face of the tooth is depressed, and presents a general slight excavation, with a very prominent acute tubercle antero-interiorly, and a general raised border; this portion is applied against the similar depressed back part of the anterior lower molar. The back upper premolar is essentially triangular in contour, but with a bulge of the postero internal border, which nearly gives it a trapezoidal shape. It consists of the outer deep portion, made up of a single very prominent acute cusp, connected by a trenchant edge with a smaller posterior cusp, which ends the tooth behind, and of an inner low portion presenting a general slightly excavated surface marked with a slight central prominence and bounded by a well-developed sharp edge. The great development of this inner moiety along the whole of the tooth is the strongest dental character of the species in comparison with *L. vulgaris*. The cuspidate part of the anterior lower molar abuts against this portion. The next upper premolar is a stout two-rooted conical cusp, with a cingulum and well-developed heel fore and aft, and, in addition, a postero-internal depressed part, against which the apex of the posterior lower premolar is apposed. The next premolar is altogether similar, but smaller. The anterior premolar is single-rooted, very small, and in peculiar position, altogether internal to the canine, with which it is in close apposition; both of the anterior premolars, in fact, are in close relation with the canine; the first one being thrown entirely to one side of the general dental axis. This small tooth not seldom aborts on one side; but I have not happened to find it absent altogether. The upper canines are not peculiar; they are perhaps shorter and stouter for their length than usual in this family. The lateral pair of incisors moderately surpass the rest in size; the others are on a par with each other; the ends of all are obtusely rounded, without obvious lobation.

The back lower molar is small and circular, as usual; it shows no special points. The front lower molar consists of an anterior tricuspidate half and posterior depressed portion. The three cusps are very prominent, subequal in size, forming a triangle, with one angle anterior and median, the two others posterior and lateral; the postero-internal cusp is rather smaller than the two others, the ridge connecting which forms the trenchant edge of the tooth. The back part of the tooth is a simple depression, with raised periphery, which, at its outer part, is twice nicked, with slight marginal cusps as a consequence. The posterior premolar is a stout conical cusp, well heeled fore and aft, with a secondary cusp half-way up its back border, as in *Taxidea*. The next premolar is smaller, but similar, except in lacking the secondary cusp. The front premolar is again similar, but smaller still, and without an anterior heel, being closely apposed to the canine. The latter is short, very stout, and much curved. The inferior incisors are much crowded and very irregular, even more so than in *Mustelinæ*, offering an interesting approach to the condition which culminates in *Enhydris* in the disappearance of one pair. The outer pair are moderately larger than the rest; the next pair—the middle tooth on each side—set almost entirely back of the general incisor plane; they are quite deep, though little of their face appears in front. The middle pair are narrow, and closely approximated. The ends of the outer pair are lobate; of the others, not appreciably so.

Variations in the skull of the Otter.

As in other cases, I present under this head Mr. J. A. Allen's measurements and comments, extracted from the paper above cited in the synonymy:—

"Specimens of this species from northern and southern localities do not differ materially in size; skulls from Newfoundland, Maine, Lake Superior, Washington, and Georgia agreeing very closely in dimensions. In a series of eighteen (mainly from northern localities), nine attain or exceed a length of 4.25, and three reach 4.50, while two only fall as low as 4.00. Seven specimens from the vicinity of Lake Umbagog, Maine, (in Mus. Comp. Zoöl.) average 4.28 in length and 2.93 in width; two of these reach 4.50 in length and two fall slightly below 4.00 (3.96 and 3.97). Two specimens from Washington. D. C., have a length respectively of 4.45 and 4.50; one specimen from Saint Simon's Island, Georgia, is nearly as large (4.32), while a Fort Cobb specimen has a length of 4.22. These four are the only ones from very southerly points. Four other specimens, from as many localities, range from 4.05 to 4.15; while three specimens from Newfoundland range from 4.03 to 4.25. While these specimens are too few to warrant positive conclusions as to geographical variations, they seem to point to a great constancy of size throughout a wide range of latitude."

Measurements of eighteen skulls of LUTRA CANADENSIS.

Catalogue number.	Locality.	Length.	Width.
501	Newfoundland	4.20	2.75
498	......do......	4.03	2.53
500	......do......	4.15	2.57
499	......do......	4.35	2.90
555	Umbagog Lake, Maine	4.40	3.00
556	......do......	4.27	2.85
557	......do......	4.50	2.90
559	......do......	3.97	2.70
558	......do......	3.96	2.70
489	......do......	4.50	3.00
4446	Lake Superior	4.15	2.85
11839	Fort Berthold, Dak	4.25	2.82
2247	Saranac Lake, N. Y	4.05	2.57
13671	Bayfield, Wis	4.06	2.82
8097	Fort Cobb, Ind. Ter	4.22	2.87
	Washington, D. C	4.50	2.95
433	......do......	4.45	
3142	Saint Simon's Island, Georgia	4.32	2.75

History of the species.

The existence of a true Otter in North America was known to the earliest systematic writers. Thus, Buffon described an Otter from Canada, noting its larger size in comparison with the European species, and a difference in the color of the fur. But all the authors of the last century persisted in confounding it with either the European *L. vulgaris*, or with the South American Carigueibeju, Sarigovion, or Saricovienne, both totally distinct. Pennant had also a certain "Slender Otter" of North America, which became a *Lutra gracilis*[*] of authors, and may

---

[*] In establishing, in 1816, the genus *Pusa* for the Sea Otter, afterward called *Enhydra* by Fleming, Oken has two species (Lehrb. Naturg. 1816, p. 986): one of these, which he calls *Pusa orientalis*, is *Enhydra marina* of authors; in the other, *L. gracilis*, we see the old "Slender Otter" of Pennant, *Lutra gracilis* Shaw, referred by Fischer with a query to his genus *Enydris* (=*Enhydra* Flem.). Pennant's beast came from "Statenland"; Oken says "Staatenland, Insel an Amerika *bei New-York*". If he means by this what is known now as Staten Island, New York, it would make his animal to be *Lutra canadensis*; perhaps, however, his geography was at fault.

or may not have been the present species. The error of confounding the species with that of Europe was refuted before the history of our species had been disentangled from that of the Brazilian Otter, with which ours was confounded by various French, and even American, writers, until a comparatively late period.

The first binomial name I have found for this species is the *Mustela canadensis* of Turton, p. 57.* This name, which appears to have been overlooked, I consider undoubtedly based upon the North American species; it consequently anticipates the name *Lutra canadensis* bestowed in 1823 by Sabine, who is usually quoted as the authority for our species. In the same year, Fr. Cuvier is said to have separated the Canada Otter from that of South America, and to have also described as distinct a second North American species, under the name of *Lutra lataxina*, which became current with several writers. A *Lataxina mollis* was described by J. E. Gray in 1843, and his type-specimen was afterward figured by Audubon and Bachman as a variety of *L. canadensis*. But it is certain that neither of these names indicates anything different from the common North American species. Of a certain "*Mustela hudsonica* Lacépède", quoted by some authors as pertaining to our Otter, I know nothing.

Prof. Wyman, in an article on the articulation of the jaw, above cited, named our species *Lutra americana* in 1847.

---

* In quoting Sabine as the authority for the name "*canadensis*", previous compilers of the synonymy of this species appear to have altogether overlooked the much earlier "*Mustela canadensis*" of Turton's English version of the *Systema Naturæ*, p. 57. As Turton gives no references, I am uncertain whether or not he is the originator of the name, as the animal was known before his time; but this is the earliest use of the name in binomial nomenclature that I have found. Turton, like Pennant and others of his predecessors, refers to the American Otter in connection with the European species; but this "*Mustela canadensis*" of his (p. 57) is additional to his other notice of *Mustela lutra* as an inhabitant of Europe, Asia, and America. The diagnosis is merely "black; fur smooth; tail long, taper;" inhabits North America", which would do very well for the true *Mustela canadensis* or *M. pennanti* (Pekan, Fisher); but it is as pertinent as many of his diagnoses, and further fixed by its coming under his section "A. Hind feet palmate. Otters", as opposed to his "B. Feet cleft. Weasels". Under head of the latter, he has, on page 59, another "*Mustela canadensis*", which is the animal so named by Schreber, the Pekan. Turton's double employ of the same name for two entirely different animals is to be carefully noted to prevent confusion of quotations. "*Mustela canadensis*, Turton, p. 57" is *Lutra canadensis*. "*Mustela canadensis*, Turton, p. 59" is *Mustela canadensis*.

Two lately introduced names require special notice: these are *Lutra californica* Baird (nec Gray) and *Lutra destructor* Barnston. The specimen which Prof. Baird referred incorrectly to *L. californica* of Gray, taken by Dr. Newberry in the Cascade Mountains of Oregon, is now before me. The palms and soles are rather less hairy than is usual in *L. canadensis;* still they are decidedly furred between the digits of both feet; the soles show the curious callosities diagnostic of *L. canadensis*, and the characteristic large nasal pad of *L. canadensis* is well exhibited. Other Otters from the same region show as fully furred feet as any from the Eastern States, and the peculiarities of the one from the Cascade Mountains can only be regarded as those of an individual, within the normal range of variation of *L. canadensis*, to which it must unquestionably be referred. Prof. Baird indeed separated it with much evident hesitation, and mainly because it was supposed (though erroneously) to represent a species already instituted by another author (*cf. op. cit.* p. 188). The true *californica* of Gray is elsewhere discussed.

The *Lutra destructor* is represented in the National Museum by specimens received from Mr. Barnston as typical examples of his supposed species. They are rather smaller than usual, and perhaps not full-grown, even though already in breeding condition; but they possess all the essential specific characters of *L. canadensis*, to which I have not the slightest hesitation in referring them. *L. canadensis* is so strongly marked a species in certain respects, already fully detailed, that there is no difficulty in recognizing it, notwithstanding its great variability in non-essential particulars. The skin and skull of *L. destructor* exhibit nothing beyond the normal range of variation of *L. canadensis*.

Geographical distribution.

The Otter is generally distributed over North America, apparently nowhere in great abundance, yet absolutely wanting in few, if any, localities adapted to its habits. Being a shy and rather solitary animal, it is among those that decrease rapidly in numbers with the settling of a country; but its very wildness, together with its wariness and sagacity, stands between it and total extirpation, even in populous districts; while the nature of its haunts further conduces to its persistence. Writing about twenty-five years ago, Mr. Audubon speaks of the Otter as being no longer found abundantly in many parts of the country where it was formerly numerous, and as having been nearly extirpated in the Atlantic States east of Maryland. Such statement, however, seems stronger than the facts would warrant; for Mr. Allen speaks of the animal as still "not rare" in Massachusetts as late as 1869, he having known of some half dozen specimens which were taken near Springfield during the ten preceding years. The "Eastern Shore" of Maryland appears to have always been a favorite locality with the Otter; Audubon specially mentions this region, and specimens are still taken there or in other spots along the Potomac, not far from Washington City. The last one I saw from this region was brought, freshly killed to the Smithsonian Institution in 1874. Northerly the Otter extends, according to Richardson, nearly to the Arctic Ocean, along the Mackenzie and other rivers; and it also inhabits the northernmost system of lakes. In the times of the author just mentioned, some seven or eight thousand pelts* were annually exported from British America to England, and the trade does not appear to have decreased to this day, for I find among the quotations of sales of Otters within two or three years by the Hudson's Bay Company, in London, over eleven thousand set down for 1873. If the skulls, unaccompanied by skins, which I have examined from Alaska, are really of this species, the Otter is abundant in that new portion of United States territory. According to Messrs. Gibbs and Suckley, writing in 1859, the Otter, called by the Yakima Indians *nookshi*, increased in abundance in Oregon and Washington with the decline of the fur trade, and were numerous in the waters of the Cascade Range. Dr. J. S. Newberry (1857) attests the presence of the Otter "on all parts of the Pacific coast, both on the sea shore and in the inland streams and lakes. In the Cascade Mountains, where neither otter nor beaver had been much hunted, and where both were abundant, we found the beaver in the streams, but the otter in great abundance in the mountain lakes where the streams take their rise. There they subsist on the western brook-trouts and a *Coregonus* with a

---

* This statement, however, it should be observed, is widely discrepant from some others, unless only some special lines of importation are here referred to by the author. According to Bell, there were imported into England, of the skins of the North American Otter, 713,115 in 1830, 494,067 in 1831, 222,493 in 1832, though only 23,889 in 1833. "After September 1, 1833, the duty was reduced from ¼d. each to 1s. per hundred, since which I believe the importation has gradually increased."—(*British Quadrupeds*, 1837, p. 136.)

crayfish, *Astacus Klamathensis* . . . In Klamath lakes the otter is quite common . . . . their food is a large sucker (*Catastomus occidentalis*) and a species of *Gila*, both rather sluggish fish and such as would be easily caught"—unlike the very active *Salmonidæ* just mentioned. At the time to which the writer refers, the pelts were much more in demand than those of the Beaver, $2.50 being paid in goods by the Hudson's Bay Company at Vancouver, while Beaver brought only one-fifth as much.

In the muddy waters of the Missouri Basin, not overstocked with fish, the Otter seems to exist but sparingly. Audubon only "observed traces" of their presence in his journey up to the Yellowstone. Hayden includes the species among the animals observed in the Upper Missouri country, where, however, it does not appear to have come under Mr. Allen's observation. North of this area, in the region of the Red River and other streams, thence westward to the Rocky Mountains, I ascertained the general, though probably not abundant, occurrence of the species. Mr. Allen found the Otter to be, in Iowa, "common on the Raccoon Rivers, and generally more or less so throughout the State";—"occasional along the streams" of Kansas;—and "more or less frequent in Salt Lake Valley, and in the adjoining mountains". Drs. Coues and Yarrow give the species as found sparingly in various portions of the Southwestern Territories. My recent exploration of portions of Colorado did not reveal the presence of the Otter, but I do not on this account deny its existence, perhaps in abundance, in the numerous mountain lakes and streams of that State, which harbor countless Beavers, and seem in every way suited to the requirements of the Otter.*

In Audubon's time, the Otter was "still abundant in the rivers and reserve-dams of the rice fields of Carolina", and was "not rare in Georgia, Louisiana and Texas". According to Mr. Allen, it is still "abundant" in Florida, where it is little hunted, its fur being, in this southern region, of comparatively little value. But the southern limits of the distribution of the species remain to be determined. A Mexican Otter is certainly of a different species from ours, whether or not the latter also exists in that country; and I am not aware of any unquestionable citation of true *canadensis* as Mexican. I am therefore much surprised at Dr. von Frantzius's recent citation of this spe-

*Since this paragraph was penned, I have seen a specimen in Mrs. Maxwell's collection, from the vicinity of Boulder, Colorado.

cies from Costa Rica,* which is considerably beyond the usually recognized range of true *canadensis*, the actual occurrence of which so far south may possibly be still open to question. With this single exception, I do not know of, at least I do not recall at present writing, any special indication of the presence of *L. canadensis* proper south of the United States, though in a general way it has been often accredited with a range coextensive with the continent of North America, and has even been ascribed, with a query, to South America.

Habits of Otters.

Although I have observed the "seal" of the Otter and its curious "slides" in various parts of our country during the years I have been a student of our animals, I cannot truly aver that I have ever laid eyes upon a living individual; and to speak of its habits, I must give information at second hand. Presuming upon the reader's knowledge of the thoroughly aquatic and highly piscivorous nature of the animal, I turn to the various histories at our disposal in further elucidation of its habits.

According to Richardson, one of the earliest authors giving accounts of the species with precision, "the Canada Otter resembles the European species in its habits and food. In the winter season, it frequents rapids and falls, to have the advantage of open water; and when its usual haunts are frozen over, it will travel to a great distance through the snow, in search of a rapid that has resisted the severity of the weather. If seen, and pursued by hunters on these journies, it will throw itself forward on its belly, and slide through the snow for several yards, leaving a deep furrow behind it. This movement is repeated with so much rapidity, that even a swift runner on snow-shoes has much trouble in overtaking it. It also doubles on its track with much cunning, and dives under the snow to elude its pursuers. When closely pressed, it will turn and defend itself with great obstinacy. In the spring of 1826, at Great Bear Lake, the Otters frequently robbed our nets, which were set under the ice, at the distance of a few yards from a piece of open water. They generally carried off the heads of the fish, and left the bodies sticking in the net.

"The Canada Otter has one litter annually about the middle of April of from one to three young."

* Arch. für Naturg. 1869, p. 289.

In the Middle and Southern States, Audubon says they are about one month earlier.*

The sliding of the Otter, which Sir John describes, is not alone resorted to in the endeavor to avoid pursuit; and again, it is something more than simply an easy way of slipping down a wet sloping bank into the water. It seems to be a favorite amusement of this creature, "just for fun". Godman speaks of the diversion in the following terms:—"Their favorite sport is sliding, and for this purpose in winter the highest ridge of snow is selected, to the top of which the Otters scramble, where, lying on the belly with the fore-feet bent backwards, they give themselves an impulse with their hind legs and swiftly glide head-foremost down the declivity, sometimes for the distance of twenty yards. This sport they continue apparently with the keenest enjoyment until fatigue or hunger induces them to desist."

Statements of similar import are made by various writers, and accord with Audubon's personal observations, as rendered by him in the following language:—

"The otters ascend the bank at a place suitable for their diversion, and sometimes where it is very steep, so that they are obliged to make quite an effort to gain the top; they slide down in rapid succession where there are many at a sliding place. On one occasion we were resting ourself on the bank of Canoe Creek, a small stream near Henderson, which empties into the Ohio, when a pair of Otters made their appearance, and not observing our proximity, began to enjoy their sliding pastime. They glided down the soap-like muddy surface of the slide with the rapidity of an arrow from a bow, [†] and we counted each one making twenty-two slides before we disturbed their sportive occupation.

"This habit of the Otter of sliding down from elevated places to the borders of streams, is not confined to cold countries, or to slides on the snow or ice, but is pursued in the Southern States, where the earth is seldom covered with snow, or the waters frozen over. Along the reserve-dams of the rice fields of Carolina and Georgia, these slides are very common.

---

* According to Bell, the European Otter goes with young nine weeks, and produces three to five young ones in March or April (Brit. Quad. 1-37, 136). The period of gestation of our species, if different, probably remains to be ascertained.

†[A statement certainly too figurative for literal acceptation.]

From the fact that this occurs in most cases during winter, about the period of the rutting season, we are inclined to the belief that this propensity may be traced to those instincts which lead the sexes to their periodical associations."

The food of the Otter, and the manner in which it is procured, are noted by the same author in the following terms:—

"The Otter is a very expert swimmer, and can overtake almost any fish, and as it is a voracious animal, it doubtless destroys a great number of fresh water fishes annually. We are not aware of its having a preference for any particular species, although it is highly probable that it has. About twenty-five years ago we went early one autumnal morning to study the habits of the Otter at Gordon and Spring's Ferry, on the Cooper River, six miles above Charleston [S. C.], where they were represented as being quite abundant. They came down with the receding tide in groups or families of five or six together. In the space of two hours we counted forty-six. They soon separated, ascended the different creeks in the salt marshes, and engaged in capturing mullets (*Mugil*). In most cases they came to the bank with a fish in their mouth, despatching it in a minute, and then hastened back again after more prey. They returned up the river to their more secure retreats with the rising tide. In the small lakes and ponds of the interior of Carolina, there is found a favourite fish with the Otter, called the fresh-water trout (*Grystes salmoides*).

"Although the food of the Otter in general is fish, yet when hard pressed by hunger it will not reject animal food of any kind. Those we had in confinement, when no fish could be procured were fed on beef, which they always preferred boiled. During the last winter we ascertained that the skeleton and feathers of a wild duck were taken from an Otter's nest on the banks of a rice field reserve-dam. It was conjectured that the duck had either been killed or wounded by the hunters, and was in this state seized by the Otter, ...

"On throwing some live fishes into a small pond in the Zoological Gardens in London, where an Otter [presumably, however, of another species] was kept alive, it immediately plunged off the bank after them, and soon securing one, rose to the surface holding its prize in its teeth, and ascending the bank, rapidly ate it by large mouthfuls, and dived into the water again for another. This it repeated until it had caught and eaten all the fish which had been thrown into the water for its use. When thus engaged in devouring the luckless fishes the Otter bit through them, crushing the bones, which we could hear snapping under the pressure of its powerful jaws."

The nest of the *European* Otter is said to be formed of grass and other herbage, and to be usually placed in some hole of a river's bank, protected either by the overhanging bank or by the projecting roots of some tree. Its fossorial ability, and the general intelligence it displays in the construction of its retreats, have been greatly exaggerated by some writers, to judge by the more temperate language used by the distinguished author of the *History of British Quadrupeds*. "We read of its excavating a very artificial habitation," says Bell, " burrowing under ground to a considerable distance; making the aperture of its retreat always under water, and working upwards, forming here and there a lodge, or dry resting-place, till it reaches the surface of the ground at the extremity of its burrow, and making there a breathing-hole, always in the middle of a bush or thicket.[*] This statement is wholly incorrect. The Otter avails itself of any convenient excavation, particularly of the hollows beneath the overhanging roots of trees which grow on the banks of rivers, or any other secure and concealed hole near its fishing-haunt; though in some cases it fixes its retreat at some distance from the water, and when driven by a scanty supply of fish, it has been known to resort far inland, to the neighbourhood of the farm-yard, and attack lambs, sucking pigs, and poultry,—thus assuming for a time the habits of its more terrestrial congeners." I am not aware that such extravagant statements have been made, with any authority at least, respecting the American Otter; and indeed one has only to regard the general configuration of the animal, and particularly the shape of the fore limbs and condition of the claws, to become convinced that the mining operations of the animal are necessarily limited. It does not appear that the underground retreats of the Otter are constructed with the skill and ingenuity of even those of the Muskrat. A retreat examined by Audubon has been thus described by this author:—

"One morning we observed that some of these animals resorted to the neighbourhood of the root of a large tree which

---

*[The author remarks the similarity of such an account with that given by Mr. George Bennett in describing the retreats of the *Ornithorhynchus* of Australia, though the former is found in books published long prior to the discovery of the latter animal.]

stood on the side of the pond opposite to us, and with its overhanging branches shaded the water. After a fatiguing walk through the tangled cane-brake and thick under-wood which bordered the sides of this lonely place, we reached the opposite side of the pond near the large tree, and moved cautiously through the mud and water to its roots: but the hearing or sight of the Otters was attracted to us, and we saw several of them hastily make off at our approach. On sounding the tree with the butt of our gun, we discovered that it was hollow, and then having placed a large stick in a slanting position against the trunk, we succeeded in reaching the lowest bough, and thence climbed up to a broken branch from which an aperture into the upper part of the hollow enabled us to examine the interior. At the bottom there was quite a large space or chamber to which the Otters retired, but whether for security or to sleep we could not decide. Next morning we returned to the spot, accompanied by one of our neighbours, and having approached and stopped up the entrance under water as noiselessly as possible, we cut a hole in the side of the tree four or five feet from the ground, and as soon as it was large enough to admit our heads, we peeped in and discovered three Otters on a sort of bed composed of the inner bark of trees and other soft substances, such as water grasses. We continued cutting the hole we had made, larger, and when sufficiently widened, took some green saplings, split them at the but-end, and managed to fix the head of each animal firmly to the ground by passing one of these split pieces over his neck, and then pressing the stick forcibly downwards. Our companion then crept into the hollow, and soon killed the otters, with which we returned home."

Their structure being identical, the American and European Otters cannot differ in their general movements and attitudes. In speaking of the conformation of the latter species, Bell remarks that evidently every facility consistent with the preservation of its structural relations with the rest of the group is given to the Otter for the pursuit and capture of its proper food. " It swims and dives with great readiness and with peculiar ease and elegance of movement; and although its action on land is far from being awkward and difficult, yet it is certainly in the water that the beautiful adaptation of its structure to its habits is most strikingly exhibited. It swims in nearly a horizontal position, and dives instantaneously after the fish that may glide beneath it, or pursues it under water, changing its course as the fish darts in various directions to escape from it, and when the prey is secured, brings it on shore to its retreat to feed."

Yielding a pelt of great beauty and value, from the exquisite softness and rich warm color of the fur, as well as from the size of the animal, the American Otter is systematically pursued by professional trappers. I have already given some figures showing the thousands annually destroyed, and will condense from Mr. Gibson's work, already often quoted, the account of the various methods employed—for every trapper has his own notions and ways of doing things, and in the pursuit of so valuable and so wary a creature as the Otter there is room for large and varied experience. The animal seems to be taken in this country usually, if not invariably, with the steel trap, a special size and make of which, with two springs, goes by the name of "Otter trap". Searching for a "slide", or place where the animal habitually crawls from the water up the bank, the hunter sets the trap on the spot, a few inches under water. No bait is here required; and devices are used in securing the trap by which the animal may be led into deep water when caught, or lifted upward, the design in either case being to prevent the animal's escape by gnawing off the imprisoned limb. The trap may also be placed at the top of the slide, two or three feet back of the slope, in a place hollowed to receive it, and covered with snow. Under such circumstances, care is taken not to handle the trap with the bare hands. It is scented with various animal odors, and, to further insure success, a "way" is made to the trap by means of parallel logs. The trap is sometimes simply set in the beaten track made in the snow, carefully hidden; or at the entrance of the burrow; or at the base of a slanting log with one end under water, the Otter being attracted by bait or odor placed beyond on the other end; or a rock which projects over a stream is utilized in the same way. In all these methods, the utmost care is necessary to obliterate traces of the trapper's presence, as the sight and smell of the Otter are acute, and his wariness, caution, and sagacity at a very high rate. "In winter when the ponds and rivers are frozen over the otters make holes through the ice at which they come up to devour their prey. Where the water is a foot deep beneath any of these holes the trap may be set in the bottom, the chain being secured to a heavy stone. When the otter endeavors to emerge from the hole he will press his foot on the trap and thus be caught. If the water is deep enough beneath the hole the trap may be baited with a small fish attached to the pan, and then carefully lowered with its chain and stone to the bottom. For this purpose the Newhouse, No. 3, is best adapted, as the otter is in this case caught by the head." Audubon speaks of the latter method as one very commonly employed in Carolina. His figure of the Otter represents the animal as caught by the fore foot in a trap, baited with a fish *on the pan*, placed on a slanting log just out of the water. But traps baited on the pan are not set by experts in this mode of trapping. Audubon has also drawn his animal as coming *down* the log from the upper end, which the animal could not have reached without passing over the trap in the other direction. Though drawn, furthermore, "to represent the pain and terror felt by the creature when its foot is caught by the sharp saw-like teeth of the trap", the Otter is nevertheless holding its foot quietly in the trap, and resting very composedly upon the log, as if it feared to displace the trap. In reality, however, an Otter so caught would be off the log and into the water, trap and all, in a fraction of a second after the jaws snapped. In writing the text to this fancy sketch, moreover, Audubon appears to have forgotten that the trap had no "sharp saw-like teeth"; it is correctly drawn with straight-edged jaws, as usually manufactured.

For commercial purposes, the skin of the Otter is removed by a cross-slit down the hind legs, and withdrawn whole, without splitting along the belly, the tail, however, being slit its whole length along the under side. The skin is stretched with the hair inside, the tail alone being spread out flat.

The hunting of the Otter for sport does not appear to be practiced in this country, at least to any extent, and the gun is only incidentally and rarely used for its destruction. The mode of hunting the European animal has been graphically described by Bell, to whom I return for this portion of the subject:—

"Otter-hunting, formerly one of the most interesting and exciting amusements of which the English sportsman could boast, has of late years [1837] dwindled into the mere chase of extirpation. It was in other days pursued with much of the pomp and circumstance of regular sport: the Dogs were chosen for their perseverance and resolution; 'good Otter-hounds,' says an old sportsman, . . . 'will come chaunting and trailing along by the river side, and will beat every tree-root, every osier-bed, and every tuft of bulrushes:—nay, sometimes they will take the water, and beat it like a Spaniel.' The huntsman and others of the party carried Otter spears, to strike the Otter when driven within their reach; horsemen and footmen joined in the chase; and the whole company formed a cavalcade of no inconsiderable extent and importance. These scenes are now no longer witnessed, or but rarely, in England; but in Wales the chase of the Otter is still kept up with some spirit, in certain romantic districts of that romantic country . . . In beating for an Otter, it is necessary to mark the character and direction of his 'seal,' or footmark in the mud or soil, as well as the recent or older appearance of his 'spraints,' or dung. These signs of his having been either remotely or more recently on the spot will afford a tolerably certain indication whether the animal be still in the neighbourhood, or whether a further search must be made for later marks of his presence. When the Otter is found, the scene becomes exceedingly animated. He instantly takes the water, and dives, remaining a long time underneath it, and rising at a considerable distance from the place at which he dived. Then the anxious watch that is kept of his rising to 'vent,' the steady purpose with which the dogs follow and bait him as he swims, the attempts of the cunning beast to drown his assailants, by diving whilst they have fastened on him, the baying of the hounds, the cries of the hunters, and the fierce and dogged resolution with which the poor hopeless quarry holds his pursuers at bay, inflicting severe, sometimes fatal wounds, and holding on with unflinching pertinacity even to the last,—must altogether form a scene as animated and exciting as the veriest epicure in hunting could desire. The return from such a day's sport as this in the county of Carmarthen is thus described by a correspondent of the Sporting Magazine:—'Sitting near the window, I beheld approaching the bridge a cavalcade, and found it was Squire Lloyd of Glansevin, escorted by the gentlemen of the neighbourhood, returning from Otter-hunting. The gentlemen in the front rank were mounted; and next the horsemen were three men neatly dressed in scarlet coats and white trousers, with long spears, on which were suspended three huge Otters. Now the huntsman appeared with his well-disciplined hounds; and then followed the cart, with nets, spears, and other paraphernalia, and an old ballad-singer appeared in the rear, who sung the praises of the high-bred hounds and their worthy master.'"

The general intelligence of the Otter is of a high order, and his docility is such that he may not only be thoroughly tamed, but taught to work for his master. Audubon speaks of four American Otters which a gentleman had tamed so completely that they never failed to come like dogs when whistled for, crawling slowly with apparent humility toward their master; and also gives his own experience in domesticating several Otters, which became so tame that they would romp with him in his study. These, he says, were taken when quite young, and became as gentle as puppies in two or three days; they preferred milk and boiled corn-meal, refusing fish or meat till they were several months old. On this subject I shall once more quote the attractive page of Bell, and conclude this lengthy compilation with some quaint and interesting paragraphs respecting the use of the Otter as food; the actual reference being, it will be understood, to the European species :—

"That the Otter may not only be readily and easily tamed and domesticated, but taught to catch and bring home fish for its master, is a fact which is so well known, and has been so often proved, that it is surprising it should not have been more frequently acted upon. From Albertus Magnus down to the late excellent Bishop Heber, instances have been continually narrated, some of which have gone no further than the domestication of pet Otters, while in others the animal has been rendered a useful purveyor of fish for the family table. Amongst other writers who have attested similar facts, honest Izaak Walton says, 'I pray, sir, save me one [young Otter], and I'll try if I can make her tame, as I know an ingenious gentleman in Leicestershire, Mr. Nicholas Seagrave, has done; who hath not only made her tame, but to catch fish, and do many other things of much pleasure.' Albertus Magnus, Aldrovandus, Gesner, and others, had asserted it; yet Buffon, losing for once his accustomed credulity, and running to an opposite extreme, refuses to believe in the susceptibility of the Otter to be brought to a state of domesticity. The former of these writers states that, in Sweden, Otters were kept in the houses of the great for the express purpose of catching fish, which they would do at a signal from the cook, and bring home their provender to be dressed for dinner. Numerous instances have been recorded in later times, by Daniel, Bewick, Shaw, and others; in one of which an Otter had been known to take eight or ten salmon in a day: and the following passage in the journal of Bishop Heber confirms some previous statements, that one of the Asiatic species, probably *Lutra nair*, (Fr. Cuv.) may be rendered similarly useful:—'We passed, to my surprise, a row of no less than nine or ten large and very beautiful Otters, tethered with straw collars and long strings to bamboo stakes on the banks (of the Malta Colly). Some were swimming about at the full extent of their strings, or lying half in and half out of the water; others were rolling themselves in the sun on the sandy bank, uttering a shrill whistling noise, as if in play. I was told that most of the fishermen in this neighbourhood kept one or more of these animals, who were almost as tame as Dogs, and of great use in fishing; sometimes driving the shoals into the nets, sometimes bringing out the large fish with their teeth. I was much pleased and interested with the sight. It has always been a fancy of mine that the poor creatures whom we waste and persecute to death, for no cause but the gratification of our cruelty, might by reasonable treatment be made the sources of abundant amusement and advantage to us.' This interesting account justifies the conclusion drawn by the good prelate from the scene that so much delighted him, that 'the simple Hindoo shows here a better taste and judgment than half the Otter hunting and Badger baiting gentry of England.' With such instances as these before us, there seems to be no reason why this animal, so tractable and docile as it is proved to be, should not be very generally domesticated for the purposes of sport, or employed by fishermen as a means of assisting them in their calling.

"The method which has been recommended to train them for this purpose is as follows:—They should be procured as young as possible, and they are at first fed with small fish and water. Then bread and milk is to be alternated with the fish, and the proportion of the former gradually increased till they are led to live entirely on bread and milk. They are then taught to fetch and carry, exactly as Dogs are trained to the same trick; and when they are brought to do this with ease and docility, a leather fish stuffed with wool is employed for the purpose. They are afterwards exercised with a dead fish, and chastised if they disobey or attempt to tear it; and finally, they are sent into the water after living ones. In this way, although the process is somewhat tedious, it is believed that the Otter may be certainly domesticated, and rendered subservient to our use. . . . . .

"The habits of the Otter, and its rank fishy taste, have procured for it the distinction of being permitted by the Church of Rome to be eaten ón maigre days. The quiet humour of good old Izaak Walton could not rest without a sly hit at this fact:—

"*Piscator.* I pray, honest huntsman, let me ask you a pleasant question: do you hunt a beast, or a fish?

"*Hunt.* Sir, it is not in my power to resolve you; yet I leave it to be resolved by the College of *Carthusians*, who have made vows never to eat flesh. But I have heard the question hath been debated among many great clerks, and they seem to differ about it; yet most agree that *her tail is fish*, and if her body be fish too, then I may say that a fish will walk upon land (for an Otter does so), sometimes five or six or ten miles in a night.

"Now, were we to adopt the reference recommended by honest Izaak, the description of this animal would have fallen within the province of my good friend Mr. Yarrell rather than mine; for, says Pennant, 'in the kitchen of the Carthusian Convent near Dijon, we saw one preparing for the dinner of the religious of that rigid order, who, by their rules, are prohibited during their whole lives the eating of flesh.'"

Extinct species of North American Otter.

Lutra piscinaria, *Leidy.*

Lutra ? ———, *Leidy,* Contrib. Extinct. Vert. Fn. of the Western Terr. (4to Rep. U. S. Geol. Surv. vol. 1) 1873, p. 230.
Lutra piscinaria, *id. ibid.* p. 316, pl. xxxi, f. 4 (tibia, ‡ nat. size, from Idaho).

Based on a tibia submitted to Dr. Leidy's inspection by the Smithsonian Institution, procured by Clarence King on Sinker Creek, Idaho, in association with remains of *Equus excelsus* and *Mastodon mirificus.*

"The tibia pertains to a carnivore, and resembles that of an otter more than that of any other animal with which I have had an opportunity of comparing it. Its differences, excepting size, are trifling. The tubercle for insertion of the quadriceps extensor is less prominent, so as to give the head of the bone proportionally less thickness in relation with its breadth. The ridge for the attachment of the interosseous membrane at the lower part of the bone is more prominent and sharper. The distal end in front just above the articulation is flatter, and the groove for the flexor tendons behind is deeper.

	Lines.
"Length of the bone internally	59
"Width of the head	15
"Thickness at the inner condyle	$10\frac{1}{2}$
"Width of the distal end between the most prominent points	11
"Thickness at the inner malleole	8"

CHAPTER XI.

SUBFAMILY ENHYDRINÆ: SEA OTTER.

General considerations—The genus *Enhydris*—Generic characters—*Enhydris lutris*, the Sea Otter—Synonymy—Habitat—Specific characters—Description of external characters—Description of the skull and teeth—History of the species—"The Sea Otter and its hunting"—The habits of the Sea Otter.

LASTLY, we come to consider a particular modification of the Musteline type of structure, which may be regarded as an exaggeration of various features characterizing the *Lutrinæ*, with the superposition of others not elsewhere found in the family. With the general aspect of an ordinary Otter, the *Enhydrinæ* present a special modification of the limbs, more particularly of the hind limbs, which are developed into flipper-like organs, not very dissimilar to those of some Seals. There is also a special condition of the pelage. The cranium, in general, is like that of the *Lutrinæ*, but the teeth are unlike anything else seen in the family *Mustelidæ*. One pair of incisors is wanting, which makes the dental formula unique. Moreover, the whole dentelure is modified in adaptation to a piscivorous regimen. The sectorial teeth are defunctionalized as such; if the teeth of ordinary carnivorous quadrupeds be likened to fresh-chipped, sharp and angular bits of rock, those of the *Enhydrinæ* are comparable to water-worn pebbles.

The *Enhydrinæ* are represented, as far as known, by a single genus and species, inhabiting the coasts and islands of the North Pacific. It is the only thoroughly marine species of the family; it furnishes one of the most valuable of all pelts in a commercial point of view, and its chase is an important industry.

The Genus ENHYDRIS. (FLEMING.)

< Mustela sp. *Linnæus*, Syst. Nat. 1758–66.
< Phoca sp. *Pallas*, Zoog. R.-A. 1831.
< Lutra sp. of various authors.
= Pusa, *Oken*, Lehrb. Naturg. 1816. (Not of Scopoli.)
= Enhydra,* *Fleming*, Philos. Zoöl. 1822. (Also written *Enhydris, Enydris*.)
= Latax, *Gloger*, N. Act. Nat. Cur. 1827. (Not of Gray.)

---

* *Etym.*—See anteà, p. 29, for discussion of the philological bearings of this word.

GENERIC CHARACTERS.—*Dental formula:* i. $\frac{3-3}{2-2}$; c. $\frac{1-1}{1-1}$; pm. $\frac{3-3}{3-3}$; m. $\frac{1-1}{2-2} = \frac{16}{16} = 32$.* Grinding teeth very peculiar in shape, without trenchant edges or acute angles, all being bluntly tuberculous on the crown, and rounded off in contour. Molar of upper jaw irregularly oval. Last upper premolar not dissimilar in shape, and but little smaller; others abruptly less in size. Anterior under molar much the largest of the lower teeth; posterior premolar and posterior molar next in size. Skull generally as in *Lutrinæ*, in straight upper outline, very short rostrum, truncate in front and flat on top, backward production of palate, size and shape of anteorbital foramen, &c., but much broader for its length; thus, the interpterygoid emargination is, if anything, broader than deep. Sagittal and occipital crests and mastoid processes very salient. Glenoid not locking condyle. Coronoid of mandible sloping backward with convex fore and straight or slightly concave posterior border, its bluntly rounded apex in the vertical line over condyle.

General external aspect of *Lutrinæ*, but limbs modified. Fore legs short, with small paws; digits webbed; palms naked. Hind feet with elongated digits, flipper-like, webbed, and furry both sides; claws small, hidden in the fur. Habits aquatic; habitat marine.

The character of the genus is so fully exhibited in the following account of the only species that further remark is not required in this connection.

The Sea Otter.

Enhydris lutris.

PLATES XIX, XX.

Lutra marina, *Steller*, N. C. Petrop. ii. 1751, 367, pl. 26.—*Erxl.* Syst. Anim. 1777, 445 (description pertinent, but synonymy mixed with that of another species).—*Schreb.* Säug. iii. 1778, 465, pl. 128 (Steller).—*Zimm.* Geog. Gesch. ii. 1780, 313, no. 211.—*Shaw*, Gen. Zoöl. i. 1800, 444, pl. 101.—*Dem.* Mamm. i. 1820, 189, no. 291; "Nouv. Dict. xviii, 216; Ency. Méth. pl. 79, f. 3."—*Harlan*, Fn. Amer. 1825, 72.—*Gloger*, N. Act. Nat. Cur. xiii. pt. ii. 1827, 510 (proposes *Latax* as a better name than *Pusa*), 875 *seq.*; "Isis, ii. 1829, 132 *seq.*; Féruss. Bull. xv. 136 *seq.*"—*Godm.* Am. N. H. i. 1831, 228. pl. —, f. i.—*Wagn.* Arch. f. Naturg. ii. 1836, 281.
Sutra [sic] marina, *H. W. Elliott*, Amer. Sportsman, Sept. 12 and 19, 1874 (biography; under pseudonym of "Alaska").
Lutra (Enhydra) marina, *Rich.* F. B.-A. i. 1829, 59, no. 21; Zoöl. Beechey's Voy. 1839, 5.
Enhydra marina, *Flem.* Philos. Zoöl. ii. 1822, 187.—*Griff.* An. Kingd. v. 1827, 132, no. 369.—*Martin*, P. Z. S. iv. 1836, 59 (osteology).—*Aud. & Bach.* Q. N. A. iii. 1853, 170, pl. 137.—*Newb.* P. R. R. Rep. vi. 1857, 43.—*Bd.* M. N. A. 1857, 189.—*Coop. & Suck.* N. H. Wash. Terr. 1860, 115.—*Dall.*, Alaska and its Res. 1870, 489 (habits).—*H. W. Elliott*, Condition of Affairs in Alaska, 8vo ed. chap. v. 1875, pp. 54–62 (history, habits, the chase, economic and commercial relations).

---

*It is said that the *young* Sea Otter has i. $\frac{3-3}{3-3}$, like all other *Mustelidæ*. The middle pair of incisors are those that are wanting in the adult. Prof. Baird (M. N. A. 1857, 189), overlooking the peculiarity of the incisor formula of the adult, but correctly noting the one less premolar than in *Lutrinæ*, gives a wrong total of 34 teeth in all, instead of 32.

SYNONYMY OF ENHYDRIS LUTRIS.

Enydris marina, *Licht.* Darstell. Säug. 1827, 34, pls. 49, 50.—*Erman,* Reise, ——, —, pl. 11, 12.—"*Wagn.* Gelehrte Anzeigen, i. ——, 555; Suppl. Schreber, ii. 1841, 274."
Enhydris marina, *Schinz,* Syn. 1844, 357.—*Giebel.* Säug. 1855, 794.—*Gervais,* Journ. de Zool. iv. 1875, pp. 200-206 (osteology).
Latax marina, *Less.* Nouv. Tabl. R. A. 1842, 71.
Mustela lutris, *L.* S. N. i. 1758, 45, no. 1 (ex Act. Petrop. 1749, 267); 1766, 66, no. 1.—*Schreb.* Säug. iii. 1777, pl. 128 (name on plate).—*Gm.* S. N. i. 1788, 92, no. 1 (excl. var. B, which = *brasiliensis*).—*Turt.* S. N. i. 1806, 57.
Phoca lutris, *Pall.* Zoog. R.-A. i. 1831, 100, no. 34.
Lutra lutris, *Less.* Man. 1827, 155, no. 419.—*Fr. Cuv.* "Dict. Sci. Nat. xxvii. 245"; Suppl. Buff. i. 1831, 204.—*Is. Geoff.* "Dict. Class. ix. 518".—*Fisch.* Syn. 1829, 227, no. 7.
Enhydris lutris, *Gray,* P. Z. S. 1865, 136, pl. 7; Cat. Carn. Br. Mus. 1869, —.
Enhydra lutris, *De Kay,* N. Y. Zoöl. 1842, 41.—*Gray,* Cat. Mamm. Br. Mus. 1843, 72.—*Gerr.* Cat. Bones Br. Mus. 1862, 102.
Pusa orientalis, *Oken,* Lehrb. Naturg. Th. iii. Abth. ii. 1816, p. 986.
Lutra stelleri, *Less.* Man. 1827, 156, no. 423.
Enydris stelleri, *Fisch.* Syn. 1829, 229.
?Lutra gracilis, *Shaw,* G. Z. i. 1800, 447 (based on "*Slender Otter*" of Pennant, Quad. ii. 85., North America, Staten Land).
?Lutra gracilis (*sub Pusa*), *Oken,* Lehrb. Naturg. Th. iii. Abth. ii. 1806, 986.
?Enydris gracilis, *Fisch.* Syn. 1829, 229 (from Shaw, *l. c.*).
Meerotter, *Steller,* "Hamb. Mag. xi. 460, with fig."—*Müll.* Naturs. i. 1773, 259.
Seeotter, *Hallen,* "Naturg. vierf. Th. 1757, 567".
Seebiber *oder* **Seeotter,** "*Steller,* Kamtsch. 1774, 97".
Seebiber, *Müll.* "Samml. iii. ——, 244".
Kamtschatkische Bieber, *id.* "*ibid.* 529".
Sea Otter, *Penn.* Syn. Quad. 1771, 241, no. 175 (in part); Hist. Quad. 1781, —, no. 230; Arct. Zoöl. i. 1784, 88, no. 36.—*Cook,* "Third Voy. 1874, ii. 295, pl. 43; Meares's Voy. 1790, pp. 241, 260."—*Hume & Menzies,* Phil. Trans. 1796, 385.—*Scammon,* Am. Nat. iv. 1870, 65 (detailed biography); "Overland Monthly", iv. 25 (biography); Marine Mamm. 1874, chap. vi. pp. 169-175, woodcuts, pl. xxii. lower fig. (biography, etc.).
Sea Beaver, *Krasch.* "Hist. Kamts. (Grieve's transl.), 1764, 131."
Kamtschatskoi Bobr *or* **Bobr morskoi** *Russian.*
Kalko, *Aliq.*
Kalan, *Steller, l. c., Gray, l. c.*
Castor marin, *Krasch.* "Hist. Kamtsch. ——, 444".
Loutre de mer, *Cook,* "Third Voy. (French transl.), pl. 43."
Loutre marine, *Desm. l. c., Fr. Cuv. l. c.*
Loutre du Kamtchatka, *Geoffr.* "Collect. Mus. d'Hist. Nat."; *Lesson, l. c.*
Loutre marine à tête blanche, "Dict. Sc. Nat. fasc. vii. pl. 19, f. 2".—*Desm. l. c.*
Var. **L. marina with a white head,** *Harl.* op. cit. 74.
Loutre de Steller, *Less. l. c.*
Saricovienne, *Buff.* "Suppl. vi. 4to, 287" (in part).

HAB.—The North Pacific. On the American side, south to Lower California.

SPECIFIC CHARACTERS.—Hind feet broad, like a Seal's flippers, the soles furry; fore feet small, like a Cat's paws, the palms naked; tail terete, obtuse, about ¼ the length of head and body. Form massive. Color dark liver-brown, bleaching on the head, everywhere silvered with hoary ends of the longer hairs. Length over all about 4 feet, of which the tail is a foot or less; hind foot about 6 inches long by 4 broad. Girth about 2½ feet.

*Description of external characters.**

In general superficial aspects, the Sea Otter is not unlike a Seal—a resemblance increased by the flipper-like hind feet. The

---

* From No. 9457, Mus. Smiths. Inst., Alaska, *Dr. Minor.*

body is a swollen cylinder, abrupt behind, tapering before to a small globose head without notable constriction of neck; the limbs are short; the tail is short, terete-tapering, obtuse; there is a remarkable disparity in size and shape between the fore and hind feet, not seen in any other species of *Mustelidæ*. In life, the skin is remarkably loose and "rolling"; the pelt of an individual four feet long readily stretches to six feet; and when the animal is lifted up by the skin, a foot or so of "slack" gathers. The pelage is notable for the preponderance of the woolly under fur, the longer stiffer hairs being very scanty. It is of the same general character all over the body; but on the head, feet, and tail consists chiefly of a finer fur, with little or no admixture of bristly hairs. The only naked parts are the muffle and palms.

The naked muffle, an inch broad, and deeper than this, is lozenge-shaped, with acute superior and inferior angles, obtuse lateral angles, straight or slightly sinuous upper sides, the lower sides somewhat irregular for most of their length, owing to the nostrils; these open quite broadly upon the surface. The face of the nasal pad is minutely papillate, and divided part way by a vertical line of impression. The eyes, of moderate size, are high up, forming a nearly equilateral triangle with the apex of the muzzle. The ears are situated remarkably low down—far below the eyes, and in fact little above the level of the commissure of the mouth; they are very small, flat, obtusely pointed, sparsely and very shortly pilous outside, only partially furry inside. The whiskers are few, short, extremely stout and stiff, directed downward for the most part; there are a few other bristles over the eyes, but none are noted on the chin or cheeks.

The fore feet are remarkably small, giving the limb an appearance which suggests amputation at the wrist; the digits are very short and much consolidated; the very small, short, and much arched claws are almost entirely hidden in the fur. The general contour of the foot is circular in front. The palm is naked, and minutely granular, with small roughened tubercles. The baldness reaches up to the wrist on the outer side in a narrow space. The hind feet, on the contrary, are notable for their expansion and flattening into strong effective oars. The general shape is trapezoidal—the longest side exterior, the side represented by the ends of the digits next longest; the inner border shorter, while the angle represents the fourth and much the shortest side. The digits are entirely webbed by membranes stretching from tip to tip of all the toes. When widespread, the ends of the toes describe a slight curve, the inner one being a little shorter than the next, the rest regularly graduated. The claws are short, stout, arched, and rather obtuse, hidden in the dense fur, which completely invests the foot above and below. The tail is short, stout, and terete, with a slight taper throughout, at the end rather abruptly contracting to an obtuse tip.

The coloration varies greatly with age and season. When the animal is in good state, like the specimen now under particular consideration, it is deep liver-brown, about the same above and below, everywhere silvered or "frosted" with the hoary tips of the longer stiff hairs. These colorless hairs are rather more numerous below than above, giving a lighter tone to the under parts, the body of which, however, is of much the same color as the back. There are fewer or none such light-tipped hairs on the tail and limbs, which consequently appear of a more uniform liver-brown. On the fore part, just in advance of the shoulders, the color lightens rather abruptly into a grayish or light muddy brown, and the bleaching increases on the head, which is of a brownish-white. The whiskers are colorless; the muffle black; the claws dark.

Among the numerous specimens examined, including some not "in condition", great variation is exhibited in the extent to which the ground color is overlaid with the hoary. The longer hairs are sometimes so numerous and so extensively bleached that the animal appears mostly grizzly, completely bleached upon the neck and head. The light hairs, instead of being purely hoary, are frequently of a yellowish tint, as if soiled. The variations in the ground color are chiefly due to presence or absence of a "red" shade, which, in the best specimens, produces the rich liver-brown hue or chocolate-shaded color, and the absence of which leaves the brown of a plain dark character. There is often a noticeable blackish area between the fore-legs.*

The variability of this species in size, though great, is only on a par with that of its allies. The dimensions may be gathered from the measurements already given; but, though these are incomplete, they are not here supplemented with others,

---

* The tendency to special particoloration on the throat and breast is strong in *Mustelidæ*. It is fully carried out in *Gulo*, *Mustela*, *Putorius vison*, &c., and even indicated here in *Enhydris*.

since such as could be given from the dried skins before me would be only approximate. The ear is about an inch long, measured from the notch in front, and about two-thirds as wide.

Young (a very young individual, under two feet long, also collected by Dr. Minor in the Aleutian Islands).—The coat is comparatively much longer than that of the adult, loose, rather harsh, and of a peculiar fluffy character, with kinky fibre. The naked muffle is much as in the adult, but quite smooth; the ears are entirely hidden in the abundant wool of the parts; the hind feet scarcely show their proper shape; the tail is clubbed, rather thicker at the abrupt end than at the base; a decided constriction of the neck appears in the specimen as mounted. The feet are quite blackish; otherwise the animal is dull grayish-brown, everywhere strongly frosted with hoary, lighter, and more uniformly brownish-white on the head and neck, bleaching to dingy white underneath the head and before the shoulders.

To sum the salient external peculiarities of this species in comparison with *Lutra*, it is only necessary to mention the more massive form, the much shorter, more uniformly terete and obtuse tail, and wholly peculiar structure of the feet. To exhibit the characters of the species in the clearest light, I add to the foregoing technical description the following account from Meares's Voyage (1790), showing how the appearance of the animal would strike an unscientific observer:—

"The Sea Otter is furnished with a formidable set of teeth; its fore paws are like those of the River Otter, but of much larger size, and greater strength; its hind feet are skirted with a membrane, on which, as on the fore feet, there grows a thick and coarse hair. The fur varies in beauty according to the age of the animal. The young cubs, of a few months old, are covered with a long, coarse, white hair, which protects the fine down that lies beneath it. The natives often pluck off this coarse hair, when the lower fur appears like velvet, of a beautiful brown colour. As they increase in size, the long hair falls off, and the fur becomes blackish, but still remains short. When the animal is full grown, it becomes of a jet [?] black, and increases in beauty; the fur then thickens, and is thinly sprinkled with white hairs. When they are past the age of perfection, and verge towards old age, their skin [fur] changes into a dark brown, dingy colour, and of course diminishes in value. The skins of those killed in the winter are of a more beautiful black, and in every respect more perfect than those which are taken in the summer and autumn. The male Otter is beyond all comparison the more beautiful than the female and is distinguished by the superior jetty colour, as well as velvety appearance of his skin; whereas the head, throat and belly of the female, are not only covered with fur that is white, but which is also of a very coarse texture. The skins in the highest estimation are those which have the belly and throat plentifully interspersed with a kind of brilliant silver hairs, while the body is covered with a thick black fur of extreme fineness, and a silky gloss."

Among other earlier descriptions, that of Dr. Pallas, modestly styled "ad complementum Stellerianæ", in delicate compliment to the previous traveller, may be cited in illustration of some of the more infrequent variations. Pallas speaks of a specimen from Kodiak, which was yellowish-white, shaded on the back with gray (e *flavescenti-alba, medio dorso grysea nebula enumbrato*). The old animal, he says, is glossy black, with somewhat reddish under fur, and often over five feet long; the young are rather dark grayish.

*Description of the skull and teeth.** (See Plates XIX, XX.)

With a general resemblance to that of *Lutra*, the skull of the Sea Otter differs not only in its superior size, but in its massiveness, depth, breadth behind, truncation anteriorly, and several details which will appear in the sequel. There is a general condition which would suggest, in common parlance, such terms as "huge", "bulky", "misshapen", and a superficial likeness to the skulls of some of the *Pinnipedia*, with which the Sea Otter is intimately associated in its mode of life.

As evidenced by the sutures in some very young skulls before me, the disposition of the several bones is much as in *Lutra*, but there are some peculiarities. The malar bone is, as it were, shifted bodily backward; it reaches to the glenoid fossa, and, in fact, just misses a share in the articulation of the lower jaw, while in front it stops altogether short of the bridge over the anteorbital foramen, which is thus circumscribed only by a very slender rod from the maxillary. The intermaxillary bones are so short and deep as to be almost vertical; their apices merely

---

*The osteology of this species has been specially studied by Martens (P. Z. S. 1836, 59) and Gervais (Journ. de Zool. iv. 1875, 200-206). Gerrard gives the vertebral formula as C. 7; D. 14; L. 6; S. 3; Cd. 18.

meet the extremity of the nasals. More than a third of the incisive foramen is maxillary, not intermaxillary. The orbitosphenoid recedes deeply in its surroundings. Other points will appear in a topographical account of the skull.

Viewed from above, the cranium differs from that of *Lutra* in greater inflation of the cerebral walls, especially anteriorly, where the encroachment upon the temporal fossæ is decided. Supraorbital processes are not so well developed (about as in a Badger or Marten; the development in *Lutra* is exceptional). The nasal orifice is greatly foreshortened in this view, owing to the abrupt truncation of the mandible. It is difficult to say how long the rostrum is, owing to the configuration of the parts, but it may be estimated at about one-sixth of the total length of the skull. Owing to its verticality, the anteorbital foramen is scarcely seen in this view (it comes into sight inside the orbit in *Lutra*). In old specimens, there is a strong sagittal crest wanting in *Lutra;* in young ones, an irregular elevated tablet. The top of the rostrum and adjoining interorbital space is a smooth, flat tablet, as in *Lutra*. The occipital contour is much as described in *Lutra*.

In profile, the skull shows the same flatness on top as is seen in *Lutra*, with the additional feature of an almost vertical anterior truncation from the ends of the nasals, at little more than a right angle, and almost straight down to the incisors. Such contour is highly characteristic, and reminds one of the same part in a Walrus. Owing to the slight supraorbital process and little marked malar protuberance, the orbit is not well defined; not so well as it is in the other subfamilies, excepting *Mephitinæ*. The zygomatic arch rises abruptly behind. Its upper border is then about straight and horizontal to the orbit; its lower border is a strong regular curve throughout. Other matters to be noted in the profile view are much as in *Lutra*.

From below: The zygomatic width is about three-fourths the length; the intermastoid diameter but slightly less. The palate reaches back of the molars about half-way to the ends of the pterygoids. The emargination between these bones is extremely wide and shallow. Perhaps here only in the family, the depth of the emargination is no greater, or less than, its width. The recess is sometimes almost semicircular, though the sides are usually more nearly parallel, and the end transverse. In detail, the shelf of the palate is altogether irregular. The walls of the glenoid fossæ are rarely, if ever, so much developed as to lock the jaw. I have not witnessed such case. The back wall, instead of overlapping strongly at its outer angle, is regularly produced into a border all along. The inflation of the bullæ is about as in *Lutra*. The posterior foramen lacerum is a large circular hole. The articular surfaces of the condyles differ from those of *Lutra* in not being produced toward each other; they are simply oval. The great foramen is irregularly circular rather than transversely elliptical, having a strong median superior as well as inferior emargination. In the under jaw, the symphysis is shorter and apparently less solid than usual. I find the union incomplete in some middle-aged specimens. The ramus of the under jaw is deep and thick, and has a decided twist, scarcely or not recognizable in other genera, by which the lower part is exflected posteriorly. The coronoid is very broad to the rounded end; the hind border rises straight and a little obliquely backward, so as to overhang the condyle; the front border is strongly, somewhat irregularly, curved. The muscular impression on the outside of the coronoid is deep and extensive, reaching below to the very edge of the jaw, and forward to a point below the last molar.

The dentition of *Enhydris* is peculiar in several respects. As in *Lutrinæ*, but not as in any other subfamily of *Mustelidæ*, there is the same number of teeth in both jaws (16); but this equality is brought about in a curious way, loss of the upper anterior premolar being rectified, so far as preserving equality of teeth in the two jaws is concerned, by lack of one pair of inferior incisors. Thus there are four fewer teeth than in *Lutrinæ* ($\frac{1-6}{1-6}$ instead of $\frac{1-8}{1-8}$). This is the only instance in the family of less than six incisors below, or of an unequal number of incisors in the two jaws. In the presence of an equal number of premolars above and below, *Enhydris* agrees with all the other North American genera excepting *Lutra* ($\frac{4-4}{3-3}$) and normal *Conepatus* ($\frac{2-2}{3-3}$); in the presence of three premolars above and below, it agrees with all but *Lutra*, *Conepatus*, *Mustela*, and *Gulo* (the two last having $\frac{4-4}{4-4}$); in the presence of three premolars below, it agrees with all excepting *Mustela* and *Gulo*.

In the physical character of the teeth, as well as in the dental formula, *Enhydris* is peculiar in its family. All the grinders are of a singularly massive, tubercular, almost bulbous character, with no trenchant edges, acute cusps, or even angular edges. This is in evident adaptation to the piscivorous regimen of the animal. The teeth of even the youngest specimens have an appearance of being greatly worn, as is not, however, the case. In fact, there is less difference with age here than elsewhere in the family. The back upper molar is the largest tooth of all, being as wide as, and much longer than, the sectorial tooth. It is irregularly oval in shape, its long axis oblique; its face is studded with obtuse tubercles in a manner scarcely admitting of detailed description. The back upper premolar is squarish, with rounded-off angles, and presents outwardly a pair of large obtuse tubercles, whereof the anterior one is the larger, separated by a groove from an interior lower portion of the tooth occupied by a single large, blunt, conical tubercle. The next premolar is a blunt cone with a heel behind. The anterior premolar is entirely similar, but much smaller, and crowded inward from the general axis of dentition. It has but one fang; the tooth behind it is two-rooted; the sectorial tooth roots by three fangs, two external, one internal; the upper molar is set in three irregular shallow sockets. The back lower molar is transversely elliptical rather than circular; its face is smooth and flattened, with a crosswise central depression. The anterior lower molar is completely and bluntly tuberculous, showing only traces of its likeness to the same tooth elsewhere in the family in a slightly elevated, tri-tuberculous, anterior part, and a flattish, depressed hind part. The back lower premolar is an irregular, low, blunt cone, with a secondary eminence part way up its inner aspect. The other premolars are successively smaller and simpler. The front premolar and back molar are single-rooted; the anterior molar has four roots; the next tooth three; the next two. The canines, both above and below, are rather small, comparatively; the latter is not much curved. Of the superior incisors, the lateral pair are moderately larger than the rest, and taper somewhat toward the end from an elbow near the base. The others are smaller, especially narrow, and somewhat club-shaped; none are obviously lobate. Of the inferior incisors, it is seen to be the median pair that are missing, for the next pair (here the middle pair) have the backward set, which usually distinguishes them in other genera. These incisors are all strongly clubbed at their extremities, which are irregularly nicked.

History of the species.

The history of this species may be considered to have begun in the middle of the last century. One of the earliest accounts, if not the first one of any scientific pretensions, was that of the celebrated navigator Steller, who described the animal, in 1751, under the name of *Lutra marina*, a term not yet wholly obsolete, though untenable under the rules of nomenclature. This may have been the first introduction of the species to the notice of civilized, or at least of scientific, men, though the animal had, of course, long been known to the natives of the countries along the shores of which it was found. It was known to the Russians as the Sea or Kamtschatkan Beaver (Bobr morskoi and Kamtschatskoi Bobr), and to the Kamtschatkans themselves as the Kalan; while other barbarous nations had their own equivalent terms, or several such, to indicate different ages or states of pelage. Notwithstanding the accuracy of Steller's account, which is quoted and sometimes consulted to the present day, and in spite of the numerous striking peculiarities which the animal offers upon the most casual inspection, the compilers of various systematic treatises soon suffered under a confusion of ideas, and perpetrated blunders that were not for many years eradicated. Linnæus confounded it with the Saricovienne or Brazilian Otter, *Lutra brasiliensis;* and the same mistake was even made by several much more accomplished therologists, like Brisson and Pennant. It would be presumed that its remarkable features would have prevented this; instead, however, we find that the singular construction of the hind feet, general aspect, and mode of life have caused it to be classed among the Seals—Pallas indeed, an eminent naturalist and observing traveller, calls it a *Phoca;* and in the latest publication upon the subject, Capt. C. M. Scammon's *Marine Mammals* (1874), it is located again in the midst of *Pinnipedia*. It is, of course, unnecessary to seriously discuss a procedure which, like this, is indefensible upon any but the most superficial and unscientific considerations, drawn from the aquatic habits of the animal, and the modifications required for this end. Its relationships with the Pinnipeds are entirely those of analogy.

Linnæus was right, according to the terms of classification of his day, in placing it in the genus *Mustela*; a group nearly equivalent to the family *Mustelidæ* as now understood. Overlooking or ignoring Steller's name of *Lutra marina*, which, though binomial in the letter, was merely a Latin translation of a vernacular term, and not binomial upon any system, he called the species *Mustela lutris*, a name the specific portion of which must stand, even though, as already intimated, it includes an altogether different animal, *Lutra brasiliensis;* for the Stellerian name *marina* was not used by any binomial writer until after Linnæus had applied *lutris.* Steller's more obviously appropriate designation of *marina* was, nevertheless, adopted by Erxleben, Schreber, Desmarest, and other distinguished naturalists of various countries, and became generally current. In consequence, doubtless, of the very marked characters which the species affords, only two or three nominal species have been based upon it. The first of these, instituted by Oken, the famous anatomist and naturalist, is, in fact, scarcely a nominal species in the usual acceptation of the term, being merely, like the *Mustela lutris* of Linnæus, a renaming of the well-known animal, without intention of separating from it a second species. Oken called it *Pusa orientalis,* in 1816,* in the work above cited, apparently inventing both the generic and specific term, in this application at least. R.-P. Lesson is responsible for another synonym, having, in 1827, renamed the species *Lutra stelleri,* a compliment to the distinguished navigator who gave us the early account, but one which the rules of nomenclature forbid us to adopt, however we might incline to such course. Lesson appears to have fancied that the Kamtschatkan Otter, *Lutra* or *Mustela lutris* of authors, and *Lutra marina* of Erxleben, was a true Land Otter, different from Steller's animal, and, in fact, such was *partly* the case. We have yet to consider a very problematical animal, the Slender Otter of Pennant, which became the *Lutra gracilis* of Shaw, the *Enydris gracilis* of Fischer, and is mentioned under *Pusa* by Oken, said to be from Staaten-Land, Nord-Amerika. It is impossible to determine what this is, owing to the imperfection of the description, but it was probably based upon a Sea Otter; Pennant himself appears to have given it up, as it does not figure in his later work, "Arctic Zoölogy". Oken speaks of "Staaten-Land, *bei New-York*", evidently having what is now known as Staten Island in view; but it is safer to presume upon a geographical error here than to refer the animal to *Lutra canadensis,* which, as is well known, is the only Otter of the Eastern United States, where the Sea Otter certainly does not occur.

These specific names are the only ones I have come upon in

---

* De Blainville gives the date of the name as 1814, but I have not been able to trace it back of 1816.

searching the literature of the species; but we have still to consider the several designations resulting from their combination with various generic designations, some of which are old, and belong to other groups, while others were newly invented for this particular species. The former are, in the order of their successive use, *Lutra*, *Mustela*, and *Phoca*, after Steller, Linnæus, and Pallas respectively; these need not detain us. It was three-quarters of a century, nearly, from its original introduction to the system, before the strongly marked characters of the species were made typical of a new genus—*Pusa* of Oken, already mentioned, being the first-named of this sort. *Pusa* had, however, already been used by another writer in connection with a genus of Seals now commonly known as *Halichœrus*, but in such a peculiar way as to raise one of those technical questions of synonymy which authors interpret differently, in absence of fixed rule. Scopoli based his *Pusa* upon a figure of Salomon Müller's, recognizable with certainty as *Halichœrus*, and gave characters utterly irreconcilable with those of this animal. This is the whole case. Now it may be argued that there being no such animal whatever as Scopoli says his *Pusa* was, his name drops out of the system, and *Pusa* of Oken, virtually an entirely new term, is tenable for something else, namely, for the Sea Otter. On the other hand, Scopoli's quotations show exactly what he meant, in spite of his inept diagnosis; his name *Pusa* therefore holds, and cannot be subsequently used by Oken in a different connection. This is the view I take in this and all similar cases, when a name can be identified by any means whatsoever, intrinsic or circumstantial, no matter how wide of the mark the ascribed characters may be. And even if it be not the first *tenable* name of a genus—in other words, if it be only a synonym of a prior name—it cannot be used again as a tenable name for a different genus. This name *Pusa* thus disposed of, another to be similarly treated is *Latax* of Gloger. Though applied by some authors, particularly J. E. Gray (*more suo*, with little regard for the obvious requirements of the case), to species of *Lutra* proper, *Latax* was nevertheless based by its founder upon the Sea Otter, *Lutra marina*, in the xiiith vol. (1827) of the N. Act. Nat. Curios. p. 511 (reprinted in the Isis for 1829, and in Férussac's Bulletin). This well-identified name[*] is, how-

---

[*] It is, however, doubtful whether *Latax* can be considered as established at all; for Gloger, treating of the Sea Otter under the name *Lutra marina*, simply takes occasion to criticise the fitness of Oken's term *Pusa*, and to suggest that *Latax* might be a more apt designation.

ever, an unquestionable synonym of *Enhydra* of Fleming, instituted in 1822, in that author's "Philosophy of Zoölogy", and which, under its various forms of *Enhydris*, *Enydris*, and *Enhydra*, has been most generally employed of late years.

Besides the technical accounts of very numerous authors who never saw the animal alive, there are many other notices of more general interest, in unscientific works, giving information upon its habits and manners, and various figures, more or less true to life, are extant. The famous navigator Cook treats of the Sea Otter, and gives a fair representation. The description from Meares's Voyage, accurate, though untechnical, is frequently quoted. Menzies's article in the Philosophical Transactions for 1796 may be noted in this connection. Pennant, as usual, has an extended biographical notice. Probably the first anatomical article of any note is Martins's, upon the osteology of the species; that of M. Gervais is specially important. In late times, detached notices of its habits have multiplied, from the pens of a number of naturalists who have visited the northwest coast, and largely contributed to a complete history. Capt. Scammon's several articles above quoted, all to much the same effect, are specially noteworthy, though certain points may require to be scrutinized and checked by the observations of others. The author last mentioned also reproduces the figure by Wolf, which accompanied Dr. J. E. Gray's paper on the *Mustelidæ*, in Zoölogical Society's Proceedings for 1865; this is probably the most life-like representation of the species extant. J. W. Audubon's plate, published in the work of his father and Dr. Bachman, is a finished drawing of unmistakable character, probably the best one generally accessible to American students. Neither Sir John Richardson nor Audubon had met with the species alive, and their biographies, the principal ones which until lately had appeared in works upon American Mammals, are necessarily at second hand. The only American biographies, indeed, at all approaching completeness are those of Mr. Elliott and Capt. Scammon, already cited.

"*The Sea-Otter and its Hunting.*"[*]
[By H. W. Elliott.]

"The sea-otter, like the fur-seal, is another illustration of an animal long known and highly prized in the commercial world,

---

[*] Having no original information to offer respecting the commercial history, the chase, or the habits of the Sea Otter, I extract an account which there is reason to believe to be the most complete, accurate, and reliable at

yet respecting the habits and life of which nothing definite has been ascertained or published. The reason for this is obvious, for, save the natives who hunt them, no one properly qualified has ever had an opportunity of seeing the sea-otter so as to study it in a state of nature, for, of all the shy, sensitive beasts, upon the capture of which man sets any value, this creature is the most keenly on the alert and difficult to obtain; and, like the fur-seal in this Territory, it possesses the enhancing value of being principally confined to our country. A truthful account of the strange, vigilant life of the sea-otter, and of the hardships and perils encountered by its hunters, would surpass in novelty and interest the most attractive work of fiction.

"When the Russian traders opened up the Aleutian Islands, they found the natives commonly wearing sea-otter cloaks, which they parted with at first for a trifle, not placing any especial value on the animal, as they did the hair-seal and the sea-lion, the hair and skin of which were vastly more palatable and serviceable to them; but the offers of the greedy traders soon set the natives after them. During the first few years the numbers of these animals taken all along the Aleutian Chain, and down the whole northwest coast as far as Oregon, were

---

our service. The following matter constitutes Chap. V, pp. 54-62, of Mr. Henry W. Elliott's "Report on the Condition of Affairs in the Territory of Alaska", 8vo, Washington, Government Printing Office, 1875. Mr. Elliott has proven a trustworthy observer and zealous naturalist, and had excellent opportunities of studying the whole subject during his long residence in Alaska as special agent of the Treasury Department, charged with the Government interests in the Fur Seal Fisheries.

A quotation from Sir John Richardson (Fn. Bor.-Am. p. 59), touching the early aspects of the Sea Otter business, will not be here out of place:—

"The fur of the Sea Otter being very handsome, was much esteemed by the Chinese, and, until the market at Canton was overstocked, prime skins brought extraordinarily high prices. The trade for a considerable period was in the hands of the Russians, who soon after the discovery of the northwest coast of America, by Beering [sic] and Tschirikow, sent mercantile expeditions hither. Captain Cook's third voyage drew the attention of English speculators to that quarter, and vessels were freighted both by private adventurers and the India Company, for the purpose of collecting furs and conveying them to Canton. Pennant, alluding to this traffic, says, 'What a profitable trade (with China) might not a colony carry on, were it possible to penetrate to that part of America by means of rivers and lakes.' The event that Pennant wished for soon took place. Sir Alexander Mackenzie having traversed the continent of America, and gained the coast of the Pacific, his partners in trade followed up his success, by establishing fur posts in New Caledonia, and a direct commerce with China; but the influx of furs into that market soon reduced their price."

very great, and compared with what are now captured seem perfectly fabulous; for instance, when the Prybilov Islands were first discovered, two sailors, Lukannon and Kaiekov, killed at St. Paul's Island, in the first year of occupation, *five thousand*, the next year they got less than a *thousand*, and in six years after not a single sea-otter appeared, and none have appeared since. When Shellikov's party first visited Cook's Inlet, they secured three thousand; during the second year, two thousand; in the third, only eight hundred; the season following they obtained six hundred; and finally, in 1812, less than a hundred, and since then not a tenth of that number. The first visit made by the Russians to the Gulf of Yahkutat, in 1794, two thousand sea-otters were taken, but they diminished so rapidly that in 1799 less than three hundred were taken. In 1798 a large party of Russians and Aleuts captured in Sitka Sound and neighborhood twelve hundred skins, besides those for which they traded with the natives there, fully as many more; and in the spring of 1800 a few American and English vessels came into Sitka Sound, anchored off the small Russian settlement there, and traded with the natives for over two thousand skins, getting the trade of the Indians by giving fire-arms and powder, ball, &c., which the Russians did not dare to do, living then, as they were, in the country. In one of the early years of the Russian American Company, 1804. Baranov went to the Okotsk from Alaska with fifteen thousand sea-otter skins, that were worth as much then as they are now, viz, fully $1,000,000.

"The result of this warfare upon the sea-otters, with ten hunters then where there is one to-day, was not long delayed. Everywhere throughout the whole coast-line frequented by them the diminution set in, and it became difficult to get to places where a thousand have been as easily obtained as twenty-five or thirty. A Russian chronicler says: 'The numbers of several kinds of animals are growing very much less in the present as compared with past times; for instance the Company here (Ounalashka) regularly killed more than a thousand sea-otters annually; now (1835) from seventy to a hundred and fifty are taken; and there was a time, in 1826, when the returns from the whole Ounalashka district (the Aleutian Islands) were only *fifteen skins*.'

"It is also a fact coincident with this diminution of sea-otters, that the population of the Aleutian Islands fell off almost in the same proportion. The Russians regarded the lives of these people as they did those of dogs, and treated them accordingly; they took, under Baranov and his subordinates, hunting-parties of five hundred to a thousand picked Aleuts, eleven or twelve hundred miles to the eastward of their homes, in skin-baidars and bidarkies, or kyacks, traversing one of the wildest and roughest of coasts, and used them not only for the severe drudgery of otter-hunting, but to fight the Koloshians and other savages all the way up and down the coast; this soon destroyed them, and few ever got back alive.

"When the Territory came in our possession, the Russians were taking between four and five hundred sea-otters from the Aleutian Islands and south of the peninsula of Alaska, with perhaps a hundred and fifty more from Kenai, Yahkutat, and the Sitkan district; the Hudson's Bay Company and other traders getting about two hundred more from the coast of Queen Charlotte's and Vancouver's Islands, and off Gray's Harbor, Washington Territory.

"Now, during the last season, 1873, instead of less than seven hundred skins, as obtained by the Russians, our traders secured not much less than *four thousand skins*. This immense difference is not due to the fact of there being a proportionate increase of sea-otters, but to the organization of hunting-parties in the same spirit and fashion, as in the early days above mentioned. The keen competition of our traders will ruin the business in a comparatively short time if some action is not taken by the Government; and to the credit of these traders let it be said, that while they cannot desist, for if they do others will step in and profit at their expense, yet they are anxious that some prohibition should be laid upon the business. This can be easily done, and in such manner as to perpetuate the sea-otter, not only for themselves, but for the natives, who are dependent upon its hunting for a living which makes them superior to savages.

"Over two-thirds of all the sea-otters taken in Alaska are secured in two small areas of water, little rocky islets and reefs around the island of Saanach and the Chernobours, which proves that these animals, in spite of the incessant hunting all the year round on this ground, seem to have some particular preference for it to the practical exclusion of nearly all the rest of the coast in the Territory. This may be due to its better adaptation as a breeding-ground. It is also noteworthy that all the sea-otters taken below the Straits of Fuca are shot by the Indians and white hunters off the beach in the surf at Gray's Harbor, a stretch of less than twenty miles; here some fifty to a hundred are taken every year, while not half that number can be obtained from all the rest of the Washington and Oregon coast-line; there is nothing in the external appearance of this reach to cause its selection by the sea-otters, except perhaps that it may be a little less rocky.

"As matters are now conducted by the hunting-parties, the sea-otters at Saanach and the Chernobours do not have a day's rest during the whole year. Parties relieve each other in succession, and a continuous warfare is maintained. This persistence is stimulated by the traders, and is rendered still more deadly to the sea-otter by the use of rifles of the best make, which, in the hands of the young and ambitious natives, in spite of the warnings of the old men, must result in the extermination of these animals, as no authority exists in the land to prevent it. These same old men, in order to successfully compete with their rivals, have to drop their bone-spears and arrows, and take up fire-arms in self-defense. So the bad work goes on rapidly, though a majority of the natives and the traders deprecate it. With a view to check this evil and to perpetuate the life of the sea-otter in the Territory, I offer the following suggestions to the Department:

"1st. Prohibit the use of fire-arms of any description in the hunting of the sea-otter in the Territory of Alaska.

"2d. Make it unlawful for any party or parties to hunt this animal during the months of June, July, and August, fixing a suitable penalty, fine, or punishment.

"The first proposition gives the sea-otter a chance to live; and, with the second, may possibly promote an increase in the number of this valuable animal.

"The enforcement by the Government of this prohibition will not be difficult, as it is desired by a great majority of the natives and all the traders having any real interest in the perpetuation of the business. A good deputy attached to the customs, whose salary and expenses might be more than paid by a trifling tax on each otter-skin, say $1, could, if provided with a sound whale-boat, make his headquarters at Saanach and Belcovski and carry the law into effect. The trade of the Kodiak district centers at the village of that name, and the presence of the collector or his deputy will exert authority, and cause the old native hunters and many of the younger who have reflection to comply with his demands. The collector then being provided with the small revenue-steamer spoken of in my chapter upon the duty of the Government toward the Territory, can insure compliance with the instructions given him, and punish violations.

"This proposed action on the part of the Government is urgent and humane, for upon the successful hunting of the sea-otter some five thousand christianized natives are entirely dependent for the means to live in a condition superior to barbarism.

"*The habits of the sea-otter.* (*Enhydra marina.*)

"I have had a number of interesting interviews with several very intelligent traders, and an English hunter who had spent an entire winter on Saanach Island, shooting sea-otters, and enduring, while there, bitter privation and hardship; and chiefly from their accounts, aided by my own observation, I submit the following:

"*Saanach Island, Islets,* and *Reefs,* is the great sea-otter ground of this country. The island itself is small, with a coast-line circuit of about eighteen miles. Spots of sand-beach are found here and there, but the major portion of it is composed of enormous water-worn bowlders piled up by the surf. The interior is low and rolling, with a ridge rising into three hills, the middle one some 800 feet in height. There is no timber on it, but abundant grass, moss, &c., with a score of little fresh-water lakes, in which multitudes of ducks and geese are found every spring and fall. The natives do not live upon the island, because the making of fires and scattering of food-refuse alarms the otters, driving them off to sea; so that it is only camped upon, and fires are never built unless the wind is from the southward, for no sea-otters are ever found to be north of the island. The sufferings to which the native hunters subject themselves every winter on this island, going for many weeks without fires, even for cooking, with the thermometer down to zero, in a northerly gale of wind, is better imagined than described.

"To the southward and westward, and stretching directly out to sea, some five to eight miles from Saanach Island, is a succession of small islets, bare, most of them, at low water, but with numerous reefs and rocky shoals, beds of kelp, &c. This is the great sea-otter ground of Alaska, together with the

Chernobour Islets, to the eastward about thirty miles, which are similar to it. The sea-otter rarely lands upon the main island, but it is found just out of water on the reef-rocks and islets above mentioned, in certain seasons, and at a little distance at sea during calm and pleasant weather. The adult sea-otter is an animal that will measure from three and a half to four feet at most, from nose to tip of tail, which is short and stumpy. The general contour of the body is closely like that of the beaver, with the skin lying in loose folds, so that when taken hold of in lifting the body out from the water, it is as slack and draws up like the hide on the nape of a young dog. This skin, which is taken from the body with but one cut made in it at the posteriors, is turned inside out, and air-dried, and stretched, so that it then gives the erroneous impression of an animal at least six feet in length, with girth and shape of a weasel or mink. There is no sexual dissimilarity in color or size, and both manifest the same intense shyness and aversion to man, coupled with the greatest solicitude for their young, which they bring into existence at all seasons of the year, for the natives get young pups every month in the year. As the natives have never caught the mothers bringing forth their offspring on the rocks, they are disposed to believe that the birth takes place on kelp-beds, in pleasant or not over-rough weather. The female has a single pup, born about fifteen inches in length, and provided during the first month or two with a coat of coarse, brownish, grizzled fur, head and nape grizzled, grayish, rufous white, with the roots of the hair growing darker toward the skin. The feet, as in the adult, are very short, webbed, with nails like a dog, fore-paws exceedingly feeble and small, all covered with a short, fine, dark, bister-brown hair or fur. From this poor condition of fur they improve as they grow older, shading darker, finer, thicker, and softer, and by the time they are two years of age they are 'prime,' though the animal is not full-grown until its fourth or fifth year. The white nose and mustache of the pup are not changed in the adult. The whiskers are white, short, and fine. The female has two teats, resembling those of a cat, placed between the hind limbs on the abdomen, and no signs of more; the pup sucks a year at least, and longer if its mother has no other; the mother lies upon her back in the water or upon the rocks, as the case may be, and when she is surprised, she protects her young by clasping it in her fore-paws and turning her back to the danger they shed their fur just as the hair of man grows and falls out; the reason is evident, for they must be ready for the water at all times.

"The sea-otter mother sleeps in the water on her back, with her young clasped between her fore-paws. The pup cannot live without its mother, though frequent attempts have been made by the natives to raise them, as they often capture them alive, but, like some other species of wild animals, it seems to be so deeply imbued with fear of man that it invariably dies from self-imposed starvation.

"Their food, as might be inferred from the flat molars of dentition, is almost entirely composed of clams, muscles, and sea-urchins, of which they are very fond, and which they break by striking the shells together, held in each fore-paw, sucking out the contents as they are fractured by these efforts: they also undoubtedly eat crabs, and the juicy tender fronds of kelp or sea-weed, and fish.

"They are not polygamous, and more than an individual is seldom seen at a time when out at sea. The flesh is very unpalatable, highly charged with a rank smell and flavor.

"They are playful, it would seem, for I am assured by several old hunters that they have watched the sea-otter for half an hour as it lay upon its back in the water and tossed a piece of sea-weed up in the air from paw to paw, apparently taking great delight in catching it before it could fall into the water. It will also play with its young for hours.

"The quick hearing and acute smell possessed by the sea-otter are not equaled by any other creatures in the Territory. They will take alarm and leave from the effects of a small fire, four or five miles to the windward of them; and the footstep of man must be washed by many tides before its trace ceases to alarm the animal and drive it from landing there should it approach for that purpose.

"There are four principal methods of capturing the sea-otter, viz, by *surf-shooting*, by *spearing-surrounds*, by *clubbing*, and by *nets*.

"The surf-shooting is the common method, but has only been in vogue among the natives a short time. The young men have nearly all been supplied with rifles, with which they patrol the shores of the island and inlets, and whenever a sea-otter's head is seen in the surf, a thousand yards out even, they fire, the great distance and the noise of the surf preventing the sea-otter from taking alarm until it is hit; and in nine times out of ten, when it is hit, in the head, which is all that is exposed, the shot is fatal, and the hunter waits until the surf brings his quarry in, if it is too rough for him to venture out in his 'bidarkie.' This shooting is kept up now the whole year round.

"The spearing-surround is the orthodox native system of capture, and reflects the highest credit upon them as bold, hardy watermen. A party of fifteen or twenty bidarkies, with two men in each, as a rule, all under the control of a chief elected by common consent, start out in pleasant weather, or when it is not too rough, and spread themselves out in a long line, slowly paddling over the waters where sea-otters are most usually found. When any one of them discovers an otter, asleep, most likely, in the water, he makes a quiet signal, and there is not a word spoken or a paddle splashed while they are on the hunt. He darts toward the animal, but generally the alarm is taken by the sensitive object, which instantly dives before the Aleut can get near enough to throw his spear. The hunter, however, keeps right on, and stops his canoe directly over the spot where the otter disappeared. The others, taking note of the position, all deploy and scatter in a circle of half a mile wide around the mark of departure thus made, and patiently wait for the re-appearance of the otter, which must take place within fifteen or thirty minutes, for breath; and as soon as this happens the nearest one to it darts forward in the same manner as his predecessor, when all hands shout and throw their spears, to make the animal dive again as quickly as possible, thus giving it scarcely an instant to recover itself. A sentry is placed over its second diving-wake as before, and the circle is drawn anew; and the surprise is often repeated, sometimes for two or three hours, until the sea-otter, from interrupted respiration, becomes so filled with air or gases that he cannot sink, and becomes at once an easy victim.

"The coolness with which these Aleuts will go far out to sea in their cockle-shell kyacks, and risk the approach of gales that are as apt to be against them as not, with a mere handful of food and less water, is remarkable. They are certainly as hardy a set of hunters, patient and energetic, as can be found in the world.

"The clubbing is only done in the winter-season, and then at infrequent intervals, which occur when tremendous gales of wind from the northward, sweeping down over Saanach, have about blown themselves out. The natives, the very boldest of them, set out from Saanach, and skud down on the tail of the gale to the far outlying rocks, just sticking out above surf-wash, where they creep up from the leeward to the sea-otters found there at such times, with their heads stuck into the beds of kelp to avoid the wind. The noise of the gale is greater than that made by the stealthy movements of the hunters, who, armed each with a short, heavy wooden club, dispatch the animals, one after another, without alarming the whole body, and in this way two Aleuts, brothers, were known to have slain seventy-eight in less than an hour and a half.

"There is no driving these animals out upon land. They are fierce and courageous, and when surprised by a man between themselves and the water, they will make for the sea, straight, without any regard for the hunter, their progress, by a succession of short leaps, being very rapid for a small distance. The greatest care is taken by the sea-otter hunters on Saanack. They have lived in the dead of a severe winter six weeks at a time without kindling a fire, and with certain winds they never light one. They do not smoke, nor do they scatter or empty food-refuse on the beaches. Of all this I am assured by one who is perhaps the first white eye-witness of this winter-hunting, as he lived on the island through that of 1872–'73, and could not be induced to repeat it.

"The hunting by use of nets calls up the strange dissimilarity existing now, as it has in all time past, between the practice of the Atka and Attou Aleuts and that of those of Ounalashka and the eastward, as given above. These people capture the sea-otter in nets, from 16 to 18 feet long and 6 to 10 feet wide, with coarse meshes made nowadays of twine, but formerly of sinew.

"On the kelp-beds these nets are spread out, and the natives withdraw and watch. The otters come to sleep or rest on those places, and get entangled in the meshes of the nets, seeming to make little or no effort to escape, paralyzed as it were by fear, and fall in this way easily into the hands of the trappers, who tell me that they have caught as many as six at one time in one of these small nets, and frequently get three. They also watch for surf-holes or caves in the bluffs, and, when one is found to which a sea-otter is in the habit of resorting, they set this net by spreading it over the entrance, and usually capture the animal.

"No injury whatever is done to these frail nets by the sea-otters, strong animals as they are; only stray sea-lions destroy them. The Atka people have never been known to hunt sea-otters without nets, while the people of Ounalashka and the eastward have never been known to use them. The salt-water and kelp seem to act as a disinfectant to the net, so that the smell of it does not repel or alarm the shy animal."

www.ingramcontent.com/pod-product-compliance
Lightning Source LLC
Chambersburg PA
CBHW020227240426
43672CB00006B/444